VOLUME NINETY TWO

ADVANCES IN
PROTEIN CHEMISTRY AND
STRUCTURAL BIOLOGY

Dynamics of Proteins and Nucleic Acids

VOLUME NINETY TWO

ADVANCES IN
PROTEIN CHEMISTRY AND
STRUCTURAL BIOLOGY
Dynamics of Proteins and Nucleic Acids

Edited by

TATYANA KARABENCHEVA-CHRISTOVA

Department of Applied Sciences, Faculty of Health and Life Sciences, Northumbria University, Newcastle upon Tyne, United Kingdom

AMSTERDAM • BOSTON • HEIDELBERG • LONDON
NEW YORK • OXFORD • PARIS • SAN DIEGO
SAN FRANCISCO • SINGAPORE • SYDNEY • TOKYO
Academic Press is an imprint of Elsevier

Academic Press is an imprint of Elsevier
The Boulevard, Langford Lane, Kidlington, Oxford, OX5 1GB, UK
32 Jamestown Road, London NW1 7BY, UK
Radarweg 29, PO Box 211, 1000 AE Amsterdam, The Netherlands
225 Wyman Street, Waltham, MA 02451, USA
525 B Street, Suite 1800, San Diego, CA 92101-4495, USA

First edition 2013

Notice
No responsibility is assumed by the publisher for any injury and/or damage to persons or
property as a matter of products liability, negligence or otherwise, or from any use or
operation of any methods, products, instructions or ideas contained in the material herein.
Because of rapid advances in the medical sciences, in particular, independent verification of
diagnoses and drug dosages should be made.

ISBN: 978-0-12-411636-8
ISSN: 1876-1623

For information on all Academic Press publications
visit our website at store.elsevier.com

Printed and bound by CPI Group (UK) Ltd, Croydon, CR0 4YY
Transferred to digital print 2012

Working together
to grow libraries in
developing countries

www.elsevier.com • www.bookaid.org

CONTENTS

INTRODUCTION TO DYNAMICS OF PROTEINS AND NUCLEIC ACIDS

Tatyana Karabencheva-Christova

Department of Applied Sciences, Faculty of Health and Life Sciences, Northumbria University, Newcastle upon Tyne, United Kingdom

Proteins and nucleic acids are biological macromolecules and their interactions underline all molecular processes in life (Berg, Tymoczko, & Stryer, 2002). Their ability to participate in highly specific, effective, and regulated interactions is determined by their structure (Fersht, 1999). Proteins and nucleic acids are large molecules containing many degrees of freedom and therefore exhibit a broad variety of internal motions ranging from femtoseconds to milliseconds and hours (Adcock & McCammon, 2006; Karplus & McCammon, 2002). Many of those motions play important roles in processes at different levels such as ligand binding, allosteric regulation, transport, protein–protein, and protein–nucleic acids interactions (Karplus & Kuriyan, 2005). X-Ray crystallography provides tremendous insight into the structures of proteins and nucleic acids, at different conformational states (e.g., free enzymes, enzyme–substrate complexes, active and nonactive states); however, the provided information is static and represents averaged information of the crystallographic experiment (Kruschel & Zagrovic, 2009). Conformational dynamics and flexibility, however, could influence and change the static picture. Numerous experimental (e.g., NMR, variety of spectroscopic and time-resolved techniques) and computational methods (e.g., molecular dynamics (MD), Monte Carlo methods) became successfully applied to provide novel and mutually complementary information about protein and nucleic acids dynamics and interactions.

In this volume of *APCSB*, we present the state-of-the-art contributions focused on protein and nucleic acids dynamics and interactions using instrumental or modelling approaches and their synergetic combination. Chapter 1 by Arnab Mukherjee and Wilbee D. Sasikala is focused on the nature of the drug–DNA intercalation using experimental techniques and computational methods for understanding its molecular mechanism which is of a particular interest for the development of anticancer pharmaceuticals. Chapter 2 by Yuko Okamoto and coworkers introduces and reviews important generalized–ensemble algorithms for molecular simulations which can greatly enhance the conformational sampling of biomolecules. A great advance in understanding conformational dynamics of single–protein molecules by direct mechanical manipulation with force spectroscopy techniques is described in Chapter 3 by Ciro Cecconi and coauthors. Chapter 4 by Mamannamana Vijayan and coauthors is centered on the important analysis of ligand specificity and modes of oligomerization in β–prism I fold lectins. A comprehensive computational investigation on the conformational changes in proteins and DNA caused by different environmental factors such as pH, temperature, and ligand binding through MD simulations is presented in Chapter 5 by Qing-Chuan Zheng and Wen-Ting Chu. The important insight into protein dynamics studied by NMR spectroscopy with main attention on ^{15}N spin relaxation methods is the subject of Chapter 6 by Oscar Millet and coauthors. Chapter 7 by Maria Barbi and Fabien Paillusson focuses on protein–DNA electrostatics, explaining the fundamental processes of searching and recognition of DNA by some proteins involved in gene expression and regulation. Chapter 8 by Biswa Ranjan Meher and Seema Patel describes MD, mutation effects, and drug resistance of the important HIV-1 protease involved in the control of the HIV life cycle.

REFERENCES

Adcock, S. A., & McCammon, J. A. (2006). Molecular dynamics: Survey of methods for simulating the activity of proteins. *Chemical Reviews*, 6(5), 1589–1615.

Berg, J. M., Tymoczko, J. L., & Stryer, L. (2002). *Biochemistry*. New York: W. H. Freeman and Co.

Fersht, A. R. (1999). *Structure and mechanism in protein science: A guide to enzyme catalysis and protein folding*. New York: W. H. Freeman and Co.

Karplus, M., & Kuriyan, J. (2005). Molecular dynamics and protein function. *Proceedings of the National Academy of Sciences of the United States of America*, 102(19), 6679–6685.

Karplus, M., & McCammon, J. A. (2002). Molecular dynamics simulations of biomolecules. *Nature Structural Biology*, 9, 646–652.

Kruschel, D., & Zagrovic, B. (2009). Conformational averaging in structural biology: Issues, challenges and computational solutions. *Molecular Biosystems*, 5(12), 1606–1616.

CHAPTER ONE

Drug–DNA Intercalation: From Discovery to the Molecular Mechanism

Arnab Mukherjee[1], Wilbee D. Sasikala

Chemistry Department, Indian Institute of Science Education and Research, Pune, India
[1]Corresponding author: e-mail address: arnab.mukherjee@iiserpune.ac.in

Contents

Abstract

The ability of small molecules to perturb the natural structure and dynamics of nucleic acids is intriguing and has potential applications in cancer therapeutics. Intercalation is a special binding mode where the planar aromatic moiety of a small molecule is inserted between a pair of base pairs, causing structural changes in the DNA and leading to its functional arrest. Enormous progress has been made to understand the nature of the intercalation process since its idealistic conception five decades ago. However, the biological functions were detected even earlier. In this review, we focus mainly on the acridine and anthracycline types of drugs and provide a brief overview of the development in the field through various experimental methods that led to our present understanding of the subject. Subsequently, we discuss the molecular mechanism of the

Advances in Protein Chemistry and Structural Biology, Volume 92
ISSN 1876-1623
http://dx.doi.org/10.1016/B978-0-12-411636-8.00001-8
1

intercalation process, free-energy landscapes, and kinetics that was revealed recently through detailed and rigorous computational studies.

1. INTRODUCTION

Small molecules interact with nucleic acids through various covalent and noncovalent interactions (Waring, 1981) and interrupt their natural biological functions. Of the different binding modes, intercalation of small molecules to nucleic acids is a unique noncovalent interaction where a small planar aromatic moiety is inserted between the adjacent base pairs of DNA (Lerman, 1961). Intercalation generally causes stabilization, local unwinding, lengthening, and some other structural changes in the DNA. While the overall B-form of the DNA is maintained, alteration in various biological functions results from intercalation (Lerman, 1961; Li & Crothers, 1969; Waring, 1970). The intercalation process has tremendous therapeutic implication and the intercalators are often used as drugs in cancer treatments, and also to treat microbial and parasitic infections (Martínez & Chacón-García, 2005). Moreover, intercalating agents are also used as a probe to study the structure and dynamics of nucleic acids (Syed et al., 2013; Wilson, 1999).

1.1. History of intercalators

The therapeutic action of Cinchona bark in the treatment of malaria was identified in the early 1630s. In 1820, Pelletier and Caventou isolated quinine from Cinchona bark (Schulemann, 1932). Because quinine was in short supply and its chemical structure was unknown, the German chemist William Henry Perkin came across the first coal tar dye aniline purple during the search for the synthetic form of quinine in 1856 (Meyers, 2007). The synthetic dye industry, based on the acridine ring system, started after this serendipitous discovery of aniline purple. During the 1890s, Paul Ehrlich, the founder of chemotherapy, identified that the uptake of dyes into different cell types of body is different. He used dyes to distinguish between different cells such as mast cells and blood cells. The concept of staining different kinds of cell was then used for staining bacteria and this selective staining of microbes by dyes triggered the use of dyes to kill microbes at reasonably high doses. Ehrlich in 1908 proposed the concept of "magic bullet" where the molecules can selectively kill microorganisms causing different diseases at sufficiently high doses without affecting the host

(Strebhardt & Ullrich, 2008; Witkop, 1999). After the discovery of the first magic bullet salvarsan used for the treatment of syphilis (Witkop, 1999), there was a massive use of heteroaromatic dyes as antibacterial agents in the late nineteenth century (Wainwright, 2001). In 1912, Ehrlich and Benda introduced trypaflavin and acriflavine against trypanosomiasis. Later on, Browning et al. reported proflavine and acriflavine as wound antiseptics (Browning, Gulbransen, & Thornton, 1917; Gulbransen & Browning, 1921; Wainwright, 2001). Following this, several acridine derivatives emerged as antiseptic and antibacterial agents with mutagenic and cytotoxic properties. The discovery of penicillin in 1928 by Alexander Fleming led to the decline of the use of acridine derivatives as antibacterial agents. But mepacrine and pyronardine were used as antimalarial agents in the 1940s (Denny, 2002; Wainwright, 2001). Actinomycin D, the first antibiotic discovered in the 1940s from soil microbe *Streptomyces* species, has anticancer property (Hollstein, 1974). In the 1950s, the first anthracycline antibiotic duanorubicin (daunomycin) was isolated from *Streptomyces* species present in the soil and was found to be active against cancer (Mross, Massing, & Kratz, 2006). This further led to the emergence of several anthracycline antibiotics such as doxorubicin, idarubicin, and pirarubicin (Mross et al., 2006). Even though the acridines and anthracyclines were used as antibacterial and anticancer agents, the mechanism of action of these compounds was unknown till 1950s. Oster's (1951) work on fluorescence quenching of acriflavine by the presence of small amount of nucleic acid suggested that acriflavine interacts with nucleic acids and this interaction was predicted as the reason for the antibacterial activity of acridines. From temperature and viscosity measurement, he found that the nature of fluorescence quenching of acriflavine by nucleic acids is very different from the quenching induced by KI and suggested that acriflavine may fit between the rings of purine and pyrimidine with its quaternary nitrogen in close proximity with the phosphate group. Note that this study (Oster, 1951) appeared 10 years before the intercalation hypothesis of Lerman (1961). In 1955, Heilweil illustrated that acriflavine has different binding affinities toward the different binding site of DNA. Based on the Watson and Crick DNA model, he also suggested that the planar ring of the acriflavine can slip between the parallel planes made by the purine and pyramidine bases. The quaternary amine group of the acriflavine may face the DNA backbone to satisfy the electrostatic interaction without many changes in the DNA configuration (Heilweil & Winkle, 1955). Peacocke and Skerrett used spectrophotometric and equilibrium dialysis approach to study the interaction of proflavine with DNA and found that initially single proflavine interacts strongly with DNA and later aggregate form of proflavine interacts weakly with DNA (Peacocke & Skerrett, 1956). This strong and weak interaction indicated therefore an intercalative and an external binding mode. In 1961, Lerman proposed that the mode of interaction of acridine, proflavine, and acridine orange to DNA is by the insertion between base pairs. He introduced the term *intercalation* for this process for the first time. The increase in viscosity, the decrease in sedimentation coefficient, and the loss of ordering in the long axis of the DNA led Lerman to propose the intercalation model (Lerman, 1961). This review further discusses the biological significance, structural classification, and thereafter experimental and theoretical characterization of both thermodynamic and kinetic aspects and thus furnishes our present understanding of the topic.

1.2. Biological significance of intercalation process

Most of the intercalating agents are used clinically as the antibacterial, antiparasitic, and antitumor agents (Berman & Young, 1981; Denny, 2002; Waring, 1981), as intercalation inhibits DNA replication and transcription process thereby inhibiting the cell division and growth (Lenglet & David-Cordonnier, 2010). Since cancer cells are more prone to growth, intercalation typically affects cancerous cells, although other normal cells also get affected. Here, we briefly discuss various clinical applications of intercalators and the mechanism of their biological action.

1.2.1 Clinical use of intercalators

Intercalators are often used clinically as drugs to various diseases. The uses of intercalators in some of the diseases are mentioned below.

1.2.1.1 Anticancer compounds

Actinomycin D is used as an anticancer antibiotic in the treatment of different types of cancers such as sarcomas (Olweny et al., 1974), Wilms tumor (Malogolowkin et al., 2008), germ cell cancers (Bradof, Hakes, Ochoa, & Golbey, 1982), and melanoma (Giermasz et al., 2002). An anthracycline antibiotic such as daunorubicin is used in the treatment of acute myeloid leukemia (AML) (Brunnberg et al., 2012), neuroblastoma (Samuels, Newton, & Heyn, 1971), and chronic myelogenous leukemia (Kantarjian et al., 1992). Doxorubicin is mainly used in the treatment of Hodgkin's lymphoma (Minuk et al., 2012) and different types of other cancers also. Idarubicin is an anthracycline analog used in the treatment of AML

(Harousseau et al., 1989). Mitoxantrone is an anthracenedione antineoplastic agent used in the treatment of non–Hodgkin's lymphoma and Hodgkin's disease (Silver et al., 1991). Ellipticine, a natural compound isolated from Australian evergreen tree of the Apocynaceae family, is a potent anticancer compound (Canals, Purciolas, Aymami, & Coll, 2005). Amsacrine (m–AMSA and acridinyl anisidide) is an antineoplastic agent which has been used in the treatment of acute lymphoblastic leukemia (Horstmann et al., 2005). Even though proflavine has anticancer property, it is not used clinically because of its nonselective nature and toxicity. So efforts were made to modify proflavine to make it more selective, less toxic, and potent anticancer drug (Baruah & Bierbach, 2003; Bazzicalupi et al., 2008).

1.2.1.2 Antiparasitic compounds

Mepacrine or quinacrine is an acridine derivative used in the treatment of malaria (Handfield-Jones, 1949). Other acridine derivatives also show antimalarial activity (Valdés, 2011). Acridine derivatives such as mepacrine, proflavine, and ethidium bromide are used in the treatment of trypanosomiasis (Figgitt, Denny, Chavalitshewinkoon, Wilairat, & Ralph, 1992).

1.2.1.3 Antimicrobial agents

Proflavine was used as disinfectant and antibacterial agent (Browning, Gulbransen, Kennaway, & Thornton, 1917). Currently, it is used as a topical antiseptic (Wainwright, 2001). Proflavine is also a promising candidate in HIV treatment due to its recently discovered intercalating ability into viral RNA (DeJong, Chang, Gilson, & Marino, 2003). Trapaflavine, ethacridine, and acriflavine were also used as antibacterial agents (Wainwright, 2001). Berberine, an alkaloid obtained from the plant *Berberis*, has got antifungal (Xu et al., 2009), antibacterial (Yu et al., 2005), and antiviral properties.

1.2.2 Biological consequence

Some of the believed mechanisms of the biological function that follows the intercalation process are given below.

1.2.2.1 Inhibition of DNA-dependent enzymes

Generally, intercalators cause enzyme inhibition by blocking the DNA sites competitively so that the enzyme cannot interact with the DNA, thereby affecting the DNA replication and transcription. Anthracycline antibiotics and amsacrine are topoisomerases II inhibitors, whereas ellipticine and its analogs inhibit not only topoisomerases II but also DNA polymerase and

RNA methylase (Carmen Avendano, 2008). Proflavine and actinomycin D are DNA polymerase inhibitors (Hurwitz, Furth, Malamy, & Alexander, 1962). Ethidium bromide, proflavine, and actinomycin D are found to be RNA polymerase inhibitors (Cavalieri & Nemchin, 1964; Waring, 1965a, 1965b).

1.2.2.2 Frame-shift mutations
The deletion or insertion of one or more nucleotide bases in DNA is called frame-shift mutations (Richards & Hawley, 2010). The codons are composed of three nucleo-bases and the mutations lead to improper codon reading during the mRNA synthesis. Since the mRNA contains the wrong codon, it codes for a wrong amino acid and leads to a mutated protein (Cummings, 2011). Ethidium bromide, proflavine, and acridine orange are known to cause frame-shift mutations (Clark & Pazdernik, 2012; Müller, 2008).

1.2.2.3 DNA damage
Doxorubicin and its analogs (Capranico, Soranzo, & Zunino, 1986), actinomycin D, and ellipticine (Ross & Bradley, 1981) cause single-strand breakage of the DNA along with DNA intercalation and further inhibits DNA replication and transcription.

1.3. Classification
In this part, we look into structural details of the intercalators that lead to following three categories: mono–intercalators, bis–intercalators, and threading intercalators. Mono–intercalators are small organic molecules with one planar moiety that intercalate between base pairs of DNA from either the major or the minor groove side. Daunomycin (Quigley et al., 1980) and proflavine (Neidle & Jones, 1975) are examples of mono–intercalators. Bis–intercalators are formed from two mono–intercalators joined together through a covalent linkage. The linkage between the two mono–intercalators should be optimal in length to allow the two mono–intercalators to insert and form stacking interactions with the base pairs. Bis-daunomycin (Hu et al., 1997), echinomycin (Pfoh, Cuesta-Seijo, & Sheldrick, 2009), triostin (Wang et al., 1984), ditercalinium (Gao et al., 1991), TOTO (Spielmann, Wemmer, & Jacobsen, 1995), YOYO (Günther, Mertig, & Seidel, 2010), etc., belong to this category. Here, the linker can remain in the minor groove or the major groove. Third category is the threading intercalators where side chains are present at both sides of the intercalating moiety.

Therefore, during intercalation process, one of the side chains must pass through the base pairs and be present on one groove of DNA and second chain on the other groove. Nogalamycin, napthalene diimide (Torigoe et al., 2002), and bisantrene (Wilson, Ratmeyer, Zhao, Strekowski, & Boykin, 1993) fall in this category. The chemical structures of different inter-calators are shown in Fig. 1.1.

However, classification, such as parallel and perpendicular intercalators, is also made based on the arrangement of the long axis of the planar inter-calating moiety with respect to the base pairs. For the parallel intercalators (such as acridine orange and proflavine), there is no bulky substituent along the long axis of intercalators, which is parallel to the intercalating base pairs as shown in Fig. 1.2A. Here, unwinding of DNA happens at the intercalation

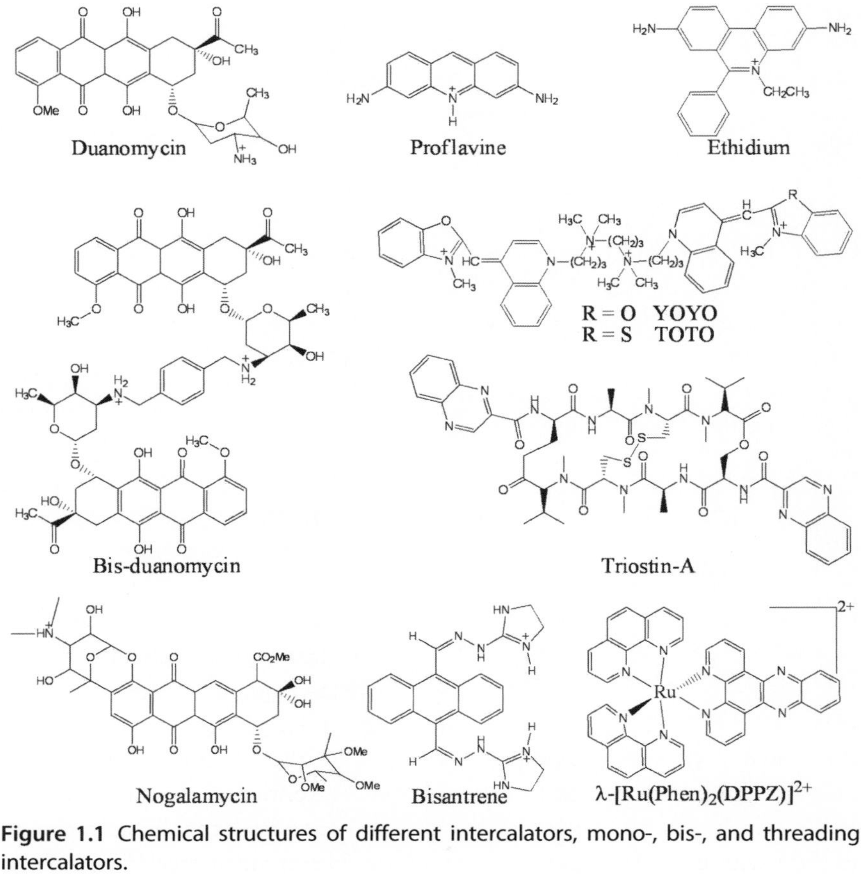

Figure 1.1 Chemical structures of different intercalators, mono-, bis-, and threading intercalators.

A   B

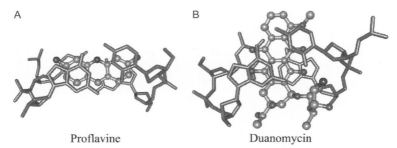

Proflavine   Duanomycin

Figure 1.2 (A) Parallel and (B) perpendicular arrangement of the ring system of inter-calators with respect to the long axis of the base pairs (blue). The view is along the DNA axis. The figure is made in UCSF Chimera (Pettersen et al., 2004). (For interpretation of the references to color in this figure legend, the reader is referred to the web version of this article.)

site (Williams, Egli, Gao, & Rich, 1992). Unwinding increases the stacking interaction between the long axis of intercalator and the flanking base pairs (Alden & Arnott, 1975). In most of the parallel intercalators, the flanking bases are flat because of the stronger stacking interaction with the inter-calators (Comba, 2010; Williams et al., 1992). Perpendicular intercalators (daunomycin, nogalamycine, etc.) contain bulky substituents on one or both sides of their long axis and orient perpendicular to the average of flanking base pairs as shown in Fig. 1.2B. No unwinding takes place at the interca-lation site because the extent of stacking interaction possible with the long axis of intercalator is less. The flanking base pairs curve around the perpen-dicular intercalator to maximize the stacking interaction. Here, the first base pair shows negative buckling and the second base pair shows positive buck-ling (Comba, 2010; Williams et al., 1992).

2. STRUCTURE AND DYNAMICS OF INTERCALATION

2.1. Structural characterization

Due to the special binding mode and specific properties of the intercalators, intercalation is possible to be characterized by various biophysical methods, some of which are briefly mentioned below.

2.1.1 Viscosity enhancement

Since intercalation increases the base pair separation by almost double (~3.4–6.8 Å) (Berman et al., 2000; Neidle & Jones, 1975), overall length of the DNA increases making the DNA rigid (Lerman, 1961). This reflects in the increased viscosity of the solvent as was first shown in the case of acridine, proflavine, and ethidium bromide (Cohen & Eisenberg, 1969; Lerman, 1961). Viscometric study by Leith (1963) also showed that acridine orange and acriflavine intercalation increases the viscosity and the increase is also seen in the case of anthracycline antibiotics (Kersten, Kersten, Szybalski, & Fiandt, 1966). However, the increase in viscosity was not significant for another intercalating drug actinomycin D, where with increase in drug concentration viscosity decreased initially and then increased at very high drug concentrations (Kersten et al., 1966). Therefore, viscosity, although an indirect proof of intercalation, is not a definite proof.

2.1.2 Sedimentation coefficient
Local unwinding of the DNA during the intercalation leads to the release of the super coil in circular DNA and makes it like an open circle. The compact super coil sediments more rapidly than the open circle form. Therefore, sedimentation coefficient was also used as a measure of intercalation for several drugs where it was shown that the intercalating drugs decrease sedimentation coefficient while the nonintercalating ones do not alter it (Crawford & Waring, 1967; Waring, 1970). However, certain non-intercalating molecules such as irehdiamine A (Saucier, 1977) also remove super coil in DNA. Therefore, sedimentation coefficient is also not a definite proof of intercalation.

2.1.3 Autocardiograph
Radioactive labeling (^{3}H (tritium) or ^{125}I) of the drug is detected in the autocardiograph and thus helps to detect the change in length of DNA due to intercalation. This was used to prove that aminoacridine intercalates into DNA (Neville & Davies, 1966).

2.1.4 Dynamic light scattering
Dynamic light scattering (DLS) provides a measure of the size of the particles from the scattered light in a solution. DLS can detect the change in conformation of the DNA due to intercalation because of the change in the size distribution caused by it (Chirico, Lunelli, & Baldini, 1990; Newman, 1984). The DLS study on the intercalation of ethidium bromide to calf thymus DNA illustrated two binding mode where one is weaker and other one is stronger. Stronger binding mode was considered to be intercalation and the weaker one as the outside-bound state (Nordmeier, 1992).

2.1.5 Increase in melting temperature

Increase in temperature transforms the double-strand DNA to single strand by disrupting the hydrogen bonds between base pairs (Cooper, 1997; Xiao, Lin, & Tian, 1994). This is known as melting. Melting temperature (T_m) of DNA is defined as the temperature where half of the DNA population is melted. T_m depends on the length and sequence of the DNA (Owczarzy et al., 1997). Intercalation leads to lengthening of DNA as well as it increases the rigidity (Fritzsche, Triebel, Chaires, Dattagupta, & Crothers, 1982; Lerman, 1961), reflecting in the increased T_m. Increase of T_m was found for actinomycin binding to DNA. Since actinomycin is known to bind to GC, the effect was only seen in case of sequences containing GC but not in AT sequence (Waring, 1968). Denaturation study indicates that forces stabilizing the DNA–actinomycin complex are stronger than the forces which stabilize the DNA structure (Davidson & Cohn, 1964). Chaires studied the thermal stability of DNA on intercalation of daunomycin to DNA (Chaires, Dattagupta, & Crothers, 1982). Melting studies performed on poly(A) · poly(U) and poly(I) · poly(C) helix at increasing concentrations of ethidium bromide indicated that the double helical structure is strongly stabilized by ethidium bromide. The magnitude of stabilization is less for poly(I) · poly(C) as compared to poly(A) · poly(U) as shown in Fig. 1.3A. But the selectivity of ethidium toward the DNA helix is showed by the interaction of ethidium bromide with triplex helical polymer poly(A) · 2poly(U). Here, ethidium

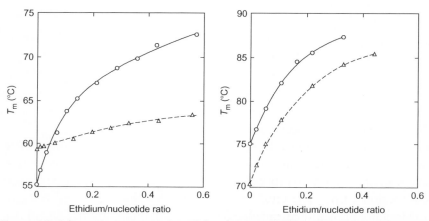

Figure 1.3 Schematic representation of the change in melting temperature of double helical DNA by ethidium bromide. (A) Poly(A) · poly(U) [○] and poly(I) · poly(C) helix [△], (B) *Escherichia coli* DNA [○] and calf thymus DNA [△]. *Reproduced with permission from Waring (1974) © the Biochemical Society.*

bromide stabilized the double helical poly(A)·poly(U) and formed coiled poly(U). So it was suggested that this differentiation between double helix and triplex happens because of the intercalation process (Waring, 1974). Figure 1.3 shows the elevation of melting temperature of double helical polymers in the presence of ethidium bromide.

2.1.6 X-ray technique

Unlike the indirect methods, fiber diffraction and X-ray crystallography probe into the molecular detail of structures. While fiber diffraction can detect various conformational aspects such as helical pitch in the DNA and winding angle, high-resolution X-ray crystallography brings out atomistic information. Double helical structure of DNA was discovered by fiber diffraction technique (Kasai, 2005). Fiber diffraction study of DNA–proflavine complex showed that the interaction of proflavine modifies the structure of DNA but maintains the rod-like nature of it with a decrease in the radius of gyration as the proflavine concentration is increased. So this gave a qualitative proof for the intercalation hypothesis proposed by Lerman in 1961 (Luzzati, Masson, & Lerman, 1961). The fiber diffraction studies of DNA with proflavine and acridine orange showed that the diffraction pattern is disordered but close to B-form pattern of DNA with an increase in the layer-line spacing of first three layer lines. The study also explained that all the bound dyes are not intercalated and that some are bound outside also (Neville & Davies, 1966).

X-ray crystallography is a method in which the atoms in the crystal diffract the beam of X-ray to specific directions. A 3D picture of density of electrons can be produced by measuring the angles and intensities of the diffracted beams which could further be used for calculating the mean positions of the atoms in the crystal as well as various other informations (Ladd & Palmer, 2003). Single crystal X-ray techniques were focused on determining the helix unwinding, sequence specificity, and neighbor exclusion of DNA intercalation process (Berman & Young, 1981). Jain and Sobell cocrystallized actinomycin D with deoxyguanosine as 1:2 complex (Jain & Sobell, 1972). The structure showed that the phenoxazone ring is in between the two deoxyguanosine which stack on alternate sides of the phenoxazone ring. The crystal structure of ethidium:5-iodouridylyl(3'-5') adenosine complex confirmed the intercalative binding mode where the interbase distance increased to 6.8 Å and led to the DNA unwinding by 6° (Tsai, Jain, & Sobell, 1975). The crystal structure of proflavine-d(CpG)$_2$ illustrated structural changes to the backbone of the dinucleotide (Neidle et al.,

1977). X-ray crystal structure of daunomycin-d(CpGpTpApCpG)$_2$ gave direct proof for daunomycin intercalation where the aglycone part of dauno-mycin intercalates between the CG base pair and the daunosamine ring stays in the minor groove covering the third base pair also. Crystal structure gave the evidence for the C–G specificity and showed that daunomycin intercalation is perpendicular, which is different from parallel intercalation of proflavine (Neidle et al., 1977; Neidle, Berman, & Shieh, 1980). This gave the evidence for other experimental observations that the selectivity of daunomycin toward the bases should be explained in terms of three base pair rather than two base pair (Chaires et al., 1982; Quigley et al., 1980). Later on, crystal structure of several other intercalators with DNA was solved and the 3D representations of some of them are shown in Fig. 1.4.

2.1.7 Photophysical study

Most of the intercalators have chromophores, which interact with the DNA base pairs by intercalation, where it changes the photophysical property of the drug. Figure 1.5 shows the change in absorbance of ethidium bromide in the presence of the DNA (Waring, 1965a). Figure 1.5A shows that on inter-action with DNA, the absorption spectrum of ethidium bromide is shifted to the higher wavelength. For all nucleic acid sequences in the study, the trend is same. Figure 1.5B shows that, by increasing the nucleic acid concentra-tion, the absorbance of ethidium bromide decreases gradually along with the shift in the absorption spectrum and the shift is maximum at the highest concentration.

Fluorescence property changes more in the presence of DNA. In certain molecules, fluorescence gets enhanced (ethidium bromide) due to the reduction in the rate of excited state proton transfer to the solvent molecules (Olmsted & Kearns, 1977), while in certain other cases, it gets quenched due to charge transfer (Thomes, Weill, & Daune, 1969). This property of drug led to understand various thermodynamics and kinetic aspects of the inter-calation process discussed next.

2.2. Thermodynamic characterization

Thermodynamic parameters of DNA intercalation process were studied extensively using different methods to understand the binding mode, bind-ing affinity, binding specificity, contribution of enthalpy and entropy of the binding process, etc. Here, we describe some of the methods and their out-comes briefly.

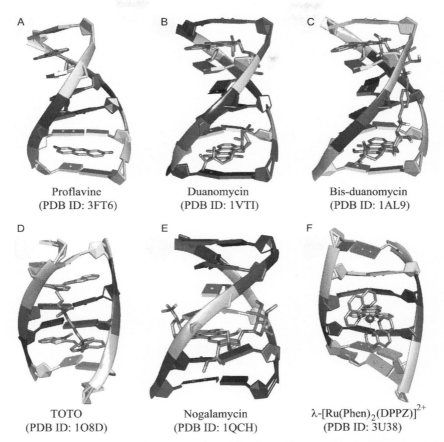

Figure 1.4 3D representations of various intercalator–DNA complexes. PDB IDs and name of the intercalators are mentioned. (A and B) Mono-intercalator, (C and D) bis-intercalators, and (E and F) threading and metallo-intercalator. (For color version of this figure, the reader is referred to the online version of this chapter.)

2.2.1 Methods

2.2.1.1 Equilibrium dialysis

Equilibrium dialysis is used to measure the amount of free and bound ligand to a macromolecule. The basic principle is that if the ligand is bound to a macromolecule which is big enough to prevent its escape though the semipermeable dialysis membrane, then the free ligand in the solution can pass though the semipermeable membrane to the dialyzate. After giving sufficient time for the equilibration to happen, the amount of free ligand present in the dialyzate can be found out by different spectroscopic techniques.

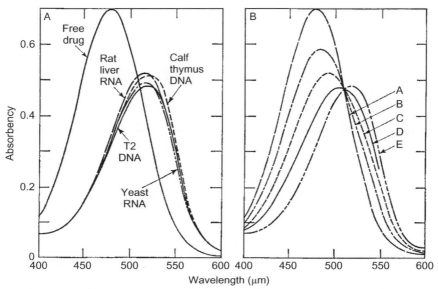

Figure 1.5 The schematic representation of the absorption spectrum of ethidium bromide in the presence of various types of nucleic acids. (A) The absorbancy of ethidium bromide decreases on addition of nucleic acid such as calf thymus DNA, T2 DNA, rat liver DNA, and yeast DNA, indicating the ethidium bromide–DNA interaction. (B) The absorbancy of ethidium bromide decreasing with increasing the T2 DNA concentration $A = 0$; $B = 1.5 \times 10^{-4}$ M; $C = 3 \times 10^{-4}$ M; $D = 5 \times 10^{-4}$ M; and $E = 1.2 \times 10^{-3}$ M. *Reprinted from publication (Waring, 1965a). Copyright (1965), with permission from Elsevier.*

The amount of bound ligand can be quantified by spectroscopic methods after cleaving the ligand–macromolecule complex present inside the dialysis membrane (Chaires et al., 1982). The percentage of ligand bound to the macromolecule is found by the following equation (George, McBurney, & Ward, 1984):

$$\%\,\text{bound} = (C_t - C_f)/C_t \times 100$$

where C_t is the concentration of the ligand in the macromolecule solution before the dialysis and C_f is the concentration of the ligand in the dialyzate after the equilibration dialysis.

The solutions obtained after the equilibrium dialysis can be used for calculating other parameters such as bound fraction of ligand and degree of binding. By varying the initial concentrations of the ligand, we can find out other different parameters such as association constant, number of binding sites, and binding capacity (Ugwu, Alcala, Bhardwaj, & Blanchard, 1996).

2.2.1.2 Spectroscopic (absorbance and fluorescence) titration analysis

Knowledge of the absorbance or fluorescence of the ligand in the presence and absence of macromolecule helps in the construction of the equilibrium-binding isotherms by changing the concentration of either the ligand or the macromolecule (Haq, 2002).

2.2.1.3 Surface plasmon resonance

This method involves the immobilization of one of the reactant such as DNA on to a sensor chip while the other reacting species is passed over the chip. If any kind of reaction takes place, refractive index of the chip will change and will increase in surface plasmon resonance (SPR) signal. After sufficient time is given for the association to happen, buffer solution is passed through the chip resulting in ligand dissociation and subsequently decrease in SPR signal. Dissociation and association rate constants thus provide a measure of equilibrium constant as shown below:

$$K_D = \frac{k_d}{k_a}$$

where k_d is the dissociation constant and k_a is the association constant (Davis & David Wilson, 2001; Haq, 2002).

2.2.2 Isothermal titration calorimetry

Isothermal titration calorimetry (ITC) is probably the most direct and widely used method to probe thermodynamic properties of binding processes. ITC equipment contains a reference cell (water or buffer solution) and sample cell containing the macromolecule solution. Initially, both the cells have the same temperature. But when the ligand is titrated to the sample cell in small portions, if the reaction is exothermic, then the heat flow to the sample cell is decreased to maintain the constant temperature as compared to the reference cell and vice versa for an endothermic reaction. It measures the binding stoi-chiometry (n) of the interaction between the macromolecule and the ligand, binding affinity (K_a), and enthalpy changes (ΔH). The Gibbs free-energy changes (ΔG) and the change in entropy (ΔS) can be calculated using the following relationship:

$$\Delta G = -RT\ln K_a = \Delta H - T\Delta S$$

where R is the gas constant and T is the absolute temperature (Caldwell & Yan, 2004).

2.2.3 Outcome

Extensive thermodynamic characterization using methods discussed earlier led to the understanding of the binding specificity and stability with different DNA sequences as well as the thermodynamic interplay of energy and entropy. Binding affinities of acriflavine to native DNA and to different synthetic polynucleotides have been studied using equilibrium dialysis and fluorescence spectroscopy to find out the amount of bound and free acriflavine. Acriflavine showed better binding affinity toward alternating polymer (RYR, R = purine, Y = pyrimidine) as compared to nonalternating ones (RR, YY) showing the heterogeneity in binding based on the DNA base composition. Binding affinities increase with the increase in the fraction of alternating nearest neighbor (NN) sequences, $\gamma = (YR + RY)/$(total NN base sequences). Among the alternating NN, $Y(3'-5')R$ has more binding affinity toward acriflavine (Baldini, Doglia, Dolci, & Sassi, 1981). Later, the analysis of various thermodynamics quantities such as $\Delta G°$, $\Delta H°$, $\Delta S°$, and $\Delta C_p°$ (Fig. 1.6) of DNA–quinacrine complex was performed at different temperatures to understand the role of binding affinity heterogeneity on the thermodynamical parameters. $\Delta G°$ has only small change in increasing the temperature from 20° to 70°. $\Delta H°$ shows a large change from ∼3 to ∼7 kcal/mol with increase in temperature and $\Delta S°$ shows a similar trend. Although the trend is same for $\Delta H°$ and $\Delta S°$ for different types of DNA, the difference in the values among the various DNAs shows the binding heterogeneity (Baldini, Castoldi, Lucchin, & Zedda, 1982).

Recently, Hossain showed that the binding of quinacrine to poly(dA)· poly(dT) is entropy driven whereas binding to other sequences such as poly(dG–dC)·poly(dG–dC), poly(dG)·poly(dC), and poly(dA–dT)·poly (dA–dT) are both enthalpically and entropically driven (Hossain & Suresh Kumar, 2009). The stability and structural studies of acridine orange indicate that major intercalative binding happens between G–C base pairs and A–T base pairs (Nafisi, Saboury, Keramat, Neault, & Tajmir-Riahi, 2007). The fluorescence-quenching experiments pointed out that intercalation of 9-amino acridine to poly[d(A–T)$_2$] and poly[d(G–C)$_2$] is energetically favorable but intercalation to poly[d(G–C)$_2$] is entropically unfavorable. For proflavine, however, intercalation to poly[d(G–C)$_2$] showed higher equilibrium constant with unfavorable entropy change compensated by favorable enthalpy change (Kwon et al., 2013).

Competition dialysis experiments demonstrated that daunomycin also shows sequence specificity toward G–C base pairs attaining comparatively larger value for drug bound per base pair with an increase in the G–C content

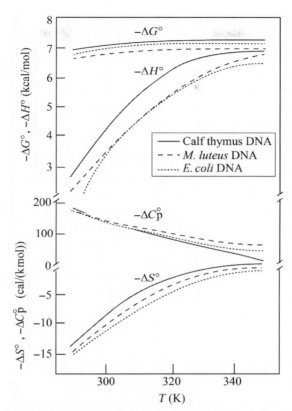

Figure 1.6 Schematic representation of thermodynamic parameters of DNA–quinacrine complex at different temperatures (Baldini et al., 1982). *Adapted from Figure 4 of Baldini et al. (1982) with kind permission of Società Italiana di Fisica.*

of DNA. This was further confirmed by CsCl density gradients (Chaires et al., 1982). Later, a more detailed study on the specificity of daunomycin binding toward different polynucleotides illustrated that the binding constants decrease in the order poly[d(A–T)]·poly[d(A–T)] > poly[d(G–C)]·poly[d(G–C)] > poly(dG)·poly(dC) > poly(dA)·poly(dT) (Chaires, 1983). The specificity of daunomycin toward the poly[d(A–T)]·poly[d(A–T)] as compared to the poly[d(G–C)]·poly[d(G–C)] was explained by Marky et al. using differential scanning calorimetry, which showed that the calculated average base–stacking enthalpy of G–C/C–G or C–G/G–C (11.9 kcal/stack) was more as compared to A–T/T–A or T–A/A–T (7.1 kcal/stack) (Marky & Breslauer, 1982). The specificity toward regular alternate polymer poly[d(A–T)]·poly[d(A–T)] was also verified by Pullman and coworkers using *ab initio* quantum chemical calculations (Chen, Gresh, & Pullman, 1985). Breslauer et al. showed that the binding affinities ($\Delta G°$) for the intercalators such as ethidium bromide, daunomycin, netropsin, and distamycin A with poly[d(A–T)]·poly [d(A–T)] and poly(dA)·poly(dT) are almost similar. But binding to alternate copolymer was found to be enthalpy driven and binding to homopolymer was found to be entropy driven. So, the comparable $\Delta G°$ indicates that there is entropy–enthalpy compensation at 25 °C (Breslauer et al., 1987). A more detailed study on the preferential binding of daunomycin to 5′ATCG and 5′ATGC sequences was performed by Chaires et al. in 1990 by footprinting titration experiments (Chaires, Herrera, & Waring, 1990) which corroborated with the theoretical studies by Pullman and coworkers where he showed that daunomycin has specificity toward mixed polynucleotides d(CGATCG)$_2$ and d(CGTACG)$_2$ (Chen et al., 1985). This is also proved by the X-ray crystal structure of daunomycin–d(CpGpTpApCpG) (Quigley et al., 1980).

Thermodynamic characterization of DNA–intercalator complex was also performed to understand the NN exclusion principle put forward by Crothers which suggested that intercalation is anticooperative at adjacent sites (Li & Crothers, 1969). To understand this in the case of daunomycin intercalation, influence of DNA concentration on daunomycin binding was studied by the dialysis of different DNA solutions against same daunomycin concentration. The daunomycin bound per base pair was found to be independent of DNA concentration, which indicated that there is no unusual cooperativity in drug binding. The following Scatchard plot (Fig. 1.7) illustrates that neighbor-exclusion parameter, $n = 3.5$ bp, and intrinsic binding constant, $K_i = 7.0 \times 10^5$ M^{-1}, gave the best fit to the plot.

The possible explanation for curvature of the Scatchard plot (Fig. 1.7) was given by the neighbor-exclusion model where the binding of one molecule influences the binding of the second molecule. The binding of the second molecule to the adjacent binding sites is excluded by the physical blockade or steric alterations of DNA. So the binding of the second daunomycin takes place 3.5 bp away from the first binding site (Chaires et al., 1982). Similarly, the thermodynamic studies on intercalation of methylene blue and quinacrine in the presence of poly(dG–dC)·poly(dG–dC), poly(dG)·poly(dC), poly(dA–dT)·poly(dA–dT), and poly(dA)·poly(dT) showed that all of them follow neighbor-exclusion principle (Hossain & Suresh Kumar, 2009). Recently, the study on the binding of [Ru(PDTA-H2)(phen)]Cl with calf thymus DNA performed using different spectroscopic methods also showed that 3 bp are involved in the binding. This also

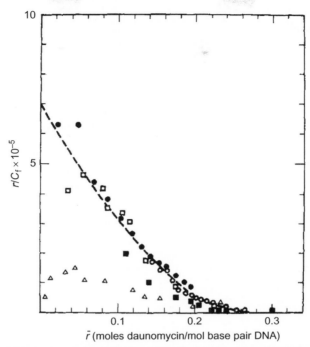

Figure 1.7 Schematic representation of Scatchard plot of DNA–daunomycin equilibrium-binding isotherm. (●) Fluorescence titration, (○) absorbance titration, (■) equilibrium dialysis with absorbance, (□) equilibrium dialysis with fluorescence, and (△) solvent partition method. The line shows the best fit to the plot with intrinsic binding constant, $K_i = 7.0 \times 10^5 \text{ M}^{-1}$, and neighbor-exclusion parameter, $n = 3.5$ bp. *Adapted with permission from Figure 3 of Chaires et al. (1982). Copyright (1982) American Chemical Society.*

indicates an exclusion character of association (Grueso, López-Pérez, Castellano, & Prado-Gotor, 2012).

Thermodynamic characterization of different analogs of the intercalating agent ethidium was performed to understand the influence of amine substituents on the intercalation process. The removal of the amine group decreased the binding affinity as well as it shifted the intercalation process from enthalpy to entropy driven (Garbett, Hammond, & Graves, 2004). The hydration changes during the intercalation of different intercalators ethidium, propidium, proflavine, daunomycin, and 7-amino actinomycin D were done using osmotic stress method. Other than ethidium–DNA complex, all other drug–DNA complex showed uptake of water during

Table 1.1 The thermodynamic parameters of intercalation process of different sequences of DNA by daunomycin, ethidium bromide, 9-amino acridine, and proflavine

Molecule	ΔG (kcal/mol)			ΔH (kcal/mol)			ΔS (cal/Kmol)		
	a	b	c	a	b	c	a	b	c
Ethidium bromide	−9.1	−7.2	−6.9	−10.0	−1.2	−8.7	−3.0	18.0	6.0
Daunomycin	−8.0	−7.9	−9.4	−12.8	−9.1	−11.0	−16.2	−4.1	−5.4
9-Amino acridine	−6.2	−6.3	−6.1	−5.1	−6.6	−4.4	3.7	−1.2	5.8
Proflavine	−6.1	−6.8	–	−5.4	−8.9	–	2.3	−7.3	–
Quinacrine	−9.1	−8.2	−8.7	−5.6	−6.0	−6.0	11.9	7.3	9.9
Methylene blue	−8.2	−8.0	−7.8	−2.4	−2.3	−2.1	19.7	19.5	19.5

Ethidium bromide: a=poly[d(A–T)]·poly[d(A–T)] (Breslauer et al., 1987), b=poly(dA)·poly(dT) (Breslauer et al., 1987), c=calf thymus DNA (Garbett et al., 2004); daunomycin: a=calf thymus DNA (Chaires et al., 1982), b=[poly(dA–dT)]₂ (Herrera & Chaires, 1989), c= poly(dA–dC)·poly (dG–dT) (Xodo, Manzini, Ruggiero, & Quadrifoglio, 1988); 9-amino acridine: a=calf thymus DNA (Kwon et al., 2013), b=poly[d(G–C)₂] (Kwon et al., 2013), c= poly[d(A–T)₂] (Kwon et al., 2013); proflavine: a=calf thymus DNA (Kwon et al., 2013), b= poly[d(G–C)₂] (Kwon et al., 2013); quinacrine and methylene blue: a=poly(dG–dC)·poly(dG–dC) (Hossain & Suresh Kumar, 2009), b=poly(dG)· poly(dC) (Hossain & Suresh Kumar, 2009), c=poly(dA–dT)·poly(dA–dT) (Hossain & Suresh Kumar, 2009).

the complex formation (Qu & Chaires, 2001). NMR studies on tilorone shows that it binds to the DNA by intercalation and it has more preference toward A–T base pairs for intercalation. The isothermal titration calorimetric experiment showed that the intercalation of tilorone is an enthalpy-driven process (Nishimura et al., 2007). Recent spectroscopic studies showed the thermodynamic parameters in the intercalation process of berberine have both van der Waals and electrostatic contributions (Li, Hu, Wang, Yu, & Yue, 2012). Table 1.1 shows the thermodynamic parameters of intercalation of different intercalators with different DNA sequences. The free-energy stability of different intercalators is within ∼10 kcal/mol.

2.3. Kinetic characterization

While thermodynamics studies provide the stability at the two end points of a reaction profile, kinetic studies point toward the *average* mechanism of the process, which could be extremely complex for intercalation. Figure 1.8 shows the different possible pathways surmised by Macgregor, Clegg, and Jovin (1987). Some pathways would lead to opening of the base pair, followed by intercalation (steps: **c** and **a**). Other pathways are either diffusion-limited association to an "outside" bound state (step: **d**) followed

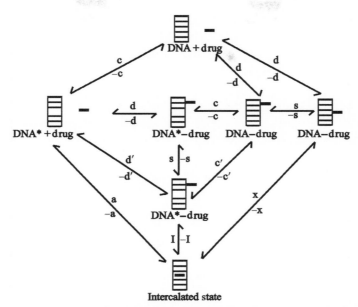

Figure 1.8 Schematic diagram of the different possible kinetic pathways of intercalation surmised by Macgregor et al. (1987). *Adapted with permission from Figure 1 of Macgregor et al. (1987). Copyright (1987) American Chemical Society.*

by synchronized (step: **x**) or drug-assisted (step: **c′**) opening of the base pair and intercalation. However, there is high probability that the drug binds in the wrong place of the DNA. In that case, it has to diffuse along the DNA to search for a proper intercalation site (steps: **d** followed by **s**).

Even though it is possible to devise chemical kinetic equation for each of the individual pathways, it will be really difficult to decouple contributions of each elementary step from the kinetic experiments. Therefore, there were efforts to combine several different methods and different competing kinetic schemes to explain and understand the possible mechanistic pathways of intercalation process, which also depends on the nature of the intercalators. Starting from the simple scheme of one-step reaction proposed by Macgregor (Macgregor, Clegg, & Jovin, 1985) in case of ethidium bromide, a complex multistep scheme was proposed for daunomycin by Rizzo, Sacchi, and Menozzi (1989). For metal intercalators, the kinetics is even more complex (Biver, Secco, & Venturini, 2008).

The most important part of the kinetic studies is the ability of the intercalators to be able to fluoresce and change the fluorescence intensity when intercalated. This gives a handle on the change of overall fluorescence intensity as the time progresses giving rise to time dependence of intensity, which is then fitted to different kinetic models. Experimental techniques to study intercalation kinetics such as temperature jump method and stopped-flow are mentioned below. The methods rely on the fact that the non-equilibrium state is created much faster than the timescale of intercalation and therefore negligible. Observation of kinetics is also limited by the experimental time resolution. Here we mention briefly the different methods used to study the kinetics of intercalation process, followed by the results and interpretations.

2.3.1 Methods
2.3.1.1 Temperature jump technique
It is used to understand the chemical kinetics of a reaction where the solution containing the complex under study is heated by pulsed laser emitting in the near infrared. The temperature of the solution is increased by a small amount in microseconds allowing to study the shift of equilibrium of reactions in milliseconds by using either absorption or fluorescence spectroscopy (Purich & Allison, 1999). The enthalpy change $\Delta H°$ of reaction is given by Van't Hoff equation (Atkins, 1997),

$$\Delta H° = RT^2 \mathrm{d}\ln K/\mathrm{d}T$$

where K is the equilibrium constant, R is the universal gas constant, and $\mathrm{d}T$ is the change in temperature.

2.3.1.2 Pressure jump technique
It involves a rapid change in pressure of the system under study and observes the return to the steady state. The shift in equilibrium of reactions is studied using absorption or fluorescence spectroscopy (Gruenewald & Knoche, 1978; Schiewek, Krumova, Hempel, & Blume, 2007). Here, the volume change is given by the following equation:

$$\Delta V° = -RT\left(\frac{\partial \ln K}{\partial P}\right)_T$$

where R is the universal gas constant, T is the temperature, P is the pressure, and K is the equilibrium constant.

2.3.1.3 Stopped-flow technique
It is one of the most widely used methods because of commercial availability of the setup at low cost. Reactants A and B are flowed into a mixing

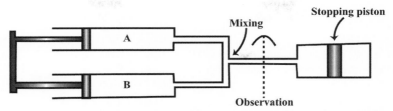

Figure 1.9 Schematic diagram of stopped-flow equipment. *Redrawn from Figure 4.11 of Connors (1990).*

chamber, and then the mixture goes to observation tube as shown in Fig. 1.9. After a few milliseconds, the flow is stopped and the observation is made as the change in fluorescence, absorbance, circular dichroism, etc. The dead-time is usually 3–5 ms (Connors, 1990; Robinson, 1975).

2.3.1.4 Dissociation using sodium dodecyl sulphate micelle

It is a detergent sequestration technique first described by Muller and Crothers (1968), where surfactant sodium dodecyl sulphate micelle acts as a hydrophobic sink for the dissociated drug from a complex. The method drives the equilibrium from the complex (DNA–drug) to the dissociated cationic drug by the absorption of the free drug into the core of the anionic micelle quantitatively. Since the micelle is negatively charged, it does not interact with the negatively charged DNA as well as do not interfere with the DNA bound drug or while it leaves the DNA. Here, the rate-limiting step is the one where the drug leaves the DNA and the absorption of the drug by the micelle is a fast process as it is a diffusion controlled process (Westerlund, Wilhelmsson, Nordén, & Lincoln, 2003).

2.3.2 Outcome

Detailed kinetic studies on the intercalation of DNA by actinomycin (Muller & Crothers, 1968) and proflavine (Li & Crothers, 1969) were performed by Crothers for the first time using a variety of association and dissociation methods. While the kinetics for actinomycin was too complex to understand, for proflavine, he proposed a simple two-step reaction kinetics (Scheme 1.1) that was obtained by fitting the two timescales to the time dependence of the fluorescence intensity.

The first step was argued to be the formation of an outside-bound state $(ID)_{out}$ and fall in the timescale of a diffusion-limited process. Second step, formation of $(ID)_{in}$, is a slow activated process, which was taken to be the intercalation process from the outside-bound state. There was another very

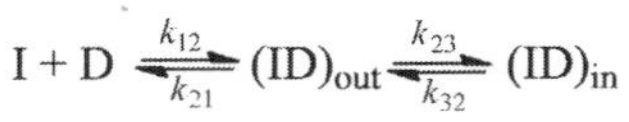

$$I + D \xrightleftharpoons[k_{21}]{k_{12}} (ID)_{out} \xrightleftharpoons[k_{32}]{k_{23}} (ID)_{in}$$

Scheme 1.1 Two step reaction kinetics (series) for intercalation of proflavine to DNA. Note: in this scheme and all the subsequent kinetic schemes, I corresponds to intercalators, D denotes DNA, and ID is the intercalator–DNA complex.

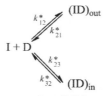

Scheme 1.2 Two step reaction kinetics (parallel) for intercalation of proflavine to DNA.

fast timescale outside the range of experimental detection. Two timescales obtained from bi-exponential fit of the time-dependent fluorescence intensity was used to find the corresponding rate constants shown by Eqs. (1.1) and (1.2):

$$\frac{1}{\tau_1} = k_{12}(C_D + C_F) + k_{21} \tag{1.1}$$

$$\frac{1}{\tau_2} = k_{32} + \frac{k_{23}(C_D + C_F)}{1/K_{12} + (C_D + C_F)'} \tag{1.2}$$

However, an alternate mechanism was also suggested as shown in Scheme 1.2.

Timescales associated with this mechanism takes the following forms shown in Eqs. (1.3) and (1.4):

$$\frac{1}{\tau_1} = k_{12}(C_D + C_F) + k_{21} \tag{1.3}$$

$$\frac{1}{\tau_2} = k_{32} + \frac{(C_D + C_F)k_{23}/k_{12}}{1/K_{12} + (C_D + C_F)} \tag{1.4}$$

The main difference between the above two reaction mechanisms is that the slow timescale τ_2 depends on the equilibrium constant between the outside-bound and free state (k_{12}). If the outside-bound state is stable, it will affect the timescale of the subsequent intercalation (τ_2). Experimental measurements, however, were not able to differentiate these two measurements (Li & Crothers, 1969). Crothers argued for the series mechanism (Scheme 1.1) over the parallel one (Scheme 1.2) because intercalation which requires structural changes could not happen directly as indicated in

Scheme 1.2. Note that proflavine being flat has the possibility to intercalate from either the major or the minor groove and the parallel mechanism is favored, that is, intercalation from major and minor grooves (Biver et al., 2008). We see later that the parallel mechanism matches the recent findings in the sense that the stability of outside-bound state affects the intercalation rate constant. However, there are more complexities involved in the process even for a simple drug like proflavine.

For the proflavine intercalation, Crothers et al. (Li & Crothers, 1969) proposed two limiting mechanisms for the intercalation process to happen (Fig. 1.10): (a) The drug waits outside the DNA which continues to fluctuate. At a certain moment, the cavity size is large enough for the drug to slip in. (b) The drug creates the cavity in the DNA and intercalates imparting some kind of induced-fit mechanism. From the indirect evidence from the experiments, he supported the second mechanism. However, at that point, it was not possible to address this question in detail in the molecular level. We will return to this discussion in the intercalation mechanism section (3.4) in the light of recent mechanistic studies of proflavine intercalation using computational methods.

For certain drugs, however, even a simpler kinetic scheme (Scheme 1.3) explains the observation. McGregor proposed this simple model for kinetics of ethidium bromide intercalation (Macgregor et al., 1985).

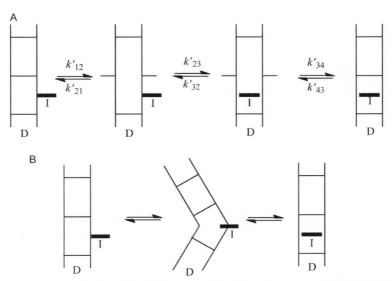

Figure 1.10 Two possible intercalation mechanisms proposed by Crothers. *Redrawn from Figure 7 of Li and Crothers (1969).*

$$I + D \underset{k_d}{\overset{k_f}{\rightleftharpoons}} ID$$

Scheme 1.3 Simple kinetic model for intercalation of ethidium bromide.

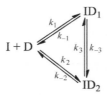

Scheme 1.4 Cyclic kinetic model for intercalation of ethidium bromide.

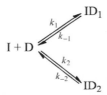

Scheme 1.5 Parallel kinetic model for intercalation of ethidium bromide.

Bresloff and Crothers (1975) proposed the cyclic mechanism (Scheme 1.4) for ethidium bromide intercalation. Two bound states form directly but they interchange amongst themselves, described as the direct transfer mechanism.

ID_1 and ID_2 are the fast-bound step and slow-bound step. However, the nature of states is in debate. While direct transfer happens between different DNA sites for substituted proflavine, it was argued to occur for proflavine in the case where intercalated state forms from major and minor groove pathways (Biver et al., 2008). However, recent studies have shown that direct transfer is present for ethidium intercalation in RNA and stabilization of double strand in the triple helix $poly(A) \cdot 2poly(U)$, where the major groove is occupied by the third strand (Waring, 1974). Therefore, it is not clear whether the direct transfer is between different grooves of DNA or between different DNA sites.

The basic difference between sequential and cyclic pathways is captured on the concentration dependence of the reaction rates. Additional information was obtained from the dependence of the rate constants on ion concentration (Wilson, Krishnamoorthy, Wang, & Smith, 1985).

The parallel mechanism shown in Scheme 1.5 is kinetically indistinguishable from the sequential mechanism. Using the dependence of the rate of intercalation on the ionic concentration, Wilson et al. (1985) also proposed the following (Scheme 1.6) two–step model for ethidium and propidium binding.

Metallo–intercalators follow a rather more complicated kinetics possibly owing to their complex structure as shown in Scheme 1.7.

This is for the binuclear Ru–based drugs (Westerlund, Wilhelmsson, Nordén, & Lincoln, 2005).

Apparently, kinetics of intercalation process depends on the nature of the drug as well. Chaires (Chaires, Dattagupta, & Crothers, 1985) performed a detailed kinetic study involving intercalation of daunomycin to DNA using stopped–flow relaxation technique and proposed a simple series mechanism (Scheme 1.8).

The first step is argued to be the outside–bound state, followed by inter-calation as the second step and the final step is the relaxation of the drug and DNA. The first bimolecular association step, which he assumed as the out-side binding of the daunomycin into DNA, has $3 \times 10^6 \, M^{-1} \, s^{-1}$ rate (k_{12}). Rizzo et al. (1989) also found three time constants for the fluorescence intensity and fitted it to a five–step process (Scheme 1.9). The main differ-ence is that the bimolecular association has two steps, one of which does not intercalate. The crystal structure of the intercalated daunomycin clearly

$$I + D \underset{k_{-1}}{\overset{k_1}{\rightleftharpoons}} ID_1 \underset{k_{-2}}{\overset{k_2}{\rightleftharpoons}} ID_2$$

Scheme 1.6 Two step kinetic model for intercalation of ethidium and propidium.

$$I + D \underset{k_{-1}}{\overset{k_1}{\rightleftharpoons}} ID_1 \underset{k_{-2}}{\overset{k_2}{\rightleftharpoons}} \underset{k_{-3}}{\overset{k_3}{\rightleftharpoons}} \cdots \cdots \underset{k_{-n}}{\overset{k_n}{\rightleftharpoons}} ID_2$$

Scheme 1.7 Multi-step kinetic scheme for metallo-intercalators.

$$I + D \underset{k_{21}}{\overset{k_{12}}{\rightleftharpoons}} ID_1 \underset{k_{32}}{\overset{k_{23}}{\rightleftharpoons}} ID_2 \underset{k_{43}}{\overset{k_{34}}{\rightleftharpoons}} ID_3$$

Scheme 1.8 Three step kinetic scheme for intercalation of daunomycine to DNA.

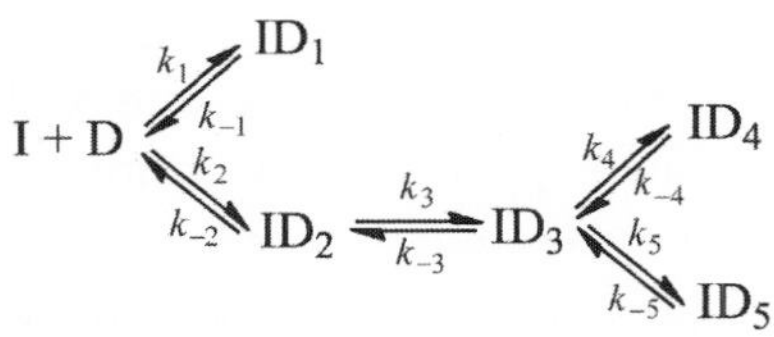

Scheme 1.9 Five step kinetic scheme for intercalation of daunomycine to DNA.

indicates that the drug must have intercalated from the minor groove side. However, the possibility of binding to the major groove is also non-negligible. Therefore, ID_1 could very well be a major groove-bound state, depending on the drug concentration, but they could not characterize it.

Another important point is that they argued the first step to be diffusion limited based on the observation that the removal of methoxyl group leads to an enhancement of the rate, however, not outside-bound state as mentioned by Chaires et al. (1985). The change in the sugar group of the drug (which does not intercalate between the pair of base pairs) also changes the bimolecular rate constants.

Forster and Stutter (1984) did temperature jump technique and they proposed only a single-step binding process. However, their main focus was on the *cooperativity* of the intercalation process. They proposed three different situations: (i) free binding site, (ii) single consecutive binding site (when one drug is already intercalated and the other one is going to intercalate), and (iii) doubly consecutive binding site (when two drugs are already intercalated and the third one is going to intercalate between the two). From theoretical point of view, we have used only one drug and it will apply to the first case of free binding to the DNA. Forster and Stutter (1984) showed that adriamycin has higher cooperativity.

2.4. Theoretical and computational evidences

Theoretical and computational effort to understand intercalation process is quite old. However, with the refinement of force-field, computational methods and increase in computer efficiency, molecular level of detail in the mechanism of the process has emerged recently. In this section, we describe the progress in the field for different aspects of the intercalation process.

2.4.1 Molecular mechanical studies

The first effort to produce a molecular model of intercalated state was by Fuller and Waring (1964) who produced a hand-built model of DNA for ethidium bromide intercalation at each base pair level. Sobell used Corey Pauling Koltun (CPK) model of actinomycin–deoxyguanosine complex to illustrate the intercalation geometry (Jain & Sobell, 1972). As an effort to model the backbone dihedral angles to avoid steric repulsion, computerized linked–atom modeling system was developed by Alden and Arnott (1975). The model was based on optimizing the stacking interactions, steric strain, and nonbonding interactions to accommodate the insertion of a molecule between the base pairs. They found that transition of two torsion angles to *trans* and sugar ring puckering to C3'-*endo* relieves the strain in the adjacent base pair during intercalation. It was assumed that intercalation of actinomycin and ethidium bromide happens from minor groove side, proflavine from major groove side, and 9-aminoacridine from either side (Jain & Sobell, 1972; Jain, Tsai, & Sobell, 1977; Sobell, Tsai, Jain, & Gilbert, 1977; Tsai, Jain, & Sobell, 1977). An empirical potential function calculation was done on dinucleotide–drug complex by Nuss, Marsh, and Kollman (1979) to predict the structure and energetics of DNA–intercalator complex. They used the potential function of the following form:

$$V = \sum_{i \neq j}^{n} \left(\frac{q_i q_j}{\epsilon R_{ij}} + B_{ij} e^{-c_{ij} R_{ij}} - \frac{A_{ij}}{R_{ij}^6} + \sum_{k=1}^{\text{ntor}} \sum_{l=1}^{3} V_l^k \cos l\varphi \right) \qquad (1.5)$$

They suggested that proflavine has to intercalate from the major groove because of steric reasons. Note that recent calculations show that indeed proflavine intercalates from the major groove, however, not for the steric reason. Energy calculations proposed that 9-aminoacridines can intercalate from the minor groove or the major groove as there is no side-chain amine groups to provide steric hindrance. The calculations on proflavine have shown that the intercalated structure predicted was closer to the X-ray crystallographic structure (Neidle et al., 1977). The calculated energy (15–30 kcal/mol) for unstacking of the base pair is qualitatively reasonable to explain the intercalation barrier and the effect of solvation–desolvation energies on the intercalation process (Nuss et al., 1979). This insightful study, however, did not consider the compensation of the base unstacking penalty by the drug and also the effect of water in the intercalation process. Dearing, Weiner, and Kollman (1981) used molecular mechanics to energy-minimize different complexes of acridine orange and proflavine with nucleotide base pairs where they used the force field (functional form of which is used current force field) of the following form:

$$E = \sum_{\text{bonds}} k_b (r - r_b)^2 + \sum_{\text{angles}} k_a (\theta - \theta_a)^2 + \sum_{\text{dihedrals}} \frac{k_d}{2} [1 + \cos(n\varphi - \gamma)]$$

$$+ \sum_{\text{nonbonded}} \left[B_{ij} r_{ij}^{-12} - A_{ij} r_{ij}^{-6} + \frac{q_i q_j}{\epsilon_{ij} r_{ij}} \right] \qquad (1.6)$$

Partial atomic charges for acridine orange and proflavine were calculated by CNDO/2 approximation. In order to incorporate the solvent effects, they used variable dielectric constants. The analysis predicted that acridine orange intercalation prefers a mixed C3'-endo (3'–5') C2'-endo sugar puckering with better stacking interactions with the base pairs, larger intercalation site, and minimum steric strain as compared to uniform C3'-endo sugar puckering. Proflavine showed uniform C3'-endo sugar puckering with less stacking interaction with the bases as compared to acridine orange but compensated by the hydrogen bonding interaction between the amine groups of proflavine and bases. First molecular dynamics simulation was performed by Singh, Pattabiraman, Langridge, and Kollman (1986) using all-atom force field for a bis–intercalator–DNA complex. He studied a bis–intercalator triostin-A between the C–G base pairs of DNA sequence d(CGTACG)$_2$ and showed that bis–intercalator stabilizes the neighboring A–T base pairs into a Hoogsteen (HG) form with van der Waals interactions between the side chain of the drug and backbone of adenine because of the reduced minor groove width in HG base pairing. Results showed that triostin-A prefers G–C-rich DNA to A–T-rich DNA. Rao and Kollman (1987) performed QM/MM studies on double intercalation of 9-amino acridines to DNA to understand the physical basis of neighbor-exclusion principle proposed by Li and Crothers (1969). The double intercalated structures of 9-amino acridine into the heptamer d(CGCGCGC)$_2$ have modeled with obeying (as shown in Fig. 1.11) and violating (as shown in Fig. 1.12) neighbor-exclusion principle using the computer graphics program CHEM (Dearing, 1981). The molecular dynamics simulations have shown that the violated structures are favored over the obeyed structures because of favorable electrostatic interactions due to the reduction in the repulsion of phosphate groups which are stretched by intercalation. But the drug–helix entropy is more favorable for the obeyed structure by 2.4 kcal/mol compared to the violated structure due to more flexibility of backbone. Therefore, violation of neighbor-exclusion principle is energetically feasible but entropically unfavorable due to the stiffening of the helical structure from double intercalation in the adjacent base pairs.

Williams et al. (1992) used two kinds of model interactions, electrostatic repulsion that wind the helix and stacking interaction that unwind it to study the exclusion principle. During intercalation, the DNA length increases but the distance between the adjacent phosphate group remains close to the normal BDNA value ($\sim$6.5 Å) by unwinding of the helix. This unwinding increases stacking interaction between the base pairs and compensate for the reduction in the stacking interaction caused by lengthening of the DNA during intercalation. So for the second intercalator to intercalate

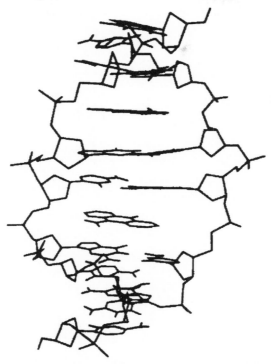

Figure 1.11 Representation of the double intercalation of 9-amino acridine to DNA obeying the neighbor-exclusion principle (structure Oby1). *Adapted from Figure 2 of Rao and Kollman (1987).*

between the NN base pair, it has to disrupt the stable stacking interaction between the already unwound bases caused during the first intercalation. So, relative stability of second intercalator in the neighborhood of first inter-calator will be less than that for the first. For intercalators with side chains such as daunomycin, the excluded volume of the side chain clearly deter-mines the neighbor exclusion for the second intercalator (Quigley et al., 1980). Neidle, Pearl, Herzyk, and Berman (1988) performed modeling study for DNA–proflavine interaction using the geometry of the intercalated state from the available crystal structure and modifying it to have a decamer with one proflavine intercalated in the center. Energetic calculations there-after showed that the conformational changes in the intercalation cause overwinding of the bases away from the intercalation site. Also the study found negative roll for base pairs involved in intercalation. The molecular dynamics simulation studies on the 2:2 proflavine complex of dCpG

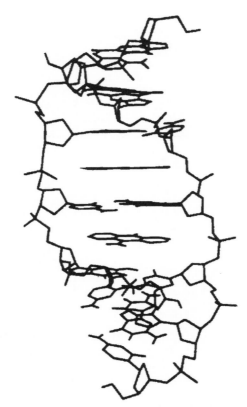

Figure 1.12 Representation of the double intercalation of 9-amino acridine to DNA violating the neighbor-exclusion principle (structure Vio1). *Adapted from Figure 3 of Rao and Kollman (1987).*

nucleotides with the hydration layer (Swaminathan, Beveridge, & Berman, 1990) using GROMOS force field (Hess, Kutzner, van der Spoel, & Lindahl, 2008) has showed the organization of water in the form of pentagonal network on the major groove and polygonal network in the minor groove as seen in the crystal structure (Neidle et al., 1980). The modeling of intercalated state of doxorubicin by molecular mechanics has shown that in the intercalated state, the sugar group remains in the minor groove and the plane of the anthraquinone ring of doxorubicin is perpendicular to the base pairs CpG (Hopfinger, Cardozo, & Kawakami, 1995). Later, crystal structure of doxorubicin in DNA has shown very similar geometry (Howerton, Nagpal, & Williams, 2003). Similar kind of study was performed by Elcock, Rodger, and Richards (1996) on 9-hydroxy ellipticine where he showed that both parallel and perpendicular orientations are favorable in terms of stacking interactions and hydrogen bonding interactions. However, the electrostatic solvation free energy was found to be more favorable for perpendicular orientation of 9-ellipticine.

Molecular dynamic simulations on the side-chain derivatives of pyrimido acridine triones in the intercalated geometry with the duplex d(GCGCGCGCGCGC)$_2$ have shown that the position and nature of side chain determine the orientation of the intercalator in the intercalation cavity (Mazerski, Antonini, & Martelli, 2000). To understand the energy of the formation of intercalation cavity by helical breathing (conformational changes in the base pair) in the case of both mono- and bis-intercalators, difference in energy between uncomplexed DNA sequence d(CGCGCGATCGCGCG)$_2$ and DNA–intercalator complex was calculated by MM–GBSA method using AMBER molecular dynamics simulation software (Case et al., 2010). The MM/GBSA simulations of DNA, DNA with mono-intercalated daunomycin, bis-intercalated daunomycin, intercalated bis-daunomycin, and intercalated 9-dehydroxy daunomycin performed by Trieb, Rauch, Wibowo, Wellenzohn, and Liedl (2004) estimated the free energy of creating one intercalation cavity to be 32.3 kcal/mol. The subsequent cavity formation was found to be less costly. Therefore, it was argued to be originated from a cooperative cavity formation, corroborated with the previous work done by Williams et al. (1992). Note that while the cost for cavity formation is estimated (Trieb, Rauch, Wibowo, et al., 2004) to be so high (32.3 kcal/mol), experimental barrier of intercalation is rather low (~15 kcal/mol) (Li & Crothers, 1969), possibly due to the favorable interaction between the drug and the DNA base pairs. A huge estimate in the rotational and translational entropy loss (~11–14 kcal/mol; Trieb, Rauch, Wibowo, et al., 2004) was attributed to the binding process, calculated using Sackur–Tetrode (Seddon & Gale, 2001) formula for entropy of ideal gases. Interestingly, this study (Trieb, Rauch, Wellenzohn, et al., 2004) pointed out the predominant formation of B_{II} conformation of the DNA at the intercalation site. Earlier study of Winger et al. (1998) showed that destacking of base pairs caused the interconversion from B_I to B_{II}. Therefore, the interconversion of the DNA forms may be an effect of unstacking itself, rather than intercalation. Similarly, ellipticine intercalation also showed a conversion of the backbone from B_I state to B_{II} state. In that case, higher entropy of the DNA–ellipticine complex originates from the sugar-phosphate group rather than from the bases (Kolar, Kubar, & Hobza, 2010).

Molecular modeling study of intercalated complexes has been performed (Cavallari, Garbesi, & Di Felice, 2009) also using G-quadruplex and porphyrin–based molecule (TMPyP), where it was shown that higher TMPyP/G4t ratio, that is, intercalator in between every tetrad dimer, deforms G4–quadruplexes and a lower ratio of 1/8 shows that G–quadruplexes are stable.

2.4.2 Quantum chemical calculation

Along with the molecular modeling studies, which probe into the overall structural and dynamical features, quantum chemical studies were applied to understand the specific drug–DNA interactions. Pack and Loew (1978) illustrated that the specificity of ethidium toward $Y(3'-5')R$ sequence compared to $R(3'-5')Y$ sequence is due to the difference in energy of unwinding of DNA during the intercalation process and not due to the stacking interaction between ethidium and base pairs. Here, the stacking interactions in both the cases remained the same. They have used semiempirical quantum chemical methods PCILO and CNDO/2 for the calculations. Pack, Hashimoto, and Loew (1981) used the same methods to study the effect of outside-bound proflavine on the induction of conformational changes of the DNA to open up the base pairs before the insertion of proflavine could happen (Li & Crothers, 1969). These calculations were done by taking the crystal structure geometry of CpG–proflavine complex (Neidle & Jones, 1975). The sequence specificity was probed by taking all different possible base pairs of DNA, by mutating the bases G and C in the crystal structure. The nucleotide base pairs with proflavine and without proflavine have shown that proflavine has a preference for dCpdG over dGpdC as it requires 23.5 kcal/mol more energy to unwind dGpdC as compared to dCpdG. Similarly, it has a preference for dApdT over dTpdA. The calculations on the BDNA and the intercalated geometry with the hydrogen bonding interaction with the outside proflavine molecule showed that the neutralization of anionic phosphate charge does not affect the sequence specificity but it lowers down the energy for opening the base pairs for intercalation to happen. This study suggested the two-step intercalation process where the outside-bound form stabilizes the open form of intercalation site so that the intercalation can occur (Pack et al., 1981). The sequence specificity of daunomycin to different polynucleotide sequences such as $d(CGTACG)_2$, $d(CGATCG)_2$, $d(CITACI)_2$, $d(TATATA)_2$, $d(CGCGCG)_2$, and $d(TACGTA)_2$ was performed by Pullman (Chen et al., 1985). Among the regular alternate sequences $d(TATATA)_2$ and $d(CGCGCG)_2$, daunomycin showed specificity toward $d(TATATA)_2$. But among the six different hexanucleotides, mixed polynucleotides $d(CGATCG)_2$

and d(CGTACG)$_2$ showed strongest affinity toward daunomycin (Chen et al., 1985) which is in agreement with the X-ray crystal structure of daunomycin–d(CpGpTpApCpG) (Quigley et al., 1980). The aglycone part of daunomycin intercalates between the C–G base pair and the daunosamine ring stays in the minor groove covering the third base pair also. These studies indicate that the selectivity of daunomycin toward the bases should be explained in terms of 3 bp rather than 2 bp (Chen et al., 1985; Quigley et al., 1980). Since most of the intercalators bear a positive charge, it was believed that the electrostatic interaction between the cationic intercalator and the negatively charged DNA plays the most important role in the stabilization of DNA–intercalator complex (Medhi, Mitchell, Price, & Tabor, 1999). Reha et al. (2002) investigated the stacking interaction energy between different intercalators and base pairs using high-level *ab initio* calculations (MP2 and dispersion corrected density functional tight-binding methods). To understand the effect of stacking energy on the vertical distance between the base pairs and intercalator, the systems chosen were AT base pair with ethidium and DAPI and GC base pair with daunomycin and ellipticine. The study showed that none of the intercalators show charge accumulation in any of the sites but they have a large polarizability and this clearly shows that the dispersion energy contributes much to the DNA–intercalator stability rather than electrostatic interactions. It also showed a dependency of Twist of base pair on the stacking energy of different intercalators by changing the Twist from 0° to 360°. This methodology was adapted by Cooper et al. by applying van der Waals density functional to study the stacking interaction between the intercalator (neutral and charged proflavine, ellipticine) and the base pair. The larger polarizabilities of all three intercalators showed that the intercalation is mostly governed by dispersion interaction. The optimization of charged and neutral proflavine in the presence of C–G base pair showed charged proflavine has stronger binding because of the presence of electrostatic interaction. Similar calculations are done for proflavine in the presence of T–A base pair and showed that the complex is less energetically stable as compared to C–G complex (Li, Cooper, Thonhauser, Lundqvist, & Langreth, 2009). *Ab initio* and semiempirical quantum chemical calculations with molecular mechanics simulations were done to understand the role of enthalpy and entropy of ethidium intercalation by applying modified Cornell et al. empirical force field. The DNA sequences used were d(GCATATATGC)$_2$ and d(GCGCGCGCGC)$_2$, and for high-level energy calculations, a minimal model composed of intercalator molecule and dimers TA or GC was used.

The hydrogen bond interaction of ethidium with DNA was found to be 10% of the total interaction energy, and most of the contribution came from dispersion energy as in the case of most of the intercalators. The interaction energies for ethidium with dimer TA and GC are comparable. The free-energy change of intercalation for ethidium was calculated to be -4.5 kcal/mol by using QM/MM methods (Reha et al., 2002). The intercalation process causes deformability to the neighborhood of intercalation site and this in turn facilitates the insertion of the second intercalator between the neighboring base pairs (Kubař, Hanus, Ryjáček, & Hobza, 2006). The stacking of phenanthroline in between the base pair A–T and in between the base pair G–C was studied by *ab initio* methods such as MP2. The analysis summarized that phenanthroline has a preference for G–C base pair and suggested that MP2/6-31+G(d,p) could be used further for studying the stacking interactions (Pankaj Hazarika, Das, Medhi, & Medhi, 2011).

3. MOLECULAR MECHANISM OF INTERCALATION

Molecular modeling and quantum chemical studies discussed earlier provide various insights into the structural and thermodynamic aspects of the intercalation process (Baginski, Fogolari, & Briggs, 1997; Trieb, Rauch, Wibowo, et al., 2004) Experimental kinetic studies using spectroscopic methods do provide the overall ensemble average picture of the mechanism, that is, timescale of the process and existence of different intermediate states, etc. (Chaires et al., 1985; Li & Crothers, 1969). However, a detail molecular picture of the process remains elusive. For example, kinetic studies depict the intercalation process of proflavine as a two-step process (Li & Crothers, 1969), ethidium bromide intercalation as a single step (Macgregor et al., 1985), while for daunomycin, both three-step (Chaires et al., 1985) and five-step process (Rizzo et al., 1989). The origin of these differences was, however, unknown, although it was understood that these differences must lie with the nature of the intercalators (Chaires et al., 1985) and minimum two-step (Li & Crothers, 1969), fast outside-bound state formation and slow activated intercalation are required for intercalation. Moreover, the limiting hypothesis of intercalation discussed earlier, that is, whether intercalation proceeds via natural fluctuation of the DNA or by the drug-induced cavity formation could not be explained from experimental results obtained (Li & Crothers, 1969).

This intriguing but difficult aim to capture the molecular detail of the sequence of events was achieved only recently. In this part, we discuss on the recent development in the area of the molecular mechanism of the inter-calation process, which will throw some light into molecular origin of the thermodynamics and kinetics of the some of the intercalation processes.

To address the above problems, we used extensive all-atom molecular dynamic simulations with accelerated sampling processes such as meta-dynamics (Alessandro & Francesco, 2008), well-tempered metadynamics (Barducci, Bussi, & Parrinello, 2008), and umbrella sampling (Torrie & Valleau, 1977) using proper collective variables (CVs) (reaction coordinates, RCs) through which the process happens. We first discuss the methods used followed by the outcome.

3.1. Methods used

Mechanism of a process is governed by the progress along the minimum free-energy path (MFEP). Therefore, here we discuss some of the methods related to the calculation of free-energy profile and MFEP. Studies of inter-calation mechanism used primarily three techniques: (a) umbrella sampling, (b) metadynamics, and (c) well-tempered metadynamics.

3.1.1 Umbrella sampling

Umbrella sampling (Torrie & Valleau, 1977) is one of the accelerated sam-pling methods in molecular dynamics simulations where the inherent barrier is overcome by employing an additional bias potential (Eq. 1.7) along a cho-sen RC or CV:

$$V_{\text{umb}}(s) = \frac{1}{2}k(s - s_0)^2 \tag{1.7}$$

where k is the force constant, s_0 is the constant CV value at which the restraint is applied, and s is dynamically changing CV. A complete free-energy profile is obtained by multiple simulations along different s_0 value to cover the entire reaction profile. Each simulation provides a biased distribution of s. Correct free energy $\left(\widetilde{F}(s)\right)$ is obtained by unbiasing (Eq. 1.7) it in the following way (Bonomi et al., 2009):

$$\widetilde{F}(s) = -\frac{1}{\beta}\ln N(s) - V_{\text{umb}}(s) \tag{1.8}$$

where $N(s)$ is the histogram of the visited configurations in the CV space, $V_{\text{umb}}(s)$ is the applied bias potential, and $\beta = \frac{1}{k_B T}$, where k_B is the Boltzmann constant and T is the temperature (K).

Thus, all the unbiased free-energy profiles obtained for different values of s are combined together using weighted histogram analysis method (Kumar, Rosenberg, Bouzida, Swendsen, & Kollman, 1992; Roux, 1995) to get the whole free-energy profile. It is assumed that all the other $3N-1$ coordinates ($N=$ no. of atoms) are faster variable and get equilibrated for a given parametric value of the RC. The choice of the RC is based on chemical intuition. Therefore, if the choice is wrong, the free energy obtained would certainly be associated with huge errors.

3.1.2 Metadynamics

Metadynamics (Alessandro & Francesco, 2008) is a relatively new, but robust, accelerated sampling method particularly for computation of multidimensional free-energy surfaces. The difference from umbrella sampling is that metadynamics is a nonequilibrium sampling; however, it provides an equilibrium free-energy landscape. Metadynamics is built with the ideology that more a system stays in a particular point in the phase space, more destabilized it should become so that it can explore the neighboring phase space and thereby the complete free-energy surface. Although one needs to describe CVs to accelerate sampling along them, the exploration is rather spontaneous. Technically it is achieved by the addition of Gaussian potential at certain time interval to the location of the CV at that point of time. Gaussian potentials are added till the free-energy surface becomes flat and thereby exploration of free energy surface (FES) is spontaneous. The accuracy and speed of sampling can be altered by changing two parameters such as Gaussian height and Gaussian width as well as changing the time interval between additions of two Gaussian functions. The metadynamics potential (Alessandro & Francesco, 2008) ($V_G(s(x), t)$) at a particular time, t, is equal to

$$V_G(s(x),t) = \omega \sum_{t'=\tau_G, 2\tau_G, 3\tau_G...t'<t} \exp\left(-\frac{(s(x) - s(x_G(t')))^2}{2\delta_s^2}\right) \qquad (1.9)$$

where ω is the height of the Gaussian, δ_s is the width of the Gaussian, τ_G is the frequency at which the Gaussian is added, $s(x)$ is the conformational space described by the CV, $s(x_G(t'))$ is the value taken by the CV at time t', and t is the sampling time.

Free energy surface $F(s)$ in the conformational space $s(x)$ is the opposite of the sum of all Gaussian potential (Alessandro & Francesco, 2008). So,

$$F(s) \sim -\lim_{t \to \infty} V_G(s(x), t) \qquad (1.10)$$

3.1.3 Well-tempered metadynamics

In normal metadynamics simulation, the free energy does not converge to a definite value leading to an average error. Continuing the simulation irreversibly pushes the system to regions of configurational space which are not physically relevant. To overcome this problem, well-tempered meta-dynamics (Barducci et al., 2008) was developed which has the advantage of controlling the regions of the free-energy surface which are physically relevant.

In case of well-tempered metadynamics, the Gaussian height is rescaled during the simulation according to the following equation (Barducci et al., 2008):

$$\omega = \omega_0 \exp - \frac{V_G(s(x),t)}{k_B \Delta T} \qquad (1.11)$$

where ω_0 is the initial Gaussian height, k_B is the Boltzmann's constant, and ΔT is a parameter with dimension of temperature.

Free energy surface is estimated as (Alessandro & Francesco, 2008; Barducci et al., 2008),

$$\widetilde{F}(s,t) = -\frac{T + \Delta T}{\Delta T} V_G(s(x),t) \qquad (1.12)$$

$T + \Delta T$ is called CV temperature, T is the simulation temperature, and $\frac{T + \Delta T}{T}$ is called bias factor.

3.2. Description of the CVs

As described earlier, CVs are chosen from the intuition of a particular process. For intercalation process, one of the CVs must be the measure of some distance between the DNA and the intercalators. However, due to the structural dissimilarity in the DNA between its major and minor grooves, Mukherjee et al. and subsequently Sasikala et al. used a vectorial distance along with other CVs to study the intercalation process of daunomycin and proflavine (chemical structures shown in Fig. 1.13) as shown in Fig. 1.14.

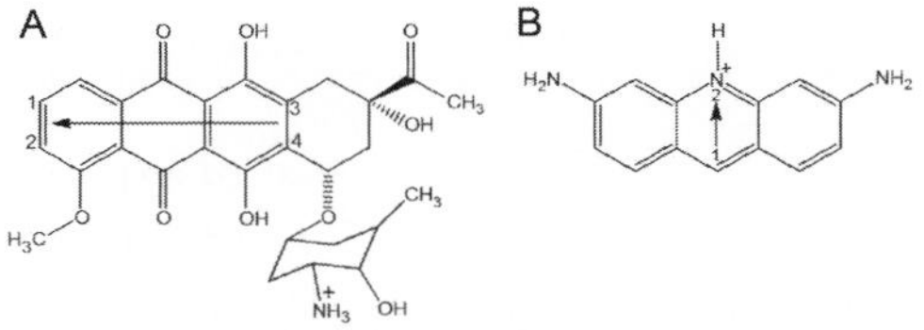

Figure 1.13 Chemical structures of (A) daunomycin and (B) proflavine.

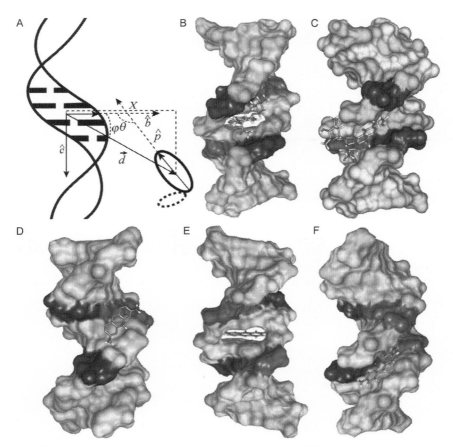

Figure 1.14 Schematic representation of collective variables and initial configurations. (A) Schematic of DNA and the drugs (daunomycin with solid oval-shaped aglycon ring and dashed oval-shaped sugar moiety; proflavine represented by the solid oval shape only). It also shows the construction of collective variables X, θ, φ, and Y. (B and C) 3D representations of the intercalated state and the minor groove-bound state of dauno-mycin. Panels (D), (E), and (F) show the major groove-bound state of proflavine, inter-calated geometry of proflavine, and minor groove-bound state of proflavine in DNA, respectively. The 3D figures are made in Chimera (Pettersen et al., 2004). (For color version of this figure, the reader is referred to the online version of this chapter.)

Figure 1.14A shows the schematic representation of DNA and drugs with the CVs description. $\vec{d}$ is the vector from the center of mass (COM) of intercalating base pair (C6, G7, C19, and G18) to COM of drug. $\hat{b}$ is the unit vector from the COM of intercalating base pairs to COM of two sugars which lie more toward the minor groove of intercalating base pairs.

$\hat{p}$ is the unit vector from the COM of two atoms (3, 4 as shown in Fig. 1.13A) of daunomycin to COM of other two atoms (1, 2 as shown in Fig. 1.13A) of daunomycin. In the case of proflavine, $\hat{p}$ is the unit vector from connecting the two atoms 1 and 2 as shown in Fig. 1.13B. $\hat{c}$ is the body-fixed vector from the COM of intercalating base pairs to the COM of the $3'$-terminal base pairs (C12 and G13).

$X = \hat{b} \cdot \vec{d}$, $\varphi = \cos^{-1}\left(\hat{b} \cdot \vec{d}\right)$, and θ corresponds to the angle between the unit vectors $\hat{b}$ and $\hat{p}$. Y is the scalar product of $\hat{c}$ and $\vec{d}$. The CV X defines the position of the drug in the minor or the major groove side of DNA. Positive X indicates that drug is in the minor groove side of the DNA and negative X shows that the drug is in the major groove side of the DNA. φ represents the position of the drug along the helical axis of DNA. φ between $0°$ and $90°$ denotes that the drug moves along the DNA in the minor groove and between $90°$ and $180°$ denotes that the drug moves along the DNA in the major groove. θ measures the angle between the vector which corresponds to the long axis of daunomycin molecule and the unit vector $\hat{b}$. In the intercalated state, θ is $90°$. In the case of proflavine, θ gives the orientation of the amine groups of proflavine with respect to DNA. If it is $\sim 180°$, then amine groups of proflavine are facing the major groove side (as in the case of crystal structure), and if it is $\sim 0°$, then amine groups are facing toward the minor groove side. The collective variable Y describes the movement of the drug in the direction of the helical axis.

3.3. Simulation protocol

For intercalated state of daunomycin (Wang, Ughetto, Quigley, & Rich, 1987) and proflavine (Maehigashi, Persil, Hud, & Williams, 2010), authors used crystal structure geometry in 12 bp BDNA after modifying the crystal structure geometry using JUMNA (Lavery, Zakrzewska, & Sklenar, 1995) program for daunomycin and NAB (Case et al., 2010) program for proflavine. To obtain the groove-bound states, docking was performed using docking softwares such as AutoDock (Morris et al., 2009) and HEX (Mustard & Ritchie, 2005). AMBER force fields were used in general; however, recent modification of parmbsc0 (Perez et al., 2007; Wang, Cieplak, & Kollman, 2000) was used in the later studies. For the intercalators proflavine and daunomycin, force-field parameters were generated using general amber force field (Wang, Wolf, Caldwell, Kollman, & Case, 2004) after optimization and the partial charge calculation in Gaussian (Frisch et al., 2003) using the basis set HF/6-31G*. Rest of the simulation protocol is similar to DNA simulations and can be found in the respective publications (Mukherjee, Lavery, Bagchi, & Hynes, 2008; Sasikala & Mukherjee, 2012, 2013; Wilhelm et al., 2012). GROMACS (Hess et al., 2008) and AMBER (Case et al., 2010) were used for performing molecular dynamics simulations, whereas in some studies, PLUMED (Bonomi et al., 2009) and GROMETA (Camilloni, Provasi, Tiana, & Broglia, 2008) were used as an additional software patch to calculate the free-energy contributions.

3.4. Mechanism of intercalation

As a first attempt on understanding the molecular mechanism of intercalation, Mukherjee et al. chose daunomycin as an intercalating drug. Crystal structure of daunomycin (Quigley et al., 1980) shows that at the intercalated state, daunosamine group of the drug stays in the minor groove. This indicates that the intercalation of daunomycin takes place from minor groove side of DNA (otherwise, the final structure would have been different in this case). Therefore, Mukherjee et al. used the CV X mentioned earlier to deintercalate the drug from the DNA using umbrella sampling method. Note here that deintercalation process is like protein unfolding where one state goes to many states, whereas intercalation is like folding where many possible states have to go to one particular native state. Therefore, as a first study, deintercalation was a natural choice to understand the intercalation mechanism, more so because the equilibrium processes should be reversible and, therefore, deintercalation process should also reflect in reverse intercalation mechanism.

In this study, daunomycin was deintercalated and finally taken to far separate state. This provided with the free-energy difference between the intercalated and separated state (-12.3 kcal/mol), in close agreement with the experimental value -9.4 kcal/mol (Chaires et al., 1996). However, this deintercalation free-energy profile could not capture the barrier observed in the intercalation process in experiment. Therefore, one more free-energy profile was calculated from the docked minor groove-bound state to the separated state. Both these free-energy profiles were calculated using X as CV. However, a final FES was created combining both X and θ (Fig. 1.14) as shown in Fig. 1.15. θ was used to help distinguish the intercalated and minor groove-bound state as by the design intercalated state corresponds to high value ($\sim150°$), whereas minor groove-bound state has low ($\sim50°$) value.

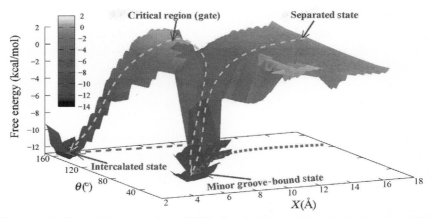

Figure 1.15 Potential of mean force (PMF) of umbrella sampling simulations along the reaction coordinate X and θ. The completely dissociated state ($X \sim 16$ Å) is considered to reference zero free-energy value. The green dashed line with an arrow (the red projections on to the X–θ plane) shows the schematic representation of the most probable path from completely separated → minor groove-bound → intercalated state. (For interpretation of the references to color in this figure legend, the reader is referred to the web version of this article.) *Adapted with permission from Mukherjee et al. (2008). Copyright (2008) American Chemical Society.*

Figure 1.15 shows the overall mechanism of the process where the daunomycin first binds to the minor groove in a downhill diffusion–limited process, although a direct intercalation could have caused only a little (~3 kcal/mol). However, once bound to the minor groove, the drug now faces a barrier of ~12 kcal/mol either for intercalation or for dissociation. This causes the slow intercalation process. This study qualitatively matches with the experimental picture where outside binding (in this case, minor groove binding) is in the regime of diffusion–limited process (Chaires et al., 1985) $(3.0 \times 10^6 \text{ M}^{-1} \text{ s}^{-1})$, whereas both the intercalative step (k_{23}, Eq. 1.2) and the dissociative step (k_{21}, Eq. 1.1) cause barrier. However, this study could not capture the following features: (i) multistep process could not be captured, (ii) barrier heights obtained were smaller than that estimated from experimental kinetics, and (iii) no direct connection between minor groove-bound state and the intercalated states were shown. FES obtained was from two different paths going from intercalated and from the minor groove-bound to the separated state. Therefore, no direct path between the minor groove-bound and the intercalated state, that is, the path of intercalation, was shown.

Subsequent work by Wilhelm et al. (2012) using both umbrella sampling and metadynamics method addressed the above shortcomings. In this study, a new coordinate, mean minimum distance between the DNA and the drug was chosen to study the deintercalation process, whereas both X and θ and Y (a CV to monitor the movement of the drug along the DNA axis) were used to study deintercalation using metadynamics method. This study captured a third state, which was originated from the dihedral rotation of the drug. Moreover, it showed the importance of physiological ion concentration on the intercalation mechanism. Lower ion concentration increased the minor groove width in case of minor groove-bound state where the dauno-mycin ring reoriented to the wider axial state and the physiological ion con-centration (150 mM NaCl) did not increase the minor groove width. This is probably the first instance where different computational methods with dif-ferent RCs provided almost similar molecular picture indicating robustness in both the methods and the calculations.

Since metadynamics allows exploration of the chosen CV depending on the nature of the FES, deintercalation toward the minor groove in the study of Wilhelm et al. (2012) indicates that indeed the pathway is through the minor groove. However, this may not be true for a flat intercalator like proflavine as both the major and the minor groove pathways may lead to the same final state. This is true for the threading intercalators also where the intercalation can happen from either the minor groove or the major groove (Scheme 1.10), as the side chains are present on both sides (Biver et al., 2008). Therefore, Sasikala and Mukherjee (2013) studied deintercalation through both grooves and intercalation of proflavine from both the minor groove and the major groove using well-tempered metadynamics simulations (Barducci et al., 2008) to obtain the complete free-energy surface of the process. This study used collective variables X, θ, and φ.

Deintercalation of proflavine starting from the initial intercalated geom-etry (Fig. 1.14E) resulted in the exit of the drug toward the major groove with a barrier of 14.7 kcal/mol, indicating that the deintercalation pathway

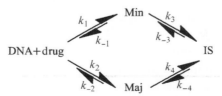

Scheme 1.10 Double pathway for intercalation of proflavine to DNA.

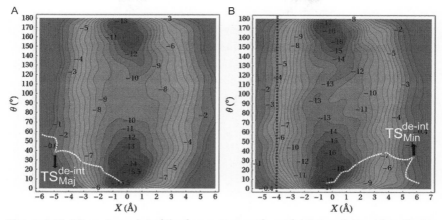

Figure 1.16 2D contour plot of the free-energy surface of deintercalation of proflavine along the collective variables X and θ. (A) Deintercalation through the major groove side and (B) deintercalation through the minor groove side. Positive value of X means the minor groove side and negative value of X means the major groove side. $\theta \sim 180°$ means the amine groups of proflavine facing the major groove side and $\theta \sim 0°$ represents the amine groups are facing toward the minor groove side. The white dotted line shows the minimum free-energy path (MFEP) showing the transition state. The blue dotted line shows the position of the configurational restraint. Each contour line represents the free-energy value in kcal/mol. (For interpretation of the references to color in this figure legend, the reader is referred to the web version of this article.) *Adapted from Sasikala and Mukherjee (2013).*

is through the major groove (Fig. 1.16A). This study also showed that proflavine can reorient within the intercalated state with a barrier of 5 kcal/mol, which decreases as the drug goes toward either groove edges. Therefore, at the groove edge (where the drug is about to enter into the DNA), the free energy is almost equal for both orientations indicating that the reason of deintercalation (or intercalation) may not be governed by the steric reason as indicated before. We discuss later about the thermodynamic reason. To obtain a barrier toward the minor groove side (see Fig. 1.16B), Sasikala et al. blocked the major groove side using a soft potential far away from the transition state and performed well-tempered metadynamics simulation which resulted in the deintercalation toward the minor groove with higher barrier (18.5 kcal/mol).

Thereafter, intercalation of proflavine was attempted from both major and minor groove-bound states. However, intercalation from a separated state to intercalated state was computationally challenging as the configurational freedom of the drug in the free state is enormously high. Therefore, a suitable configurational restraint was used away from the probable transition state and well-tempered metadynamics simulation along X and φ led to the formation of the intercalated state (Fig. 1.17C and D) and provided barriers for both the processes. Intercalation barrier from the major groove-bound state was 8.8 kcal/mol, while from the minor groove-bound state was 16.9 kcal/mol. Also Sasikala et al. obtained free-energy profile for dissociation of proflavine from major and minor groove-bound states (Fig. 1.17A and B).

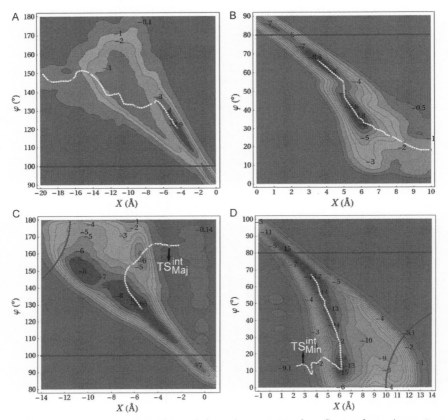

Figure 1.17 2D free-energy surface for (A) dissociation of proflavine from the major groove-bound state, (C) intercalation of proflavine from major groove side, (B) dissociation of proflavine from minor groove-bound state, and (D) intercalation of proflavine from minor groove side along the collective variables X and φ. The white dotted line represents minimum free-energy path (MFEP) and blue dotted line represents the position of the configurational restraint. The contour line represents the free energy in kcal/mol. (For interpretation of the references to color in this figure legend, the reader is referred to the web version of this article.) *Adapted from Sasikala and Mukherjee (2013).*

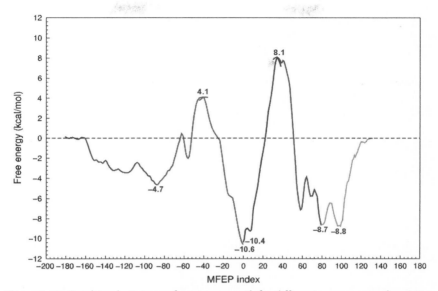

Figure 1.18 Combined minimum free-energy path for different processes such as inter-calation from the minor and the major groove, deintercalation toward the minor groove side and the major groove side and dissociation of the minor and the major groove-bound states. The MFEP index 0 shows the intercalated state, and positive MFEP index shows the minor groove side and negative MFEP shows the major groove side. The gap between each MFEP index corresponds to a radius 0.05 in (X, θ) or (X, φ). The MFEP index for each process is as follows: $132 \rightarrow 79$, separated to minor groove-bound state; $79 \rightarrow 34$, intercalation from minor groove-bound state up to transition state; $0 \rightarrow 34$, intercalated state to transition state for deintercalation toward minor groove side; $0 \rightarrow -41$, intercalated state to transition state for deintercalation toward major groove side; $-41 \rightarrow -88$, transition state for intercalation from major groove side to major groove-bound state; and $-88 \rightarrow -181$, major groove-bound state to separated state. *Adapted from Sasikala and Mukherjee (2013).* (For color version of this figure, the reader is referred to the online version of this chapter.)

Combining all these six different processes, the free-energy profile along the MFEP reveals the overall picture of the molecular mechanism of the intercalation process (Fig. 1.18).

Table 1.2 compares different calculated barriers for intercalation, de-intercalation, and dissociation of proflavine from metadynamics simulation with the experimental estimates from fluorescence kinetic data.

Deintercalation barrier through the major groove matches with numerous other experimental calculations as shown in Table 1.2 (Biver, Secco, Tinè, & Venturini, 2003; Ciatto, D'Amico, Natile, Secco, & Venturini, 1999; Corin & Jovin, 1986; Li & Crothers, 1969; Ramstein,

Table 1.2 Comparison of free-energy stabilities and barriers for intercalation, deintercalation processes, and external bound states between experimental and theoretical estimates (Sasikala & Mukherjee, 2013)

Energy (kcal/mol)	Experimental[a]	Theoretical	
		Major	Minor
$\Delta G^{\#,int}$	12.5	8.8	16.9
$\Delta G^{\#,de-int}$	14.8	14.7	18.5
ΔG_{ext}	−6.8	−4.7	−8.8
ΔG_{int}	−9.0	−10.6	−10.4

[a]$T = 25\ °C$, 0.1 M Na^+ (Ramstein et al., 1980).

Ehrenberg, & Rigler, 1980; Ramstein & Leng, 1975). However, intercalation barrier from minor groove is higher while from the major groove is lower than the experimental estimate (Biver et al., 2003; Ciatto et al., 1999; Corin & Jovin, 1986; Li & Crothers, 1969; Ramstein et al., 1980). Therefore, Sasikala et al. performed kinetic study for proflavine intercalation using the following double pathway intercalation process (Scheme 1.10) and the corresponding kinetic graph is shown in Fig. 1.19A and B.

Figure 1.19(A) shows that the major groove-bound state decays faster, while the minor groove-bound state is stable and forms within tens of nanoseconds. Figure 1.19B shows that the intercalation happens in millisecond time in agreement with experimental kinetics (Li & Crothers, 1969). However, it is not clear whether intercalation subsequently proceeds directly from minor groove-bound state or it forms via the major groove as the former would then be a cyclic intercalation and deintercalation and may violate the thermodynamic principle (Sasikala & Mukherjee, 2013).

To understand the exact pathway of intercalation, the kinetic scheme was decomposed into different pathways as shown in Fig. 1.19C and D (Sasikala & Mukherjee, 2013). Analysis of kinetic result showed that the direct intercalation from the minor groove side is a very slow process ($\gg$ms) (Fig. 1.20C, light green line), whereas Fig. 1.19D shows that the intercalation directly from the major groove side happens in microsecond timescale. However, as shown in Fig. 1.19C, intercalation through the major groove starting from the minor groove-bound state matches experimental timescale and also maintains microscopic reversibility (Sasikala & Mukherjee, 2013). Therefore, minor groove-bound state acts as a kinetic trap before releasing the drug and thereby increases the kinetics of

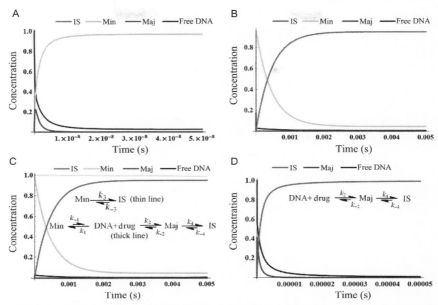

Figure 1.19 Kinetic profile of proflavine showing (A) intercalation process following the kinetic Scheme 1.10 in a shorter timescale, (B) intercalation process following kinetic Scheme 1.10 at a longer timescale, (C) intercalation from the minor groove-bound state (light green color) and intercalation from the major groove side (thick green line) through the intermediate minor groove-bound state, and (D) intercalation from the major groove-bound state without the intermediate minor groove-bound state. (For interpretation of the references to color in this figure legend, the reader is referred to the web version of this article.) *Adapted from Sasikala and Mukherjee (2013).*

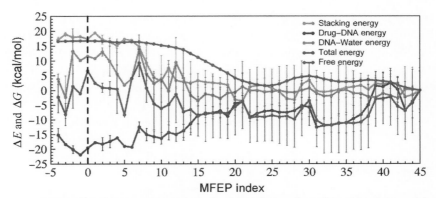

Figure 1.20 Different energy parameters for intercalation from the minor groove side. The dashed vertical line at the MFEP index 0 indicates the transition state for intercalation, whereas MFEP index 45 represents the initial minor groove-bound state. *Adapted from Sasikala and Mukherjee (2013).* (For interpretation of the references to color in this figure legend, the reader is referred to the online version of this chapter.)

intercalation (Sasikala & Mukherjee, 2013). Similar conclusions were made from the derived 2D FES of daunomycin (Mukherjee et al., 2008).

Note that this mechanism is similar to what was proposed by Crothers (1975) (Scheme 1.4) where the stability of one state affects the rate of the other. Here also the stability of the minor groove-bound state affects the rate of intercalation through the major groove side. However, there is no direct intercalation process, as Crothers rightly assumed. All pathways lead through some outside-bound state. Kinetic profile shows that this pathway through the minor groove-bound state followed by intercalation through the major groove side falls in the millisecond timescale as observed in the experiment. Direct intercalation through the major groove would result in much faster kinetic, whereas through minor groove, it would result in much slower kinetics.

Table 1.3 summarizes various thermodynamic contributions of the six processes mentioned earlier. Both the deintercalation and dissociation processes are dominated by energy the most contribution of which arises from drug–DNA interaction. The outside-bound state is found to be entropically unfavorable, similar to what is found in a recent entropy evaluation for daunomycin (Mukherjee, 2011).

For the intercalation from the minor groove, the free-energy barrier is 16.9 kcal/mol and the energy barrier is 6.7 kcal/mol. Here, as shown in Fig. 1.20, the increase in stacking energy between the base pairs is compensated by the drug–DNA interaction energy and the main contribution to the total energy (6.7 kcal/mol) is given by the desolvation energy. Rest of the free-energy barrier may be because of entropy loss. But in the case of intercalation from the major groove side, the total energy change is negligible with no desolvation energy and the free-energy barrier 8.8 kcal/mol mainly comes from the entropic contribution (-8.5 kcal/mol; Sasikala & Mukherjee, 2013). The effective barrier for intercalation, as discussed earlier, is somewhat in between the estimates for intercalation either major

Table 1.3 Thermodynamic parameters in intercalation process of proflavine (Sasikala & Mukherjee, 2013)

	Minor			Major		
Energy (kcal/mol)	$\triangle G^{\#}$	$\triangle E^{\#}$	$T \triangle S^{\#}$	$\triangle G^{\#}$	$\triangle E^{\#}$	$T \triangle S^{\#}$
Deintercalation	18.5	23.2	4.7	14.7	11.1	-3.4
Intercalation	16.9	6.7	-10.2	8.8	0.3	-8.5
Dissociation	8.8	35.3	26.5	4.7	41.5	36.5

$\Delta G^{\#}$, free-energy barrier; $\Delta E^{\#}$, energy barrier; and entropy barrier, $T\Delta S^{\#} = \Delta E^{\#} - \Delta G^{\#}$.

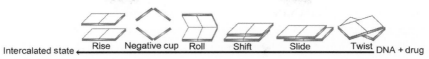

Intercalated state ← Rise Negative cup Roll Shift Slide Twist DNA + drug

Figure 1.21 Figure showing the sequence of DNA base pair step parameters during the intercalation of proflavine to DNA. (For color version of this figure, the reader is referred to the online version of this chapter.)

or minor groove. For deintercalation, the barrier toward the major groove (14.7 kcal/mol) is smaller compared to the barrier through the minor groove (18.5 kcal/mol). Therefore, the drug deintercalates through the major groove. Most of the contribution to this barrier is from the energy of destacking. Dissociation barrier, as seen in Table 1.3, also originates mainly due to interaction energy between the drug and the DNA and opposed by favorable entropy. Therefore, binding to the outside (opposite of dissociation) is entropically unfavorable. A complete entropy calculation for a different drug daunomycin corroborates the same where the entropy of daunomycin–DNA in the minor groove-bound state is −15.8 kcal/mol and in the intercalated state is 14.4 kcal/mol (Mukherjee, 2011).

Apart from the thermodynamics and kinetics from the molecular point of view, Sasikala et al. also addressed the limiting mechanism of intercalation by Crothers et al. (Li & Crothers, 1969), who suggested two possibilities: (i) the drug waits outside the DNA for a natural high fluctuation to slip in and (ii) the drug induces the cavity and intercalates. Sasikala et al. probed into this aspect by analyzing the structures collected along the MFEP which showed the sequence of events happening during the intercalation process (Fig. 1.20), where the nonstacking base pair step parameters Twist, Slide, and Shift change first and the base-stacking base pair step parameters such as Roll, Buckle, and Rise change later (Fig. 1.21). Also the intercalation follows a minimum base-stacking penalty pathway, where the increase in energy between the base pairs during the intercalation is compensated by the favorable stacking interaction between the drug and the flanking base pairs (Fig. 1.20). As the position of the drug is correlated with the changes in the above parameters, this study argued for the drug-induced cavity formation of intercalation.

4. CONCLUSION

Classic intercalators have been used as various aids for mankind in the form of dyes, therapeutic agents for different ailments, and in diagnostic applications for several hundred years. To prove the concept of intercalation of those small molecules to nucleic acids, several experimental studies were performed along with some theoretical and computational ones. Those studies were mainly focused on the thermodynamic, kinetic, and structural aspects of intercalation. As intercalators are mainly used in the treatment of cancer, search for better chemotherapeutic agents led to the discovery of more selective and sequence specific intercalators such as threading intercalators, metallo–intercalators, and finally G-quadruplex intercalators. The molecular mechanism of the intercalation process, however, received attention only recently through the computational studies of daunomycin intercalation to DNA. Recent computational studies on the proflavine intercalation mechanism strengthen the application of computational approaches to understand such a complex biological process like intercalation. The studies reveal that intercalation of small, flat intercalators like proflavine is also complex and exhibit competitive intercalating pathways originated through the interplay of interactions among drug, DNA, and water. This indicates that a far complex and richer mechanism may be possible for other drugs and is waiting to be discovered. The understanding of the molecular mechanism of intercalation will further help in better and more efficient drug design.

ACKNOWLEDGMENT

Authors acknowledge computational facility of IISER, Pune.

REFERENCES

Alden, C. J., & Arnott, S. (1975). Visualization of planar drug intercalations in B-DNA. *Nucleic Acids Research, 2*(10), 1701–1718.
Alessandro, L., & Francesco, L. G. (2008). Metadynamics: A method to simulate rare events and reconstruct the free energy in biophysics, chemistry and material science. *Reports on Progress in Physics, 71*(12), 126601.
Atkins, P. W. (1997). *Physical chemistry*. New York, NY: Macmillan Higher Education.
Baginski, M., Fogolari, F., & Briggs, J. M. (1997). Electrostatic and non–electrostatic contributions to the binding free energies of anthracycline antibiotics to DNA. *Journal of Molecular Biology, 274*(2), 253–267.
Baldini, G., Castoldi, B., Lucchin, G., & Zedda, A. (1982). Thermodynamics of the intercalation of acridines in DNA. *Il Nuovo Cimento, 1*(5), 637–647.
Baldini, G., Doglia, S., Dolci, S., & Sassi, G. (1981). Fluorescence-determined preferential binding of quinacrine to DNA. *Biophysical Journal, 36*(3), 465–477.
Barducci, A., Bussi, G., & Parrinello, M. (2008). Well-tempered metadynamics: A smoothly converging and tunable free-energy method. *Physical Review Letters, 100*(2), 020603.
Baruah, H., & Bierbach, U. (2003). Unusual intercalation of acridin-9-ylthiourea into the 5′-GA/TC DNA base step from the minor groove: Implications for the covalent DNA

adduct profile of a novel platinum–intercalator conjugate. *Nucleic Acids Research*, *31*(14), 4138–4146.

Bazzicalupi, C., Bencini, A., Bianchi, A., Biver, T., Boggioni, A., Bonacchi, S., et al. (2008). DNA binding by a new metallointercalator that contains a proflavine group bearing a hanging chelating unit. *Chemistry*, *14*(1), 184–196.

Berman, H. M., Westbrook, J., Feng, Z., Gilliland, G., Bhat, T. N., Weissig, H., et al. (2000). The protein data bank. *Nucleic Acids Research*, *28*(1), 235–242. http://dx.doi.org/10.1093/nar/28.1.235.

Berman, H. M., & Young, P. R. (1981). The intercalation of intercalating drugs with nucleic acids. *Annual Review of Biophysics and Bioengineering*, *10*, 87–114.

Biver, T., Secco, F., Tinè, M. R., & Venturini, M. (2003). Equilibria and kinetics of the intercalation of Pt-proflavine and proflavine into calf thymus DNA. *Archives of Biochemistry and Biophysics*, *418*(1), 63–70.

Biver, T., Secco, F., & Venturini, M. (2008). Mechanistic aspects of the interaction of intercalating metal complexes with nucleic acids. *Coordination Chemistry Reviews*, *252*(10–11), 1163–1177.

Bonomi, M., Branduardi, D., Bussi, G., Camilloni, C., Provasi, D., Raiteri, P., et al. (2009). PLUMED: A portable plugin for free-energy calculations with molecular dynamics. *Computer Physics Communications*, *180*(10), 1961–1972.

Bradof, J. E., Hakes, T. B., Ochoa, M., & Golbey, R. (1982). Germ cell malignancies of the ovary. Treatment with vinblastine, actinomycin D, bleomycin and cisplatin containing chemotherapy combinations. *Cancer*, *50*(6), 1070–1075.

Breslauer, K. J., Remeta, D. P., Chou, W.-Y., Ferrante, R., Curry, J., Zaunczkowski, D., et al. (1987). Enthalpy–entropy compensations in drug–DNA binding studies. *Proceedings of the National Academy of Sciences of the United States of America*, *84*(24), 8922–8926.

Bresloff, J. L., & Crothers, D. M. (1975). DNA-ethidium reaction kinetics: Demonstration of direct ligand transfer between DNA binding sites. *Journal of Molecular Biology*, *95*(1), 103–110.

Browning, C. H., Gulbransen, R., Kennaway, E. L., & Thornton, L. H. D. (1917). Flavine and Brilliant Green, powerful antiseptics with low toxicity to the tissues: Their use in the treatment of infected wounds. *British Medical Journal*, *1*, 73–78.

Browning, C. H., Gulbransen, R., & Thornton, L. H. D. (1917). The antiseptic properties of acriflavinie and proflavine, and brilliant green: With special reference to suitability for wound therapy. *British Medical Journal*, *2*, 70.

Brunnberg, U., Mohr, M., Noppeney, R., Dürk, H. A., Sauerland, M. C., Müller-Tidow, C., et al. (2012). Induction therapy of AML with ara-C plus daunorubicin versus ara-C plus gemtuzumab ozogamicin: A randomized phase II trial in elderly patients. *Annals of Oncology*, *23*(4), 990–996.

Caldwell, G. W., & Yan, Z. (2004). Isothermal titration calorimetry characterization of drug-binding energetics to blood proteins. *Optimization in Drug Discovery*, 123–149.

Camilloni, C., Provasi, D., Tiana, G., & Broglia, R. A. (2008). Exploring the protein G helix free-energy surface by solute tempering metadynamics. *Proteins: Structure, Function, and Bioinformatics*, *71*(4), 1647–1654.

Canals, A., Purciolas, M., Aymami, J., & Coll, M. (2005). The anticancer agent ellipticine unwinds DNA by intercalative binding in an orientation parallel to base pairs. *Acta Crystallographica Section D*, *61*(7), 1009–1012.

Capranico, G., Soranzo, C., & Zunino, F. (1986). Single-strand DNA breaks induced by chromophore-modified anthracyclines in P388 leukemia cells. *Cancer Research*, *46*, 5499–5503.

Carmen Avendano, J. C. M. (2008). *Medicinal chemistry of anticancer drugs* (Vol. 14). Amsterdam: Elsevier.

Case, D. A., Darden, T. A., Cheatham I, T. E., Simmerling, C. L., Wang, J., Duke, R. E., et al. (2010). *AMBER 11*. San Francisco: University of California.

Cavalieri, L. F., & Nemchin, R. G. (1964). The mode of interaction of actinomycin D with deoxyribonucleic acid. *Biochimica et Biophysica Acta (BBA)—Specialized Section on Nucleic Acids and Related Subjects, 87*(4), 641–652.

Cavallari, M., Garbesi, A., & Di Felice, R. (2009). Porphyrin intercalation in G4-DNA quadruplexes by molecular dynamics simulations. *The Journal of Physical Chemistry. B, 113*(40), 13152–13160.

Chaires, J. B. (1983). Equilibrium studies on the interaction of daunomycin with deoxypolynucleotides. *Biochemistry, 22*(18), 4204–4211.

Chaires, J. B., Dattagupta, N., & Crothers, D. M. (1982). Studies on interaction of anthracycline antibiotics and deoxyribonucleic acid: Equilibrium binding studies on interaction of daunomycin with deoxyribonucleic acid. *Biochemistry, 21*(17), 3933–3940.

Chaires, J. B., Dattagupta, N., & Crothers, D. M. (1985). Kinetics of the daunomycin–DNA interaction. *Biochemistry, 24*(2), 260–267.

Chaires, J. B., Herrera, J. E., & Waring, M. J. (1990). Preferential binding of daunomycin to 5′ATCG and 5′ATGC sequences revealed by footprinting titration experiments. *Biochemistry, 29*(26), 6145–6153.

Chaires, J. B., Satyanarayana, S., Suh, D., Fokt, I., Przewloka, T., & Priebe, W. (1996). Parsing the free energy of anthracycline antibiotic binding to DNA. *Biochemistry, 35*(7), 2047–2053.

Chen, K. X., Gresh, N., & Pullman, B. (1985). A theoetical investigation on the sequence selective binding of daunomycin to double-stranded polynucleotides. *Journal of Biomolecular Structure & Dynamics, 3*(3), 445–466.

Chirico, G., Lunelli, L., & Baldini, G. (1990). Conformation of intercalated DNA plasmids investigated by circular dichroism and dynamic light scattering. *Biophysical Chemistry, 38*(3), 201–211.

Ciatto, C., D'Amico, M. L., Natile, G., Secco, F., & Venturini, M. (1999). Intercalation of proflavine and a platinum derivative of proflavine into double-helical Poly(A). *Biophysical Journal, 77*(5), 2717–2724.

Clark, D. P., & Pazdernik, N. J. (2012). *Molecular biology*. Burlington, MA: Academic Press.

Cohen, G., & Eisenberg, H. (1969). Viscosity and sedimentation study of sonicated DNA–proflavine complexes. *Biopolymers, 8*(1), 45–55.

Comba, P. (2010). *Structure and function*. New York, NY: Springer.

Connors, K. A. (1990). *Chemical kinetics: The study of reaction rates in solution*. New York: John Wiley & Sons, Incorporated.

Cooper, G. M. (1997). *The cell: A molecular approach*. Washington, DC: ASM Press (electronic resource).

Corin, A. F., & Jovin, T. M. (1986). Proflavine binding to poly[d(A–T)] and poly[d(A–Br5U)]—Triplet-state and temperature-jump kinetics. *Biochemistry, 25*(14), 3995–4007.

Crawford, L. V., & Waring, M. J. (1967). Supercoiling of polyoma virus DNA measured by its interaction with ethidium bromide. *Journal of Molecular Biology, 25*(1), 23–30.

Cummings, M. R. (2011). *Human heredity: Principles & issues*. Belmont, CA: Brooks/Cole.

Davidson, J. J. N., & Cohn, W. E. (1964). *Progress in nucleic acid research and molecular biology*. New York, NY: Elsevier Science.

Davis, T. M., & David Wilson, W. (2001). Surface plasmon resonance biosensor analysis of RNA-small molecule interactions. In M. J. Waring & J. B. Chaires (Eds.), *Methods in enzymology. Vol. 340*. (pp. 22–51). New York, NY: Academic Press.

Dearing, A. (1981). *CHEM: A molecular display program*. San Francisco: University of California.

Dearing, A., Weiner, P., & Kollman, P. A. (1981). Molecular mechanical studies of proflavine and acridine orange intercalation. *Nucleic Acids Research, 9*(6), 1483–1498.

DeJong, E. S., Chang, C. E., Gilson, M. K., & Marino, J. P. (2003). Proflavine acts as a Rev inhibitor by targeting the high-affinity Rev binding site of the Rev responsive element of HIV-1. *Biochemistry, 42*(26), 8035–8046.

Denny, W. A. (2002). Acridine derivatives as chemotherapeutic agents. *Current Medicinal Chemistry, 9*, 1655–1665.

Elcock, A. H., Rodger, A., & Richards, W. G. (1996). Theoretical studies of the intercalation of 9-hydroxyellipticine in DNA. *Biopolymers, 39*(3), 309–326.

Figgitt, D., Denny, W., Chavalitshewinkoon, P., Wilairat, P., & Ralph, R. (1992). In vitro study of anticancer acridines as potential antitrypanosomal and antimalarial agents. *Antimicrobial Agents and Chemotherapy, 36*(8), 1644–1647.

Forster, W., & Stutter, E. (1984). Interaction of anthracycline antibiotics with bio-polymers. 9. Comparative-study of the interaction kinetics of daunomycin, adriamycin and iremycin with DNA. *International Journal of Biological Macromolecule, 6*(3), 114–124.

Frisch, M. J., Trucks, G. W., Schlegel, H. B., Scuseria, G. E., Robb, M. A., Cheeseman, J. R., et al. (2003). *Gaussian 03, Revision C.02*. Gaussian, Inc., Wallingford CT.

Fritzsche, H., Triebel, H., Chaires, J. B., Dattagupta, N., & Crothers, D. M. (1982). Studies on intercalation of anthracycline antibiotics and deoxyribonucleic acid: Geometry of intercalation of iremycin and daunomycin. *Biochemistry, 21*, 3940–3946.

Fuller, W., & Waring, M. J. (1964). A molecular model for the interaction of ethidium bromide with deoxyribonucleic acid. *Berichte der Bunsengesellschaft für physikalische Chemie, 68*(8–9), 805–808.

Gao, Q., Williams, L. D., Egli, M., Rabinovich, D., Chen, S. L., Quigley, G. J., et al. (1991). Drug-induced DNA repair: X-ray structure of a DNA–dibercalinium complex. *Proceedings of the National Academy of Sciences of the United States of America, 88*(6), 2422–2426.

Garbett, N. C., Hammond, N. B., & Graves, D. E. (2004). Influence of the amino substituents in the interaction of ethidium bromide with DNA. *Biophysical Journal, 87*(6), 3974–3981.

George, S., McBurney, A., & Ward, J. (1984). The protein binding of timegadine determined by equilibrium dialysis. *British Journal of Clinical Pharmacology, 18*, 785–790.

Giermasz, A., Makowski, M., Nowis, D., Jalili, A., Maj, M., Dabrowska, A., et al. (2002). Potentiated antitumor effects of butyrate and actinomycin D in melanoma model in mice. *Oncology Reports, 9*(1), 199.

Gruenewald, B., & Knoche, W. (1978). Pressure jump method with detection of optical rotation and circular dichroism. *The Review of Scientific Instruments, 49*(6), 797–801.

Grueso, E., López-Pérez, G., Castellano, M., & Prado-Gotor, R. (2012). Thermodynamic and structural study of phenanthroline derivative ruthenium complex/DNA interactions: Probing partial intercalation and binding properties. *Journal of Inorganic Biochemistry, 106*(1), 1–9.

Gulbransen, R., & Browning, C. H. (1921). The antiseptic potency of acriflavine, with considerations on the variability of results in testing antiseptics. *British Journal of Experimental Pathology, 2*, 95–102.

Günther, K., Mertig, M., & Seidel, R. (2010). Mechanical and structural properties of YOYO-1 complexed DNA. *Nucleic Acids Research, 38*(19), 6526–6532.

Handfield-Jones, R. P. (1949). Chloroquine, proguanil, mepacrine and quinine in the treatment of malaria caused by *Plasmodium falciparum*. *Annals of Tropical Medicine and Parasitology, 43*, 345–348.

Haq, I. (2002). Thermodynamics of drug–DNA interactions. *Archives of Biochemistry and Biophysics, 403*(1), 1–15.

Harousseau, J. L., Reiffers, J., Hurteloup, P., Milpied, N., Guy, H., Rigal-Huguet, F., et al. (1989). Treatment of relapsed acute myeloid leukemia with idarubicin and intermediate-dose cytarabine. *Journal of Clinical Oncology, 7*(1), 45–49.

Hazarika, P., Bezbaruah, B., Das, P., Medhi, O., & Medhi, C. (2011). A model study on the stacking interaction of phenanthroline ligand with nucleic acid base pairs: An ab initio, MP2 and DFT studies. *Journal of Biophysical Chemistry*, *2*(2), 152–157.

Heilweil, H. G., & Winkle, Q. V. (1955). Studies on the interaction of desoxyribonucleic acid with acriflavine. *The Journal of Physical Chemistry*, *59*(9), 939–943.

Herrera, J. E., & Chaires, J. B. (1989). A premelting conformational transition in poly(dA)-poly(dT) coupled to daunomycin binding. *Biochemistry*, *28*(5), 1993–2000.

Hess, B., Kutzner, C., van der Spoel, D., & Lindahl, E. (2008). GROMACS 4: Algorithms for highly efficient, load-balanced, and scalable molecular simulation. *Journal of Chemical Theory and Computation*, *4*(3), 435–447.

Hollstein, U. (1974). Actinomycin. Chemistry and mechanism of action. *Chemical Reviews*, *74*(6), 625–652.

Hopfinger, A. J., Cardozo, M. G., & Kawakami, Y. (1995). Molecular modelling of ligand–DNA intercalation interactions. *Journal of the Chemical Society, Faraday Transactions*, *91*(16), 2515–2524.

Horstmann, M. A., Hassenpflug, W. A., zur Stadt, U., Escherich, G., Janka, G., & Kabisch, H. (2005). Amsacrine combined with etoposide and high-dose methylprednisolone as salvage therapy in acute lymphoblastic leukemia in children. *Haematologica*, *90*(12), 1701–1703.

Hossain, M., & Suresh Kumar, G. (2009). DNA intercalation of methylene blue and quinacrine: New insights into base and sequence specificity from structural and thermodynamic studies with polynucleotides. *Molecular BioSystems*, *5*(11), 1311–1322.

Howerton, S. B., Nagpal, A., & Williams, L. D. (2003). Surprising roles of electrostatic interactions in DNA–ligand complexes. *Biopolymers*, *69*(1), 87–99.

Hu, G. G., Shui, X., Leng, F., Priebe, W., Chaires, J. B., & Williams, L. D. (1997). Structure of a DNA–bisdaunomycin complex. *Biochemistry*, *36*(20), 5940–5946.

Hurwitz, J., Furth, J. J., Malamy, M., & Alexander, M. (1962). The role of deoxyribonucleic acid in ribonucleic acid synthesis. III. The inhibition of the enzymatic synthesis of ribonucleic acid and deoxyribonucleic acid by actinomycin D and proflavine. *Proceedings of the National Academy of Sciences of the United States of America*, *48*(7), 1222–1230.

Jain, S. C., & Sobell, H. M. (1972). Stereochemistry of actinomycin binding to DNA: I. Refinement and further structural details of the actinomycin–deoxyguanosine crystalline complex. *Journal of Molecular Biology*, *68*(1), 1–20.

Jain, S. C., Tsai, C. C., & Sobell, H. M. (1977). Visualization of drug–nucleic acid interactions at atomic resolution: II. Structure of an ethidium/dinucleoside monophosphate crystalline complex, ethidium:5-iodocytidylyl (3′–5′) guanosine. *Journal of Molecular Biology*, *114*(3), 317–331.

Kantarjian, H. M., Talpaz, M., Kontoyiannis, D., Gutterman, J., Keating, M. J., Estey, E. H., et al. (1992). Treatment of chronic myelogenous leukemia in accelerated and blastic phases with daunorubicin, high-dose cytarabine, and granulocyte-macrophage colony-stimulating factor. *Journal of Clinical Oncology*, *10*(3), 398–405.

Kasai, N. (2005). *X-ray diffraction by macromolecules*. Heidelberg, New York, NY: Springer.

Kersten, W., Kersten, H., Szybalski, W., & Fiandt, M. (1966). Physicochemical properties of complexes between deoxyribonucleic acid and antibiotics which affect ribonucleic acid synthesis (actinomycin, daunomycin, cinerubin, nogalamycin, chromomycin, mithramycin, and olivomycin). *Biochemistry*, *5*(1), 236–244.

Kolar, M., Kubar, T., & Hobza, P. (2010). Sequence-dependent configurational entropy change of DNA upon intercalation. *The Journal of Physical Chemistry. B*, *114*(42), 13446–13454.

Kubař, T., Hanus, M., Ryjáček, F., & Hobza, P. (2006). Binding of cationic and neutral phenanthridine intercalators to a DNA oligomer is controlled by dispersion energy:

Quantum chemical calculations and molecular mechanics simulations. *Chemistry—A European Journal, 12*(1), 280–290.

Kumar, S., Rosenberg, J. M., Bouzida, D., Swendsen, R. H., & Kollman, P. A. (1992). The weighted histogram analysis method for free-energy calculations on biomolecules. I. The method. *Journal of Computational Chemistry, 13*(8), 1011–1021.

Kwon, J. H., Chitrapriya, N., Han, S. W., Lee, G. J., Lee, D. J., & Cho, T.-S. (2013). Effect of number and location of amine groups on the thermodynamic parameters on the acridine derivatives to DNA. *Bulletin of the Korean Chemical Society, 34*(3), 810–814.

Ladd, M. F. C., & Palmer, R. A. (2003). *Structure determination by X-ray crystallography*. New York: Kluwer Academic/Plenum Publishers.

Lavery, R., Zakrzewska, K., & Sklenar, H. (1995). JUMNA: Junction minimisation of nucleic acids. *Computer Physics Communication, 91*, 135–158.

Leith, J. D., Jr. (1963). Acridine orange and acriflavin inhibit deoxyribonuclease action. *Biochimica et Biophysica Acta (BBA)—Specialized Section on Nucleic Acids and Related Subjects, 72*, 643–644.

Lenglet, G., & David-Cordonnier, M.-H. (2010). DNA-destabilizing agents as an alternative approach for targeting DNA: Mechanisms of action and cellular consequences. *Journal of Nucleic Acids, 2010*, 1–17.

Lerman, L. S. (1961). Structural considerations in the interaction of DNA and acridines. *Journal of Molecular Biology, 3*, 18–30.

Li, S., Cooper, V. R., Thonhauser, T., Lundqvist, B. I., & Langreth, D. C. (2009). Stacking interactions and DNA intercalation. *The Journal of Physical Chemistry. B, 113*(32), 11166–11172.

Li, H. J., & Crothers, D. M. (1969). Relaxation studies of proflavine–DNA complex— Kinetics of an intercalation reaction. *Journal of Molecular Biology, 39*(3), 461–477.

Li, X.-L. , Hu, Y.-J. , Wang, H., Yu, B.-Q. , & Yue, H.-L. (2012). Molecular spectroscopy evidence of berberine binding to DNA: Comparative binding and thermodynamic profile of intercalation. *Biomacromolecules, 13*(3), 873–880.

Luzzati, V., Masson, F., & Lerman, L. S. (1961). Interaction of DNA and proflavine: A small-angle X-ray scattering study. *Journal of Molecular Biology, 3*(5), 634–639.

Macgregor, R. B., Clegg, R. M., & Jovin, T. M. (1985). Pressure-jump study of the kinetics of ethidium-bromide binding to DNA. *Biochemistry, 24*(20), 5503–5510.

Macgregor, R. B., Clegg, R. M., & Jovin, T. M. (1987). Viscosity dependence of ethidium-DNA intercalation kinetics. *Biochemistry, 26*(13), 4008–4016.

Maehigashi, T., Persil, O., Hud, N. V., & Williams, L. D. (2010). Crystal structure of proflavine in complex with a DNA hexamer duplex. Doi: http://dx.doi.org/10.2210/pdb3ft6/pdb.

Malogolowkin, M., Cotton, C. A., Green, D. M., Breslow, N. E., Perlman, E., Miser, J., et al. (2008). Treatment of Wilms tumor relapsing after initial treatment with vincristine, actinomycin D, and doxorubicin. A report from the National Wilms Tumor Study Group. *Pediatric Blood & Cancer, 50*(2), 236–241.

Marky, L. A., & Breslauer, K. J. (1982). Calorimetric determination of base-stacking enthalpies in double-helical DNA molecules. *Biopolymers, 21*(11), 2185–2194.

Martínez, R., & Chacón-García, L. (2005). The search of DNA-intercalators as antitumoral drugs: What it worked and what did not work. *Current Medicinal Chemistry, 12*, 127–151.

Mazerski, J., Antonini, I., & Martelli, S. (2000). The role of side chains in the interaction of new antitumor pyrimidoacridinetriones with DNA: Molecular dynamics simulations. *Acta Biochimica Polonica, 47*(1), 47–57.

Medhi, C., Mitchell, J. B. O., Price, S. L., & Tabor, A. B. (1999). Electrostatic factors in DNA intercalation. *Biopolymers, 52*(2), 84–93.

Meyers, M. A. (2007). *Happy accidents: Serendipity in modern medical breakthroughs*. New York, NY: Arcade Publishing.

Minuk, L. A., Monkman, K., Chin-Yee, I. H., Lazo-Langner, A., Bhagirath, V., Chin-Yee,- B. H., et al. (2012). Treatment of Hodgkin lymphoma with adriamycin, bleomycin, vin- blastine and dacarbazine without routine granulocyte-colony stimulating factor support does not increase the risk of febrile neutropenia: A prospective cohort study. *Leukemia & Lymphoma*, *53*(1), 57–63.

Morris, G. M., Huey, R., Lindstrom, W., Sanner, M. F., Belew, R. K., Goodsell, D. S., et al. (2009). AutoDock4 and AutoDockTools4: Automated docking with selective receptor flexibility. *Journal of Computational Chemistry*, *30*(16), 2785–2791.

Mross, K., Massing, U., & Kratz, F. (2006). DNA-intercalators—The anthracyclines. In H. Pinedo & C. Smorenburg (Eds.), *Drugs affecting growth of tumours* (pp. 19–81). Basel, Switzerland: Birkhäuser Basel.

Mukherjee, A. (2011). Entropy balance in the intercalation process of an anti-cancer drug daunomycin. *Journal of Physical Chemistry Letters*, *2*(24), 3021–3026.

Mukherjee, A., Lavery, R., Bagchi, B., & Hynes, J. T. (2008). On the molecular mechanism of drug intercalation into DNA: A simulation study of the intercalation pathway, free energy, and DNA structural changes. *Journal of the American Chemical Society*, *130*(30), 9747–9755.

Müller, S. (2008). *Nucleic acids from A to Z*. Weinheim: Wiley.

Muller, W., & Crothers, D. M. (1968). Studies of binding of actinomycin and related com- pounds to DNA. *Journal of Molecular Biology*, *35*(2), 251–290.

Mustard, D., & Ritchie, D. W. (2005). Docking essential dynamics eigenstructures. *Proteins*, *60*(2), 269–274.

Nafisi, S., Saboury, A. A., Keramat, N., Neault, J.-F. , & Tajmir-Riahi, H.-A. (2007). Sta- bility and structural features of DNA intercalation with ethidium bromide, acridine orange and methylene blue. *Journal of Molecular Structure*, *827*(1–3), 35–43.

Neidle, S., Achari, A., Taylor, G. L., Berman, H. M., Carrell, H. L., Glusker, J. P., et al. (1977). Structure of a dinucleoside phosphate–drug complex as model for nucleic acid–drug interaction. *Nature*, *269*(5626), 304–307.

Neidle, S., Berman, H. M., & Shieh, H. S. (1980). Highly structured water network in crys- tals of a deoxydinucleoside–drug complex. *Nature*, *288*(5787), 129–133.

Neidle, S., & Jones, T. A. (1975). Crystal structure of proflavine—DNA binding agent. *Nature*, *253*(5489), 284–285.

Neidle, S., Pearl, L. H., Herzyk, P., & Berman, H. M. (1988). A molecular model for proflavine–DNA intercalation. *Nucleic Acids Research*, *16*(18), 8999–9016.

Neville, D. M., Jr., & Davies, D. R. (1966). The interaction of acridine dyes with DNA: An X-ray diffraction and optical investigation. *Journal of Molecular Biology*, *17*(1), 57–74.

Newman, J. (1984). Dynamic light scattering as a probe of superhelical DNA–intercalating agent interaction. *Biopolymers*, *23*(6), 1113–1119.

Nishimura, T., Okobira, T., Kelly, A. M., Shimada, N., Takeda, Y., & Sakurai, K. (2007). DNA binding of tilorone: ^{1}H NMR and calorimetric studies of the intercalation. *Biochemistry*, *46*(27), 8156–8163.

Nordmeier, E. (1992). Absorption spectroscopy and dynamic and static light-scattering stud- ies of ethidium bromide binding to calf thymus DNA: Implications for outside-binding and intercalation. *The Journal of Physical Chemistry*, *96*(14), 6045–6055.

Nuss, M. E., Marsh, F. J., & Kollman, P. A. (1979). Theoretical studies of drug-dinucleotide interactions. Empirical energy function calculations on the interaction of ethidium, 9-aminoacridine, and proflavin cations with the base-paired dinucleotides GpC and CpG. *Journal of the American Chemical Society*, *101*(4), 825–833.

Olmsted 3rd, J., & Kearns, D. R. (1977). Mechanism of ethidium bromide fluorescence enhancement on binding to nucleic acids. *Biochemistry*, *16*(16), 3647–3654.

Olweny, C. L. M., Toya, T., Katongole-Mbidde, E., Lwanga, S. K., Owor, R., Kyalwazi, S., et al. (1974). Treatment of Kaposi's sarcoma by combination of actinomycin-D, vincristine and imidazole carboxamide (NSC-45388): Results of a randomized clinical trial. *International Journal of Cancer, 14*(5), 649–656.

Oster, G. (1951). Fluorescence quenching by nucleic acids. *Transactions of the Faraday Society, 47*, 660–666.

Owczarzy, R., Vallone, P. M., Gallo, F. J., Paner, T. M., Lane, M. J., & Benight, A. S. (1997). Predicting sequence-dependent melting stability of short duplex DNA oligomers. *Biopolymers, 44*(3), 217–239.

Pack, G. R., Hashimoto, G. M., & Loew, G. H. (1981). Quantum chemical calculations on the two-step mechanism of proflavine binding to DNA. *Annals of the New York Academy of Sciences, 367*(1), 240–249.

Pack, G. R., & Loew, G. (1978). Origins of the specificity in the intercalation of ethidium into nucleic acids. A theoretical analysis. *Biochimica et Biophysica Acta (BBA)—Nucleic Acids and Protein Synthesis, 519*(1), 163–172.

Peacocke, A. R., & Skerrett, N. J. H. (1956). The interaction of aminoacridines with nucleic acids. *Transactions of the Faraday Society, 52*, 261–279.

Perez, A., Marchan, I., Svozil, D., Sponer, J., Cheatham, T. E., III., Laughton, C. A., et al. (2007). Refinement of the AMBER force field for nucleic acids: Improving the description of α/γ conformers. *Biophysical Journal, 92*(11), 3817–3829.

Pettersen, E. F., Goddard, T. D., Huang, C. C., Couch, G. S., Greenblatt, D. M., Meng, E. C., et al. (2004). UCSF Chimera—A visualization system for exploratory research and analysis. *Journal of Computational Chemistry, 25*, 1605–1612.

Pfoh, R., Cuesta-Seijo, J. A., & Sheldrick, G. M. (2009). Interaction of an echinomycin–DNA complex with manganese ions. *Acta Crystallographica Section F, 65*(7), 660–664.

Purich, D. L., & Allison, R. D. (1999). *Handbook of biochemical kinetics: A guide to dynamic processes in the molecular life sciences.* San Diego, California, CA: Elsevier Science.

Qu, X., & Chaires, J. B. (2001). Hydration changes for DNA intercalation reactions. *Journal of the American Chemical Society, 123*(1), 1–7.

Quigley, G. J., Wang, A. H. J., Ughetto, G., Marel, G. V. D., Boom, J. H. V., & Rich, A. (1980). Molecular structure of an anticancer drug–DNA complex: Daunomycin plus d(CpGpTpApCpG). *Proceedings of the National Academy of Sciences of the United States of America, 77*(12), 7204–7208.

Ramstein, J., Ehrenberg, M., & Rigler, R. (1980). Fluorescence relaxation of proflavin–deoxyribonucleic acid interaction. Kinetic properties of a base-specific reaction. *Biochemistry, 19*(17), 3938–3948.

Ramstein, J., & Leng, M. (1975). Effect of DNA base composition on the intercalation of proflavine: A kinetic study. *Biophysical Chemistry, 3*(3), 234–240.

Rao, S. N., & Kollman, P. A. (1987). Molecular mechanical simulations on double intercalation of 9-amino acridine into d(CGCGCGC) × d(GCGCGCG): Analysis of the physical basis for the neighbor-exclusion principle. *Proceedings of the National Academy of Sciences of the United States of America, 84*(16), 5735–5739.

Reha, D., Kabelac, M., Ryjacek, F., Sponer, J., Sponer, J. E., Elstner, M., et al. (2002). Intercalators. 1. Nature of stacking interactions between intercalators (ethidium, daunomycin, ellipticine, and 4′,6-diaminide-2-phenylindole) and DNA base pairs. Ab initio quantum chemical, density functional theory, and empirical potential study. *Journal of the American Chemical Society, 124*(13), 3366–3376.

Richards, J. E., & Hawley, R. S. (2010). *The human genome: A user's guide.* San Diego, California, CA: Elsevier Science.

Rizzo, V., Sacchi, N., & Menozzi, M. (1989). Kinetic-studies of anthracycline-DNA interaction by fluorescence stopped flow confirm a complex association mechanism. *Biochemistry, 28*(1), 274–282.

Robinson, B. H. (1975). The stopped-flow and temperature-jump techniques—Principles and recent advances. In E. Wyn-Jones (Ed.), *Chemical and biological applications of relaxation spectrometry. Vol. 18.* (pp. 41–48). Netherlands: Springer.

Ross, W. E., & Bradley, M. O. (1981). DNA double-strand breaks in mammalian cells after exposure to intercalating agents. *Biochimica et Biophysica Acta (BBA)—Nucleic Acids and Protein Synthesis, 654*(1), 129–134.

Roux, B. (1995). The calculation of the potential of mean force using computer simulations. *Computer Physics Communications, 91,* 275–282.

Samuels, L. D., Newton, W. A., & Heyn, R. (1971). Daunorubicin therapy in advanced neuroblastoma. *Cancer, 27*(4), 831–834.

Sasikala, W. D., & Mukherjee, A. (2012). Molecular mechanism of direct proflavine–DNA intercalation: Evidence for drug-induced minimum base-stacking penalty pathway. *The Journal of Physical Chemistry. B, 116*(40), 12208–12212.

Sasikala, W. D., & Mukherjee, A. (2013). Intercalation and de-intercalation pathway of proflavine through the minor and major grooves of DNA: Roles of water and entropy. *Physical Chemistry Chemical Physics, 15*(17), 6446–6455.

Saucier, J. M. (1977). Physiocochemical studies on the interaction of irehdiamine A with bihelical DNA. *Biochemistry, 16*(26), 5879–5889.

Schiewek, M., Krumova, M., Hempel, G., & Blume, A. (2007). Pressure jump relaxation setup with IR detection and millisecond time resolution. *The Review of Scientific Instruments, 78*(4), 045101–045106.

Schulemann, W. (1932). Synthetic anti-malarial preparations. *Proceedings of the Royal Society of Medicine, 25*(6), 897–905.

Seddon, J. M., & Gale, J. D. (2001). *Thermodynamics and statistical mechanics.* Cambridge, UK: Royal Society of Chemistry.

Silver, R. T., Case, D. C., Wheeler, R. H., Miller, T. P., Stein, R. S., Stuart, J. J., et al. (1991). Multicenter clinical trial of mitoxantrone in non–Hodgkin's lymphoma and Hodgkin's disease. *Journal of Clinical Oncology, 9*(5), 754–761.

Singh, U. C., Pattabiraman, N., Langridge, R., & Kollman, P. A. (1986). Molecular mechanical studies of d(CGTACG)2: Complex of triostin A with the middle A–T base pairs in either Hoogsteen or Watson-Crick pairing. *Proceedings of the National Academy of Sciences of the United States of America, 83*(17), 6402–6406.

Sobell, H. M., Tsai, C.-C. , Jain, S. C., & Gilbert, S. G. (1977). Visualization of drug-nucleic acid interactions at atomic resolution: III. Unifying structural concepts in understanding drug-DNA interactions and their broader implications in understanding protein-DNA interactions. *Journal of Molecular Biology, 114*(3), 333–365.

Spielmann, H. P., Wemmer, D. E., & Jacobsen, J. P. (1995). Solution structure of a DNA complex with the fluorescent bis-intercalator TOTO determined by NMR spectroscopy. *Biochemistry, 34*(27), 8542–8553.

Strebhardt, K., & Ullrich, A. (2008). Paul Ehrlich's magic bullet concept: 100 years of progress. *Nature Reviews. Cancer, 8*(6), 473–480.

Swaminathan, S., Beveridge, D. L., & Berman, H. M. (1990). Molecular dynamics simulation of a deoxydinucleoside-drug intercalation complex: dCpG/proflavin. *The Journal of Physical Chemistry, 94*(11), 4660–4665.

Syed, S. N., Schulze, H., Macdonald, D., Crain, J., Mount, A. R., & Bachmann, T. T. (2013). Cyclic denaturation and renaturation of double-stranded DNA by redox-state switching of DNA intercalators. *Journal of the American Chemical Society, 135,* 5399–5407.

Thomes, J. C., Weill, G., & Daune, M. (1969). Fluorescence of proflavine–DNA complexes: Heterogeneity of binding sites. *Biopolymers, 8*(5), 647–659.

Torigoe, H., Sato, S., Yamashita, K., Obika, S., Imanishi, T., & Takenaka, S. (2002). Binding of threading intercalator to nucleic acids: Thermodynamic analyses. *Nucleic Acids Symposium Series, 2*(1), 55–56.

Torrie, G. M., & Valleau, J. P. (1977). Nonphysical sampling distributions in Monte Carlo free-energy estimation: Umbrella sampling. *Journal of Computational Physics, 23,* 187–199.

Trieb, M., Rauch, C., Wellenzohn, B., Wibowo, F., Loerting, T., Mayer, E., et al. (2004). Daunomycin intercalation stabilizes distinct backbone conformations of DNA. *Journal of Biomolecular Structure & Dynamics, 21*(5), 713–724.

Trieb, M., Rauch, C., Wibowo, F. R., Wellenzohn, B., & Liedl, K. R. (2004). Cooperative effects on the formation of intercalation sites. *Nucleic Acids Research, 32*(15), 4696–4703.

Tsai, C. C., Jain, S. C., & Sobell, H. M. (1975). X-ray crystallographic visualization of drug-nucleic acid intercalative binding: Structure of an ethidium-dinucleoside monophosphate crystalline complex, ethidium:5-iodouridylyl (3′-5′) adenosine. *Proceedings of the National Academy of Sciences of the United States of America, 72*(2), 628–632.

Tsai, C. C., Jain, S. C., & Sobell, H. M. (1977). Visualization of drug-nucleic acid interactions at atomic resolution: I. Structure of an ethidium/dinucleoside monophosphate crystalline complex, ethidium:5-iodouridylyl (3′-5′) adenosine. *Journal of Molecular Biology, 114*(3), 301–315.

Ugwu, S., Alcala, M., Bhardwaj, R., & Blanchard, J. (1996). The application of equilibrium dialysis to the determination of drug–cyclodextrin stability constants. *Journal of Inclusion Phenomena and Molecular Recognition in Chemistry, 25*(1–3), 173–176.

Valdés, A. F.-C. (2011). Acridine and acridinones: Old and new structures with antimalarial activity. *The Open Medicinal Chemistry Journal, 5,* 11–20.

Wainwright, M. (2001). Acridine—A neglected antibacterial chromophore. *The Journal of Antimicrobial Chemotherapy, 47,* 1–13.

Wang, J., Cieplak, P., & Kollman, P. A. (2000). How well does a restrained electrostatic potential (RESP) model perform in calculating conformational energies of organic and biological molecules? *Journal of Computational Chemistry, 21,* 1049–1074.

Wang, A. H., Ughetto, G., Quigley, G. J., Hakoshima, T., van der Marel, G. A., van Boom, J. H., et al. (1984). The molecular structure of a DNA-triostin A complex. *Science, 225*(4667), 1115–1121.

Wang, A. H.-J. , Ughetto, G., Quigley, G. J., & Rich, A. (1987). Interactions between an anthracycline antibiotic and DNA: Molecular structure of daunomycin complexed to d(CpGpTpApCpG) at 1.2-A resolution. *Biochemistry, 26*(4), 1152–1163.

Wang, J., Wolf, R. M., Caldwell, J. W., Kollman, P. A., & Case, D. A. (2004). Development and testing of a general AMBER force field. *Journal of Computational Chemistry, 25,* 1157–1174.

Waring, M. J. (1965a). Complex formation between ethidium bromide and nucleic acids. *Journal of Molecular Biology, 13*(1), 269–282.

Waring, M. J. (1965b). The effects of antimicrobial agents on ribonucleic acid polymerase. *Molecular Pharmacology, 1*(1), 1–13.

Waring, M. J. (1968). Drugs which affect structure and function of DNA. *Nature, 219*(5161), 1320–1325.

Waring, M. J. (1970). Variation of the supercoils in closed circular DNA by binding of antibiotics and drugs: Evidence for molecular models involving intercalation. *Journal of Molecular Biology, 54,* 247–279.

Waring, M. J. (1974). Stabilization of two-stranded ribohomopolymer helices and destabilization of a three-stranded helix by ethidium bromide. *The Biochemical Journal, 143*(2), 483–486.

Waring, M. J. (1981). DNA modification and cancer. *Annual Review of Biochemistry, 50,* 159–192.

Westerlund, F., Wilhelmsson, L. M., Nordén, B., & Lincoln, P. (2003). Micelle-sequestered dissociation of cationic DNA-intercalated drugs: Unexpected surfactant-induced rate enhancement. *Journal of the American Chemical Society, 125*(13), 3773–3779.

Westerlund, F., Wilhelmsson, L. M., Nordén, B., & Lincoln, P. (2005). Monitoring the DNA binding kinetics of a binuclear ruthenium complex by energy transfer: Evidence for slow shuffling. *The Journal of Physical Chemistry B, 109*(44), 21140–21144.

Wilhelm, M., Mukherjee, A., Bouvier, B., Zakrzewska, K., Hynes, J. T., & Lavery, R. (2012). Multistep drug intercalation: Molecular dynamics and free energy studies of the binding of daunomycin to DNA. *Journal of the American Chemical Society, 134*(20), 8588–8596.

Williams, L. D., Egli, M., Gao, Q., & Rich, A. (1992). DNA intercalation: Helix unwinding and neighbor-exclusion. *Structure and Function, 1,* 107.

Wilson, W. D. (Ed.), (1999). In *DNA and RNA intercalators: DNA aspects of molecular biology* (7). Amsterdam: Elsevier.

Wilson, W. D., Krishnamoorthy, C. R., Wang, Y. H., & Smith, J. C. (1985). Mechanism of intercalation—Ion effects on the equilibrium and kinetic constants for the interaction of propidium and ethidium with DNA. *Biopolymers, 24*(10), 1941–1961.

Wilson, W. D., Ratmeyer, L., Zhao, M., Strekowski, L., & Boykin, D. (1993). The search for structure-specific nucleic acid-interactive drugs: Effects of compound structure on RNA versus DNA interaction strength. *Biochemistry, 32*(15), 4098–4104.

Winger, R. H., Liedl, K. R., Rüdisser, S., Pichler, A., Hallbrucker, A., & Mayer, E. (1998). B-DNA's BI → BII conformer substrate dynamics is coupled with water migration. *The Journal of Physical Chemistry. B, 102*(44), 8934–8940.

Witkop, B. (1999). Paul Ehrlich and his magic bullets—Revisited. *Proceedings of the American Philosophical Society, 143*(4), 540–557.

Xiao, J. X., Lin, J. T., & Tian, B. G. (1994). Denaturation temperature of DNA. *Physical Review E, 50*(6), 5039–5042.

Xodo, L. E., Manzini, G., Ruggiero, J., & Quadrifoglio, F. (1988). On the interaction of daunomycin with synthetic alternating DNAs: Sequence specificity and polyelectrolyte effects on the intercalation equilibrium. *Biopolymers, 27*(11), 1839–1857.

Xu, Y., Wang, Y., Yan, L., Liang, R.-M. , Dai, B.-D. , Tang, R.-J. , et al. (2009). Proteomic analysis reveals a synergistic mechanism of fluconazole and berberine against fluconazole-resistant *Candida albicans*: Endogenous ROS augmentation. *Journal of Proteome Research, 8*(11), 5296–5304.

Yu, H.-H. , Cha, J.-D. , Kim, H.-K. , Lee, Y.-E. , Choi, N.-Y. , & You, Y.-O. (2005). Antimicrobial activity of berberine alone and in combination with ampicillin or oxacillin against methicillin–resistant *Staphylococcus aureus*. *Journal of Medicinal Food, 8*(4), 454–461.

CHAPTER TWO

Ligand Docking Simulations by Generalized-Ensemble Algorithms

Yuko Okamoto[*,†,‡,§,1]**, Hironori Kokubo**[¶]**, Toshimasa Tanaka**[¶]
[*]Department of Physics, Graduate School of Science, Nagoya University, Nagoya, Japan
[†]Structural Biology Research Center, Graduate School of Science, Nagoya University, Nagoya, Japan
[‡]Center for Computational Science, Graduate School of Engineering, Nagoya University, Nagoya, Japan
[§]Information Technology Center, Nagoya University, Nagoya, Japan
[¶]Pharmaceutical Research Division, Takeda Pharmaceutical Co., Ltd., Fujisawa, Japan
[1]Corresponding author: e-mail address: okamoto@phys.nagoya-u.ac.jp

Contents

Abstract

In protein chemistry and structural biology, conventional simulations in physical statistical mechanical ensembles, such as the canonical ensemble with fixed temperature and isobaric–isothermal ensemble with fixed temperature and pressure, face a great difficulty. This is because there exist a huge number of local-minimum-energy states in the system and the conventional simulations tend to get trapped in these states, giving wrong results. Generalized-ensemble algorithms are based on artificial unphysical ensembles and overcome the above difficulty by performing random walks in potential energy, volume, and other physical quantities or their corresponding conjugate parameters such as temperature and pressure. The advantage of generalized-ensemble simulations lies in the fact that they not only avoid getting trapped in states of energy local minima but also allow the calculations of physical quantities as functions of temperature or other parameters from a single simulation run. In this chapter, we review the generalized-ensemble algorithms. Some of their specific examples such as replica-exchange molecular dynamics and replica-exchange umbrella sampling are described in detail. Examples of their applications to drug design are presented.

Advances in Protein Chemistry and Structural Biology, Volume 92
ISSN 1876-1623
http://dx.doi.org/10.1016/B978-0-12-411636-8.00002-X

1. INTRODUCTION

A large number of experimentally determined protein structures are now available from the Protein Data Bank (PDB), and these structures are often used for structure-based drug design. There are many docking software packages available, such as GOLD (Jones, Willett, Glen, Leach, & Taylor, 1997), GLIDE (Friesner et al., 2006), and FlexX (Kramer, Rarey, & Lengauer, 1999). These software packages are often routinely used for computational drug design. However, most existing docking software to predict ligand structures binding with proteins employs empirical scoring functions. Docking scoring functions usually neglect some of the important factors such as the desolvation free energy of a receptor and a ligand, receptor flexibility, strain energy, and entropy.

In order to have more quantitative predictions than the above docking software, molecular simulation methods such as Monte Carlo (MC) and molecular dynamics (MD) are often used to calculate the free energy of ligand binding (Gallicchio & Levy, 2011; Gilson & Zhou, 2007; Grant, Gorfe, & McCammon, 2010; Jiang, Hodoscek, & Roux, 2009; Meng, Dashti, & Roitberg, 2011). However, molecular simulation techniques are not free from difficulties, either. They require much more computational power than docking software, and conventional canonical simulations at physically relevant temperatures tend to get trapped in states of energy local minima, giving wrong results. A class of simulation methods, which are referred to as the *generalized-ensemble algorithms*, overcome the latter difficulty (for reviews, see, e.g., Hansmann & Okamoto, 1999; Kokubo & Okamoto, 2006; Mitsutake, Sugita, & Okamoto, 2001; Okamoto, 2004; Sugita & Okamoto, 2002). In the generalized-ensemble algorithm, each state is weighted by an artificial, non-Boltzmann probability weight factor so that random walks in potential energy, volume, and other physical quantities or their corresponding conjugate parameters such as temperature and pressure may be realized. The random walks allow the simulation to escape from any energy barrier and to sample much wider conformational space than by conventional methods.

One of the effective generalized-ensemble algorithms for molecular simulations is the *replica-exchange method* (REM) (Hukushima & Nemoto, 1996) (the method is also referred to as parallel tempering; Marinari, Parisi, & Ruiz-Lorenzo, 1997). In this method, a number of noninteracting copies (or, replicas) of the original system at different temperatures are simulated independently and exchanged with a specified transition probability. The details of MD algorithm for REM, which is referred to as the *replica-exchange molecular dynamics* (REMD), have been worked out in Sugita and Okamoto (1999), and this has led to a wide application of REMD in the protein and other biomolecular systems. One is naturally led to a multidimensional (or, multivariable) extension of REM, which we refer to as the *multidimensional replica-exchange method* (MREM) (Sugita, Kitao, & Okamoto, 2000), which is also referred to as Hamiltonian REM (Fukunishi, Watanabe, & Takada, 2002). A special realization of MREM is *replica-exchange umbrella sampling* (REUS) (Sugita et al., 2000), which combines the conventional umbrella sampling method (Torrie & Valleau, 1997) and REM, and it is particularly useful in free-energy calculations for pharmaceutical design. General formulations for multidimensional generalized-ensemble algorithms have also been worked out (Mitsutake, 2009; Mitsutake & Okamoto, 2009), and special versions for isobaric–isothermal ensemble have been developed (Mori & Okamoto, 2010a; Okumura & Okamoto, 2006; Sugita & Okamoto, 2002).

In this chapter, we describe the generalized-ensemble algorithms mentioned above. We review the three methods: REM, MREM, and REUS. Examples of the results where these methods were applied to the predictions of protein–ligand binding structures are then presented.

2. METHODS

2.1. Replica-exchange method

Let us consider a system of N atoms of mass m_k ($k=1,\ldots,N$) with their coordinate vectors and momentum vectors denoted by $q=(\boldsymbol{q}_1,\ldots,\boldsymbol{q}_N)$ and $p=(\boldsymbol{p}_1,\ldots,\boldsymbol{p}_N)$, respectively. The Hamiltonian $H(q,p)$ of the system is the sum of the kinetic energy $K(p)$ and the potential energy $E(q)$:

$$H(q,p) = K(p) + E(q), \tag{2.1}$$

where

$$K(p) = \sum_{k=1}^{N} \frac{\boldsymbol{p}_k^2}{2m_k}. \tag{2.2}$$

In the canonical ensemble at temperature T each state $x \equiv (q,p)$ with the Hamiltonian $H(q,p)$ is weighted by the Boltzmann factor:

$$W_B(x; T) = \exp(-\beta H(q,p)), \tag{2.3}$$

where the inverse temperature β is defined by $\beta = 1/k_B T$ (k_B is the Boltzmann constant). The average kinetic energy at temperature T is then given by

$$\langle K(p) \rangle_T = \left\langle \sum_{k=1}^{N} \frac{p_k^2}{2m_k} \right\rangle_T = \frac{3}{2} N k_B T. \tag{2.4}$$

Because the coordinates q and momenta p are decoupled in Eq. (2.1), we can suppress the kinetic energy part and can write the Boltzmann factor as

$$W_B(x; T) = W_B(E; T) = \exp(-\beta E). \tag{2.5}$$

The canonical probability distribution of potential energy $P_{NVT}(E;T)$ is then given by the product of the density of states $n(E)$ and the Boltzmann weight factor $W_B(E;T)$:

$$P_{NVT}(E; T) \propto n(E) W_B(E; T). \tag{2.6}$$

Because $n(E)$ is a rapidly increasing function and the Boltzmann factor decreases exponentially, the canonical ensemble yields a bell-shaped distribution of potential energy which has a maximum around the average energy at temperature T. The conventional MC or MD simulations at constant temperature are expected to yield $P_{NVT}(E;T)$. An MC simulation based on the Metropolis algorithm (Metropolis, Rosenbluth, Rosenbluth, Teller, & Teller, 1953) is performed with the following transition probability from a state x of potential energy E to a state x' of potential energy E':

$$w(x \rightarrow x') = \min\left(1, \frac{W_B(E'; T)}{W_B(E; T)}\right) = \min(1, \exp(-\beta \Delta E)), \tag{2.7}$$

where

$$\Delta E = E' - E. \tag{2.8}$$

An MD simulation, on the other hand, is based on the following Newton equations of motion:

$$\dot{q}_k = \frac{p_k}{m_k}, \tag{2.9}$$

$$\dot{p}_k = -\frac{\partial E}{\partial q_k} = f_k, \tag{2.10}$$

where f_k is the force acting on the kth atom ($k = 1, \ldots, N$). This set of equations actually yield the microcanonical ensemble, however, and we have to add a thermostat in order to obtain the canonical ensemble at temperature T. Here, we just follow Nosé's prescription (Nosé, 1984a, 1984b), and we have

$$\dot{q}_k = \frac{p_k}{m_k}, \tag{2.11}$$

$$\dot{p}_k = -\frac{\partial E}{\partial q_k} - \frac{\dot{s}}{s} p_k = f_k - \frac{\dot{s}}{s} p_k, \tag{2.12}$$

$$\dot{s} = s\frac{P_s}{Q}, \tag{2.13}$$

$$\dot{P}_s = \sum_{k=1}^{N} \frac{p_k^2}{m_k} - 3Nk_B T = 3Nk_B(T(t) - T), \tag{2.14}$$

where s is Nosé's scaling parameter, P_s is its conjugate momentum, Q is its mass, and the "instantaneous temperature" $T(t)$ is defined by

$$T(t) = \frac{1}{3Nk_B} \sum_{k=1}^{N} \frac{p_k(t)^2}{m_k}. \tag{2.15}$$

However, in practice, it is very difficult to obtain accurate canonical distributions of complex systems at low temperatures by conventional MC or MD simulation methods. This is because simulations at low temperatures tend to get trapped in one or a few of local–minimum–energy states. This difficulty is overcome by, for instance, the generalized–ensemble algorithms, which greatly enhance conformational sampling.

The REM is one of the effective generalized–ensemble algorithms. The system for REM consists of M *noninteracting* copies (or, replicas) of the original system in the canonical ensemble at M different temperatures T_m $(m=1,\ldots,M)$. We arrange the replicas so that there is always exactly one replica at each temperature. Then there exists a one–to–one correspondence between replicas and temperatures; the label i $(=1,\ldots,M)$ for replicas is a permutation of the label m $(=1,\ldots,M)$ for temperatures, and vice versa:

$$\begin{cases} i = i(m) \equiv f(m), \\ m = m(i) \equiv f^{-1}(i), \end{cases} \tag{2.16}$$

where $f(m)$ is a permutation function of m and $f^{-1}(i)$ is its inverse.

Let $X = \{x_1^{[i(1)]}, \ldots, x_M^{[i(M)]}\} = \{x_{m(1)}^{[1]}, \ldots, x_{m(M)}^{[M]}\}$ stand for a "state" in this generalized ensemble. Each "substate" $x_m^{[i]}$ is specified by the coordinates $q^{[i]}$ and momenta $p^{[i]}$ of N atoms in replica i at temperature T_m:

$$x_m^{[i]} \equiv \left(q^{[i]}, p^{[i]} \right)_m. \tag{2.17}$$

Because the replicas are noninteracting, the weight factor for the state X in this generalized ensemble is given by the product of Boltzmann factors for each replica (or at each temperature):

$$
\begin{aligned}
W_{\mathrm{REM}}(X) &= \prod_{i=1}^{M} \exp\left\{ -\beta_{m(i)} H\left(q^{[i]}, p^{[i]} \right) \right\} = \prod_{m=1}^{M} \exp\left\{ -\beta_m H\left(q^{[i(m)]}, p^{[i(m)]} \right) \right\} \\
&= \exp\left\{ -\sum_{i=1}^{M} \beta_{m(i)} H\left(q^{[i]}, p^{[i]} \right) \right\} = \exp\left\{ -\sum_{m=1}^{M} \beta_m H\left(q^{[i(m)]}, p^{[i(m)]} \right) \right\},
\end{aligned}
\tag{2.18}
$$

where $i(m)$ and $m(i)$ are the permutation functions in Eq. (2.16).

We now consider exchanging a pair of replicas in this ensemble. Suppose we exchange replicas i and j which are at temperatures T_m and T_n, respectively:

$$X = \left\{ \ldots, x_m^{[i]}, \ldots, x_n^{[j]}, \ldots \right\} \to X' = \left\{ \ldots, x_m^{[j]\prime}, \ldots, x_n^{[i]\prime}, \ldots \right\}. \tag{2.19}$$

The exchange of replicas can be written in more detail as

$$
\begin{cases}
x_m^{[i]} \equiv \left(q^{[i]}, p^{[i]} \right)_m \to x_m^{[j]\prime} \equiv \left(q^{[j]}, p^{[j]\prime} \right)_m, \\
x_n^{[j]} \equiv \left(q^{[j]}, p^{[j]} \right)_n \to x_n^{[i]\prime} \equiv \left(q^{[i]}, p^{[i]\prime} \right)_n,
\end{cases}
\tag{2.20}
$$

where the definitions for $p^{[i]\prime}$ and $p^{[j]\prime}$ will be given below.

In the original implementation of the REM (Hukushima & Nemoto, 1996), MC algorithm was used, and only the coordinates q (and the potential energy function $E(q)$) had to be taken into account. In MD algorithm, on the other hand, we also have to deal with the momenta p. We proposed the following momentum assignment in Eq. (2.20) (Sugita & Okamoto, 1999):

$$
\begin{cases}
p^{[i]\prime} \equiv \sqrt{\dfrac{T_n}{T_m}} p^{[i]}, \\
p^{[j]\prime} \equiv \sqrt{\dfrac{T_m}{T_n}} p^{[j]},
\end{cases}
\tag{2.21}
$$

which we believe is the simplest and the most natural. This assignment means that we just rescale uniformly the velocities of all the atoms in the replicas by the square root of the ratio of the two temperatures so that the temperature condition in Eq. (2.4) may be satisfied immediately after replica exchange is accepted. We remark that similar momentum rescaling formulae for various constant temperature algorithms have been worked out in Mori and Okamoto (2010b).

The transition probability of this replica-exchange process is given by the usual Metropolis criterion:

$$w(X \rightarrow X') \equiv w\left(x_m^{[i]}|x_n^{[j]}\right) = \min\left(1, \frac{W_{\text{REM}}(X')}{W_{\text{REM}}(X)}\right) = \min(1, \exp(-\Delta)),$$

(2.22)

where in the second expression (i.e., $w(x_m^{[i]} \mid x_n^{[j]})$), we explicitly wrote the pair of replicas (and temperatures) to be exchanged. From Eqs. (2.1), (2.2), (2.18), and (2.21), we have

$$\Delta = \beta_m\left(E\left(q^{[j]}\right) - E\left(q^{[i]}\right)\right) - \beta_n\left(E\left(q^{[j]}\right) - E\left(q^{[i]}\right)\right)$$

(2.23)

$$= (\beta_m - \beta_n)\left(E\left(q^{[j]}\right) - E\left(q^{[i]}\right)\right).$$

(2.24)

Note that after introducing the momentum rescaling in Eq. (2.21), we have the same Metropolis criterion for replica exchanges, that is, Eqs. (2.22) and (2.24), for both MC and MD versions.

Without loss of generality, we can assume that $T_1 < T_2 < \cdots < T_M$. The lowest temperature T_1 should be sufficiently low so that the simulation can explore the experimentally relevant temperature region, and the highest temperature T_M should be sufficiently high so that no trapping in an energy-local-minimum state occurs. An REM simulation is then realized by alternately performing the following two steps:

1. Each replica in canonical ensemble of the fixed temperature is simulated *simultaneously* and *independently* for certain MC or MD steps.
2. A pair of replicas at neighboring temperatures, say, $x_m^{[i]}$ and $x_{m+1}^{[j]}$, are exchanged with the probability $w(x_m^{[i]} \mid x_{m+1}^{[j]})$ in Eq. (2.22).

A random walk in "temperature space" is realized for each replica, which in turn induces a random walk in potential energy space. This alleviates the problem of getting trapped in states of energy local minima.

After a long production run of a replica-exchange simulation, the canonical expectation value of a physical quantity A at temperature $T_m (m = 1, \ldots, M)$ can be calculated by the usual arithmetic mean:

$$\langle A \rangle_{T_m} = \frac{1}{n_m} \sum_{k=1}^{n_m} A(x_m(k)), \qquad (2.25)$$

where $x_m(k)$ $(k=1,\ldots,n_m)$ are the configurations obtained at temperature T_m and n_m is the total number of measurements made at $T=T_m$. The expectation value at any intermediate temperature T $(=1/k_B\beta)$ can also be obtained as follows:

$$\langle A \rangle_T = \frac{\sum_E A(E)P_{\text{NVT}}(E; T)}{\sum_E P_{\text{NVT}}(E; T)} = \frac{\sum_E A(E)n(E)\exp(-\beta E)}{\sum_E n(E)\exp(-\beta E)}. \qquad (2.26)$$

The density of states $n(E)$ in Eq. (2.26) is given by the multiple-histogram reweighting techniques or the weighted histogram analysis method (WHAM) (Ferrenberg & Swendsen, 1989; Kumar, Bouzida, Swendsen, Kollman, & Rosenberg, 1992; see also Mitsutake, Sugita, & Okamoto, 2003) as follows. Let $N_m(E)$ and n_m be, respectively, the potential energy histogram and the total number of samples obtained at temperature $T_m=1/k_B\beta_m$ $(m=1,\ldots,M)$. The best estimate of the density of states is then given by

$$n(E) = \frac{\sum_{m=1}^{M} N_m(E)}{\sum_{m=1}^{M} n_m \exp(f_m - \beta_m E)}, \qquad (2.27)$$

where we have for each m $(=1,\ldots,M)$

$$\exp(-f_m) = \sum_E n(E)\exp(-\beta_m E). \qquad (2.28)$$

Note that Eqs. (2.27) and (2.28) are solved self-consistently by iteration (Ferrenberg & Swendsen, 1989; Kumar et al., 1992) to obtain the density of states $n(E)$ and the dimensionless Helmholtz free energy f_m. Therefore, we can set all the f_m $(m=1,\ldots,M)$ to, for example, zero initially. We then use Eq. (2.27) to obtain $n(E)$, which is substituted into Eq. (2.28) to obtain next values of f_m, and so on.

2.2. Multidimensional REM

We now present the multidimensional/multivariable extension of the REM, which we refer to as the MREM (Sugita et al., 2000). Let us consider a generalized potential energy function $E_\lambda(x)$, which depends on L

parameters $\boldsymbol{\lambda} = (\lambda^{(1)}, \ldots, \lambda^{(L)})$, of a system in state x. Although $E_{\boldsymbol{\lambda}}(x)$ can be any function of $\boldsymbol{\lambda}$, we consider the following specific generalized potential energy function for simplicity:

$$E_{\boldsymbol{\lambda}}(x) = E_0(x) + \sum_{l=1}^{L} \lambda^{(l)} V_l(x). \tag{2.29}$$

Here, there are $L+1$ energy terms, $E_0(x)$ and $V_l(x)$ $(l = 1, \ldots, L)$, and $\lambda^{(l)}$ are the corresponding coupling constants for $V_l(x)$ (we collectively write $\boldsymbol{\lambda} = (\lambda^{(1)}, \ldots, \lambda^{(L)})$).

The crucial observation that led to MREM is: As long as we have M *noninteracting* replicas of the original system, the Hamiltonian $H(q, p)$ of the system does not have to be identical among the replicas and it can depend on a parameter with different parameter values for different replicas. The system for MREM consists of M noninteracting replicas of the original system in the "canonical ensemble" with $M (= M_0 \times M_1 \times \cdots \times M_L)$ different parameter sets $\boldsymbol{\Lambda}_m (m = 1, \ldots, M)$, where $\boldsymbol{\Lambda}_m \equiv (T_{m_0}, \boldsymbol{\lambda}_m) \equiv \left(T_{m_0}, \lambda_{m_1}^{(1)}, \ldots, \lambda_{m_L}^{(L)} \right)$ with $m_0 = 1, \ldots, M_0$, $m_l = 1, \ldots, M_l (l = 1, \ldots, L)$. Because the replicas are noninteracting, the weight factor is given by the product of Boltzmann–like factors for each replica:

$$W_{\text{MREM}} \equiv \prod_{m_0=1}^{M_0} \prod_{m_1=1}^{M_1} \cdots \prod_{m_L=1}^{M_L} \exp\left(-\beta_{m_0} E_{\boldsymbol{\lambda}_m} \right). \tag{2.30}$$

Without loss of generality, we can order the parameters so that $T_1 < T_2 < \cdots < T_{M_0}$ and $\lambda_1^{(l)} < \lambda_2^{(l)} < \cdots < \lambda_{M_l}^{(l)}$ (for each $l = 1, \ldots, L$). An MREM simulation is realized by alternately performing the following two steps:

1. For each replica, a "canonical" MC or MD simulation at the fixed parameter set is carried out *simultaneously* and *independently* for a few steps.
2. We exchange a pair of replicas i and j which are at the parameter sets $\boldsymbol{\Lambda}_m$ and $\boldsymbol{\Lambda}_{m+1}$, respectively. The transition probability for this replica-exchange process is given by

$$w(\boldsymbol{\Lambda}_m \leftrightarrow \boldsymbol{\Lambda}_{m+1}) = \min(1, \exp(-\Delta)), \tag{2.31}$$

where we have

$$\Delta = \left(\beta_{m_0} - \beta_{m_0+1} \right) \left(E_{\boldsymbol{\lambda}_m}\left(q^{[j]} \right) - E_{\boldsymbol{\lambda}_m}\left(q^{[i]} \right) \right), \tag{2.32}$$

for T-exchange, and

$$\Delta = \beta_{m_0} \left[\left(E_{\boldsymbol{\lambda}_{m_l}} \left(q^{[j]} \right) - E_{\boldsymbol{\lambda}_{m_l}} \left(q^{[i]} \right) \right) - \left(E_{\boldsymbol{\lambda}_{m_l+1}} \left(q^{[j]} \right) - E_{\boldsymbol{\lambda}_{m_l+1}} \left(q^{[i]} \right) \right) \right],$$

(2.33)

for $\lambda^{(l)}$-exchange (for one of $l=1,\ldots,L$). Here, $q^{[i]}$ and $q^{[j]}$ stand for configuration variables for replicas i and j, respectively, before the replica exchange.

Suppose we have made a single run of a short MREM simulation with $M(=M_0 \times M_1 \times \cdots \times M_L)$ replicas that correspond to M different parameter sets $\boldsymbol{\Lambda}_m\,(m=1,\ldots,M)$. Let $N_{m_0,m_1,\ldots,m_L}(E_0, V_1,\ldots,V_L)$ and $n_{m_0,m_1,\ldots,m_L}$ be, respectively, the $(L+1)$-dimensional potential energy histogram and the total number of samples obtained for the mth parameter set $\boldsymbol{\Lambda}_m = \left(T_{m_0},\lambda_{m_1}^{(1)},\ldots,\lambda_{m_L}^{(L)} \right)$. The generalized WHAM equations are then given by

$$n(E_0, V_1,\ldots,V_L) = \frac{\displaystyle\sum_{m_0,m_1,\ldots,m_L} N_{m_0,m_1,\ldots,m_L}(E_0, V_1,\ldots,V_L)}{\displaystyle\sum_{m_0,m_1,\ldots,m_L} n_{m_0,m_1,\ldots,m_L} \exp\left(f_{m_0,m_1,\ldots,m_L} - \beta_{m_0} E_{\boldsymbol{\lambda}_m}\right)}, \quad (2.34)$$

and

$$\exp\left(-f_{m_0,m_1,\ldots,m_L}\right) = \sum_{E_0, V_1,\ldots,V_L} n(E_0, V_1,\ldots,V_L) \exp\left(-\beta_{m_0} E_{\boldsymbol{\lambda}_m}\right). \quad (2.35)$$

The density of states $n(E_0, V_1,\ldots,V_L)$ and the dimensionless free energy $f_{m_0,m_1,\ldots,m_L}$ are obtained by solving Eqs. (2.34) and (2.35) self-consistently by iteration.

We now present the equations for calculating ensemble averages of physical quantities with any temperature T and any parameter λ values. After a long production run of MREM simulations, the canonical expectation value of a physical quantity A with the parameter values $\boldsymbol{\Lambda}_m\,(m=1,\ldots,M)$, where $\boldsymbol{\Lambda}_m \equiv (T_{m_0},\boldsymbol{\lambda}_m) \equiv \left(T_{m_0},\lambda_{m_1}^{(1)},\ldots,\lambda_{m_L}^{(L)} \right)$ with $m_0=1,\ldots,M_0,\; m_l=1,\ldots,$ $M_l(l=1,\ldots,L)$, and $M(=M_0 \times M_1 \times \cdots \times M_L)$, can be calculated by the usual arithmetic mean:

$$\langle A \rangle_{T_{m_0},\boldsymbol{\lambda}_m} = \frac{1}{n_m} \sum_{k=1}^{n_m} A(x_m(k)), \quad (2.36)$$

where $x_m(k)\,(k=1,\ldots,n_m)$ are the configurations obtained with the parameter values $\boldsymbol{\Lambda}_m\,(m=1,\ldots,M)$, and n_m is the total number of measurements made with these parameter values. The expectation values of A at any intermediate T $(=1/k_B\beta)$ and any $\boldsymbol{\lambda}$ can also be obtained from

$$\langle A \rangle_{T,\boldsymbol{\lambda}} = \frac{\displaystyle\sum_{E_0, V_1,\dots, V_L} A(E_0, V_1,\dots, V_L) n(E_0, V_1,\dots, V_L) \exp(-\beta E_{\boldsymbol{\lambda}})}{\displaystyle\sum_{E_0, V_1,\dots, V_L} n(E_0, V_1,\dots, V_L) \exp(-\beta E_{\boldsymbol{\lambda}})}, \quad (2.37)$$

where the density of states $n(E_0, V_1,\dots, V_L)$ is obtained from the multiple-histogram reweighting techniques.

We now describe a free-energy calculation method based on MREM, which we refer to as REUS (Sugita et al., 2000). In Eq. (2.34), we consider that $E_0(q)$ is the original unbiased potential and that $V_l(q)$ $(l=1,\dots,L)$ are the biasing (umbrella) potentials with $\lambda^{(l)}$ being the corresponding coupling constants $(\boldsymbol{\lambda} = (\lambda^{(1)},\dots,\lambda^{(L)}))$. Introducing a "reaction coordinate" ξ, the umbrella potentials are usually written as harmonic restraints:

$$V_l(q) = k_l(\xi(q) - d_l)^2, \quad (l=1,\dots,L), \quad (2.38)$$

where d_l are the midpoints and k_l are the strengths of the restraining potentials. We prepare M replicas with M different values of the parameters $\Lambda_m = (T_{m_0}, \boldsymbol{\lambda}_m)$, and the replica-exchange simulation is performed. Because the umbrella potentials $V_l(q)$ in Eq. (2.38) are all functions of the reaction coordinate ξ only, we can take the histogram $N_m(E_0,\xi)$ instead of $N_m(E_0, V_1,\dots, V_L)$. The WHAM equations can then be written as

$$P_{T,\boldsymbol{\lambda}}(E_0, \xi) = \left[\frac{\displaystyle\sum_{m=1}^{M} N_m(E_0, \xi)}{\displaystyle\sum_{m=1}^{M} n_m \exp\left(f_m - \beta_{m_0} E_{\boldsymbol{\lambda}_m}\right)} \right] \exp(-\beta E_{\boldsymbol{\lambda}}), \quad (2.39)$$

and

$$\exp(-f_m) = \sum_{E_0,\xi} P_{T_{m_0},\boldsymbol{\lambda}_m}(E_0, \xi). \quad (2.40)$$

The expectation value of a physical quantity A is now given by

$$\langle A \rangle_{T,\boldsymbol{\lambda}} = \frac{\displaystyle\sum_{E_0,\xi} A(E_0, \xi) P_{T,\boldsymbol{\lambda}}(E_0, \xi)}{\displaystyle\sum_{E_0,\xi} P_{T,\boldsymbol{\lambda}}(E_0, \xi)}. \quad (2.41)$$

The potential of mean force (PMF), or free energy as a function of the reaction coordinate, of the original, unbiased system at temperature T is given by

$$W_{T,\boldsymbol{\lambda}=\{0\}}(\xi) = -k_\mathrm{B} T \ln \left[\sum_{E_0} P_{T,\boldsymbol{\lambda}=\{0\}}(E_0, \xi) \right], \qquad (2.42)$$

where $\{0\} = (0,\dots,0)$.

2.3. Principal component analysis

We obtain the predicted binding structures of the ligand molecules, employing the principal component analysis (PCA) (Abagyan & Argos, 1992; Amadei, Linssen, & Berendsen, 1993; Garcia, 1992; Kitao, Hirata, & Go, 1991; Teeter & Case, 1990). PCA is an effective method for classifying the conformations of a molecule. First, n conformations, which have the reaction coordinate ξ that is close to the value corresponding to the global minimum of the PMF $W_{\boldsymbol{\lambda}=\{0\}}(\xi)$ and which have the potential energy (excluding the umbrella potential) close to the average (unbiased) potential energy, are extracted from the REUS simulations. We calculated the following variance–covariance matrix C_{ij}:

$$C_{ij} = \left\langle (q_i - \langle q_i \rangle)(q_j - \langle q_j \rangle) \right\rangle, \qquad (2.43)$$

where $\boldsymbol{q} = (q_1, q_2, q_3, \dots, q_{3N_L-2}, q_{3N_L-1}, q_{3N_L}) = (x_1, y_1, z_1, \dots, x_{N_L}, y_{N_L}, z_{N_L})$, $\langle q_i \rangle = \sum_{i=1}^{n} q_i / n$, and x_i, y_i, z_i are the Cartesian coordinates of the ith atom, N_L is the total number of ligand atoms, and n is the total number of MD trajectory samples. Note that we do not superimpose each structure onto a reference structure because our purpose is to predict ligand binding structures and we need to analyze differences not only in the ligand conformations but also in the ligand positions and orientations. This symmetric matrix is diagonalized, and the eigenvalues and eigenvectors are obtained. The first and the second principal component axes are defined as the eigenvectors with the largest and the second-largest eigenvalues, respectively. The ith principal component μ_i of each sampled structure is defined by the inner product:

$$\mu_i = \boldsymbol{v}_i \cdot (\boldsymbol{q} - \langle \boldsymbol{q} \rangle), \qquad (2.44)$$

where $\boldsymbol{v}_i$ is the ith eigenvector.

We analyzed the free-energy landscapes for the first and the second principal component axes and used the following equation to calculate the free energy as a function of two principal components μ_1 and μ_2:

$$F(\mu_1, \mu_2) = -k_B T \ln P(\mu_1, \mu_2), \qquad (2.45)$$

where $P(\mu_1, \mu_2)$ is the probability of finding the structure with the principal component values μ_1 and μ_2. We select a typical structure that exists in the global-minimum free-energy state on the free-energy landscape. This is our predicted binding structure.

3. RESULTS

We now present some examples of the simulation results by the algorithms described in Section 2.

These generalized-ensemble algorithms are especially useful for drug design. We examined in detail the dependency of free-energy landscapes on different charge parameters and solvent models by using six ligands as a test set (Okamoto, Tanaka, & Kokubo, 2010). The charge parameters obtained from different ligand conformations by AM1-BCC (Jakalian, Bush, Jack, & Bayly, 2000) and RESP (Bayly, Cieplak, Cornell, & Kollman, 1993; Cieplak, Cornell, Bayly, & Kollman, 1995) models were compared, and the free-energy landscapes obtained by using these charge parameters were examined in detail. Three different solvent conditions, vacuum, generalized Born (GB) implicit solvent model (Feig et al., 2004; Onufriev, Bashford, & Case, 2004), and TIP3P explicit solvent model (Jorgensen, Chandrasekhar, Madura, Impey, & Klein, 1983), were applied to explore the dependency on solvent conditions. We also examined where bioactive conformations, defined as the conformations of protein-bound ligands, were on free-energy landscapes.

We selected six pharmaceutical ligand molecules: imatinib, atazanavir, rivaroxaban, sildenafil, topiramate, and biotin, for all of which crystal structures of protein complexes have been reported. Figure 2.1 shows the structures of the six ligand molecules complexed to proteins, as registered in the PDB.

We first extracted ligand conformations from the PDB files and determined the RESP charges, which were calculated by *ab initio* molecular orbital programs at the HF/6-31G* level and RESP utilities in the Antechamber program suite (Wang, Wang, Kollman, & Case, 2006). The molecular mechanics parameters by general AMBER force field (GAFF) (Wang, Wang,

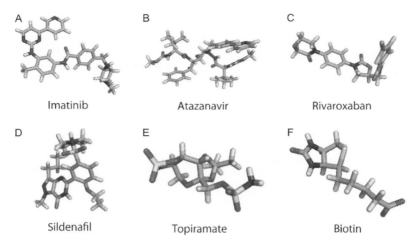

Figure 2.1 Ligand molecule conformations complexed to proteins (from PDB). PDB IDs are (A) 1XBB, (B) 2AQU, (C) 2W26, (D) 2H42, (E) 1EOU, and (F) 1STP. The figures were created with PyMOL (De Lano, 2002). *Reprinted from Okamoto et al. (2010) with kind permission of Springer (2010).* (For color version of this figure, the reader is referred to the online version of this chapter.)

Caldwell, Kollman, & Case, 2006; Wang, Wang, Kollman, et al., 2006) were employed for bond, angle, torsion, and van der Waals (vdW) parameters.

Missing force-field parameters were assigned with the parmcheck utility in AMBER (Case et al., 2005). After performing energy minimization by steepest–descent and conjugate gradient methods (in 500 steps for each minimization), we carried out an equilibration simulation for 1.0 ns at a temperature of 300 K. We performed five independent simulated annealing simulations (Kirkpatrick, Gelatt, & Vecchi, 1983) from 1000 to 200 K over 10 ns. The conformations thus obtained are referred to as the simulated annealing conformations. We obtained five independent simulated annealing conformations for each ligand. Two ligand conformations out of five were selected. The first one was the conformation with the largest root-mean-square deviation (RMSD) from the PDB conformation. The second one was the conformation which was judged to be dissimilar to both the PDB and the first conformations by visual inspection. Thus, we ultimately prepared three different conformations for each ligand: the native conformation from the PDB, simulated annealing conformation 1, and simulated annealing conformation 2.

Using these three conformations for each ligand, we again determined the atomic charge parameters by two different methods: AM1-BCC and

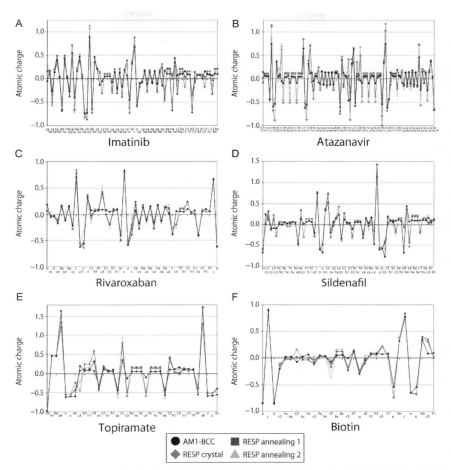

Figure 2.2 Atomic charges of six ligand molecules. The black circle shows the AM1-BCC charge and the other three symbols show the RESP charge on each atom based on the PDB conformations (red diamond, see Fig. 2.1) and conformations based on simulated annealing (blue square and green triangle). The two simulated annealing conformations are those with the largest RMSD values with respect to the PDB conformations out of five independent simulated annealing runs. The AM1-BCC charge model did not show conformational dependency; thus only the charges for the PDB conformations are shown. (For interpretation of the references to color in this figure legend, the reader is referred to the web version of this article.) *Reprinted from Okamoto et al. (2010) with kind permission of Springer (2010).*

RESP to examine the conformational dependency of charge parameters. Thus, six different charge parameters were obtained for each ligand. Figure 2.2 shows the obtained atomic charges for three ligand conformations and the two charge models. We show only the results for the PDB conformations in Fig. 2.2 because AM1–BCC charges were almost identical for the different conformations. On the other hand, we see that RESP charges show some conformational dependency. AM1–BCC was designed to yield charge sets of comparable quality to RESP charges and we confirmed that AM1–BCC well reproduces the overall trend of the RESP model in each case. However, when we look in detail, large deviations between AM1–BCC and RESP charges are observed for some atoms.

In order to examine how much these charge differences affect the free-energy landscapes, we performed REMD simulations. We used three different solvent conditions: vacuum, GB, and TIP3P. Thus, 18 independent simulations for each ligand were performed, corresponding to six charge parameter sets for each of three solvent conditions. For TIP3P conditions, we immersed each ligand in a TIP3P water cube with at least 8 Å solvation from the ligand surface, and the systems were neutralized by adding one sodium ion (for biotin) or one chloride ion (for imatinib and sildenafil) for charged ligands.

The following eight temperatures (eight replicas) were used for the REMD simulations in vacuum and GB solution: 260.5, 300.0, 345.5, 397.9, 458.3, 527.7, 607.8, and 700.0 K. On the other hand, 32 temperatures (32 replicas) were used for the REMD simulations in TIP3P water model: 300.0, 304.9, 310.2, 315.8, 321.7, 327.9, 334.3, 340.8, 347.6, 354.6, 361.7, 369.1, 376.7, 384.6, 392.7, 401.3, 410.0, 419.2, 428.8, 438.8, 449.3, 460.3, 471.8, 483.8, 496.4, 509.6, 523.3, 537.6, 552.5, 567.9, 583.7, and 600.0 K.

We performed REMD simulations of 20 ns (per replica) for all the systems and collected 20,000 snapshots at even intervals for each replica. We examined the influence on free-energy landscapes by the conformational differences in charge determination. Figures 2.3 and 2.4 show the free-energy landscapes for RESP charges determined from the PDB conformation and the simulated annealing conformations. Figure 2.3 shows the case of atazanavir in vacuum. Comparing (A-2) with (B-2), the conformations with a radius of gyration of 5.25 Å and RMSD of 4.0 Å are stable in (B-2), but these were rarely sampled and were unstable in (A-2). This implies that stable conformations can change greatly due to conformational differences in charge determination, even when the same RESP method is used. Figure 2.4 shows the case of TIP3P solution instead of vacuum with the same molecule and charge parameters as in Fig. 2.3. We see that the free-energy landscapes are very similar to each other even though the same charge parameters were used as in Fig. 2.3. We also observed a similar trend in other ligand molecules, namely that the influence of differences in charge parameters becomes small or disappears in GB and TIP3P solution compared

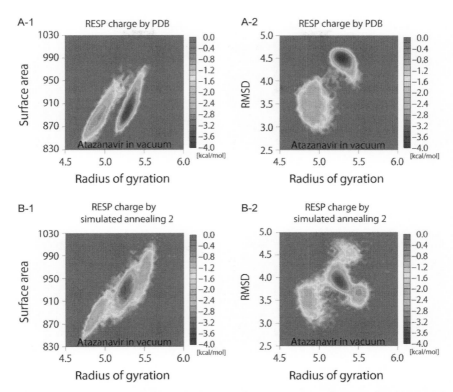

Figure 2.3 Free-energy landscapes of atazanavir in vacuum with RESP charge parameters determined using (A-1 and A-2) the PDB conformation (PDB ID: 2AQU) and (B-1 and B-2) the conformation obtained by simulated annealing. (A-1) and (B-1) The landscapes for the radius of gyration (in Å) and surface area (in Å^2), and (A-2) and (B-2) those for the radius of gyration (in Å) and RMSD (in Å). Conformations at 300 K from the REMD simulations in vacuum were analyzed to draw the free-energy landscape. *Reprinted from Okamoto et al. (2010) with kind permission of Springer (2010).* (For color version of this figure, the reader is referred to the online version of this chapter.)

to the results in vacuum. These observations are reasonable because both the GB and TIP3P models have some screening effect on internal ligand interactions. We also found that the GB and TIP3P models produced similar free-energy landscapes (data not shown).

Figure 2.5 shows the free-energy landscapes along the first and second principal component axes for all of the ligand molecules studied here. The black circles show where the bioactive conformations complexed to proteins (from the PDB) exist on these landscapes. We found that none of the protein–bound conformations exists in the global–minimum

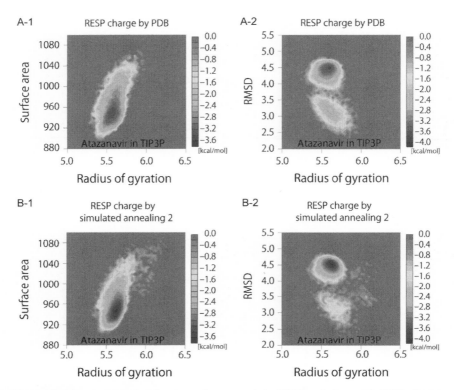

Figure 2.4 Free-energy landscapes of atazanavir in TIP3P solution with RESP charge parameters determined using (A) the PDB conformation (PDB ID: 2AQU) and (B) the conformation obtained by simulated annealing. (A-1) and (B-1) The landscapes for the radius of gyration (in Å) and surface area (in Å²), and (A-2) and (B-2) those for the radius of gyration (in Å) and RMSD (in Å). Conformations at 300 K from the REMD simulations in TIP3P water were analyzed to draw the free-energy landscape. *Reprinted from Okamoto et al. (2010) with kind permission of Springer (2010).* (For color version of this figure, the reader is referred to the online version of this chapter.)

free-energy states except in the case of imatinib, but that they exist near local–minimum states (Fig. 2.5B, E, and F) or as intermediates between stable states (Fig. 2.5C and D). Although we see that ligands rarely bind in their lowest energy conformations, protein–bound ligand conformations were sampled in water in the absence of protein with small free-energy penalties (<2.5 kcal/mol for all six ligand molecules). However, these protein–bound ligand conformations were not found in the free-energy landscapes under vacuum conditions for most ligands (data not shown).

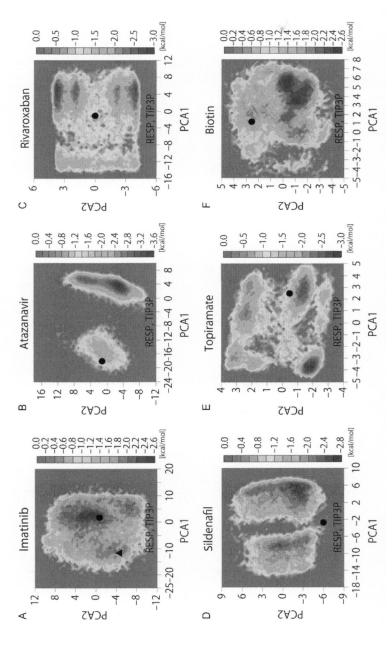

Figure 2.5 Free-energy landscapes with respect to the two major principal component axes for the six ligand molecules studied. The black circles (and triangle for imatinib) show where the protein-bound conformations (from PDB) lie on these landscapes. Conformations at 300 K from the REMD simulations were analyzed to draw the free-energy landscape. *Reprinted from Okamoto et al. (2010) with kind permission of Springer (2010).* (For color version of this figure, the reader is referred to the online version of this chapter.)

We next examine how our docking method based on REUS does for the prediction of protein–ligand binding structures (Kokubo, Tanaka, & Okamoto, 2011). The reaction coordinate ξ is the distance between the ligand and the binding site in the protein. To be more specific, it is the distance between the center of mass of the ligand molecule and the center of mass of the backbone heavy atoms of two selected residues of the protein. Two residues of the protein near the ligand binding site are arbitrarily selected.

We selected five protein–ligand complexes from PDB: 1T4E, 1SNC, 1OF6, 1ROB, and 2JJC, which correspond to the protein and ligand pairs, MDM2 and benzodiazepine, staphylococcal nuclease and PD*TP, aldolase and tyrosine, ribonuclease A and 2'-2CMP, and heat shock protein and pyrimidine-2-amine, respectively.

Figure 2.6 shows the structures of the five ligand molecules complexed to proteins, as registered in the PDB. The charge parameters of ligand molecules were determined by the RESP method (Bayly et al., 1993; Cieplak et al., 1995). The RESP charges were calculated by *ab initio* molecular orbital program at the HF/6-31G* level and RESP utilities in the antechamber program suite (Wang, Wang, Kollman, et al., 2006). The molecular mechanics parameters by GAFF (Wang, Wang, Caldwell, et al., 2006; Wang, Wang, Kollman, et al., 2006) were employed for bond, angle, torsion, and vdW parameters for ligand molecules. Missing force-field parameters were assigned with the

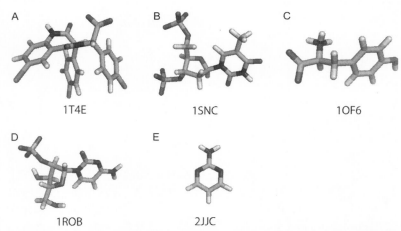

Figure 2.6 Ligand molecule conformations complexed to proteins (from PDB). PDB IDs are (A) 1T4E, (B) 1SNC, (C) 1OF6, (D) 1ROB, and (E) 2JJC. The figures were created with PyMOL (De Lano, 2002). *Reprinted from Kokubo et al. (2011) with kind permission of Wiley (2011).* (For color version of this figure, the reader is referred to the online version of this chapter.)

parmcheck utility in AMBER (Case et al., 2005). We employed AMBER ff99SB force-field parameters (Hornak et al., 2006) for proteins and the TIP3P model (Jorgensen et al., 1983) for water.

We next immersed each protein–ligand complex in a TIP3P water box with at least 10.0 Å solvation from the protein surface. The total number of water molecules was 5971 for the 1T4E system, 7561 for the 1SNC system, 14,670 for the 1OF6 system, 7093 for the 1ROB system, and 8812 for the 2JJC system. The systems were neutralized by adding sodium or chloride ions for charged complexes. For the 1T4E system, we added four chloride ions; for the 1SNC system, we added seven chloride ions; for the 1OF6 system, we added four chloride ions; for the 1ROB system, we added six chloride ions; and for the 2JJC system, we added seven sodium ions. Periodic boundary conditions and the particle mesh Ewald method (Darden, York, & Pederson, 1993) were employed for all the simulations. The Berendsen thermostat was employed for simulations for temperature control at 300 K, and the pressure was controlled by the weak–coupling method at 1.0 atm (Berrendsen, Postma, van Gunsteren, Dinola, & Haak, 1984). The SHAKE method (Ryckaert, Ciccotti, & Berendsen, 1977) was employed to constrain bond lengths involving hydrogen atoms and the time step was 2.0 fs.

The following 24 umbrella potentials were prepared: $k_m(\xi(q) - d_m)^2$, where the midpoint values were $d_m = 5.0$, 5.5, 6.0, 6.5, 7.0, 7.5, 8.0, 8.5, 9.0, 9.5, 10.0, 10.5, 11.0, 12.0, 13.0, 14.0, 15.0, 16.0, 17.5, 19.0, 20.5, 22.0, 23.5, and 25.0 Å, and the strengths of the restraining potentials were $k_m = 1.0$ kcal/(mol Å^2) for $d_m \leq 13.0$ Å and $k_m = 0.5$ kcal/(mol Å^2) for $d_m > 13.0$ Å.

Figure 2.7 shows typical snapshots from the umbrella sampling simulations with the largest midpoint value of 25 Å (22 Å only for 1OF6). We see that ligand molecules are away from the proteins and are completely outside the protein pockets.

The production runs were performed for 110–200 ns and the last 100 ns data were used for the analyses.

Figure 2.8 shows the time series of various quantities that we obtained from the REUS simulations of 1T4E as a representative system. Figure 2.8A shows the time series of the reaction coordinate (the distance of the ligand from the pocket) through umbrella potential exchange for some of the replicas: Replicas 1 (black), 9 (blue), 13 (orange), 16 (green), and 19 (pink). We observe that the reaction coordinate goes up and down between the lowest value and the highest one and that sufficient sampling has been achieved. Other replicas behaved similarly.

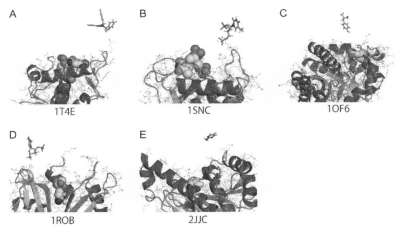

A   1T4E   B   1SNC   C   1OF6

D   1ROB   E   2JJC

Figure 2.7 Snapshots from the umbrella sampling simulations with the largest mid-point value of $d_m = 25$ Å (22 Å only for 1OF6). PDB IDs are shown below each figure. The space-filled molecules, which do not exist in these simulations actually, show the correct ligand binding positions (from PDB) as references. Water molecules are not shown to see the protein–ligand conformations clearly. The figures were created with PyMOL (De Lano, 2002). *Reprinted from Kokubo et al. (2011) with kind permission of Wiley (2011).* (For color version of this figure, the reader is referred to the online version of this chapter.)

Figure 2.8B shows the time series of the distances of the center positions of umbrella potentials corresponding to Fig. 2.8A. When the center position of umbrella position is high, the reaction coordinate is high, and when the former is low, so is the latter. There is a strong correlation between Fig. 2.8A and B, as is to be expected. The red horizontal lines at 6 Å in Fig. 2.8A and B show the correct binding distances in PDB. When we compare (A) and (B) in detail, we observe that actual distances in (A) do not necessarily follow the distances of umbrella potential in (B) especially in the case of short distances. It is reasonable because the ligand will have discrete stable regions inside the pocket. In addition, the fact that a ligand never approaches 5.0 Å distance suggests the steric exclusion. On the other hand, we observed that ligands in long distance tended to follow umbrella potentials well for all the systems.

Figure 2.8C shows the time series of RMSD of the ligand from PDB structure. We see that the RMSD goes up and down between the lowest and highest values found and that wide conformational space along the reaction coordinate has indeed been sampled during the simulations. Comparing Fig. 2.8A and C, we see that Replica 1 (black) has the distance close to the

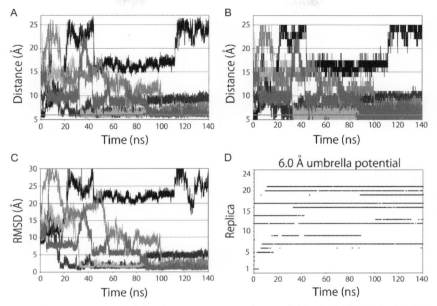

Figure 2.8 Time series of (A) the reaction coordinate ξ (distance from the protein pocket), (B) the center position of umbrella potential from the protein pocket, (C) RMSD (in Å) from the correct binding mode (a blue ligand in Fig. 2.10A), and (D) replica exchange for the third umbrella potential ($k_3 = 1.0$ kcal/(mol Å^2) and $d_3 = 6.0$ Å), obtained from the REUS simulation of 1T4E system. Time series of Replicas 1 (black), 9 (blue), 13 (orange), 16 (green), and 19 (pink) are shown for (A), (B), and (C) as representatives. The bold red lines at 6.0 Å in (A) and (B) show the correct binding distance from PDB. (For interpretation of the references to color in this figure legend, the reader is referred to the web version of this article.) *Reprinted from Kokubo et al. (2011) with kind permission of Wiley (2011).*

correct binding distance in the beginning, but it has actually a very different conformation and has a large RMSD from the experimental binding structure in PDB. However, we observe that the correct binding structures were found independently by different replicas, while the REUS simulation proceeded.

Figure 2.8D is the time series of replica exchange. The time series of replica number at an umbrella potential with the midpoint of 6.0 Å is shown. We see that many replicas experienced the umbrella potential with the midpoint value of 6.0 Å. These figures show that the replica–exchange simulations have been properly performed. Thus far, we examined how well the REUS simulation worked in the case of 1T4E system. We also confirmed that REUS simulations have been performed properly in all other systems.

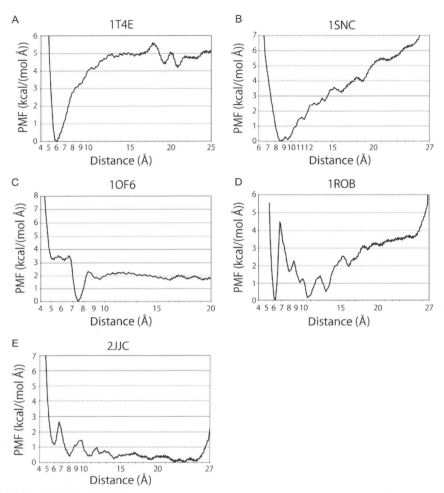

Figure 2.9 Potential of mean force profiles along the reaction coordinate ξ (distance from the protein pocket) for all the five systems (A) 1T4E, (B) 1SNC, (C) 1OF6, (D) 1ROB, and (E) 2JJC. *Reprinted from Kokubo et al. (2011) with kind permission of Wiley (2011).*

Figure 2.9 shows the PMF for the five systems that we examined in this study. We collected about 50,000 ligand structures with the global-minimum PMF inside the pocket for each system. The global-minimum PMF inside the pocket was also the global-minimum PMF over all distance range for (A) 1T4E, (B) 1SNC, (C) 1OF6, and (D) 1ROB systems. On the other hand, we found that the structures with the global-minimum PMF for (E) 2JJC system were not inside the pockets. Therefore, we collected the structures

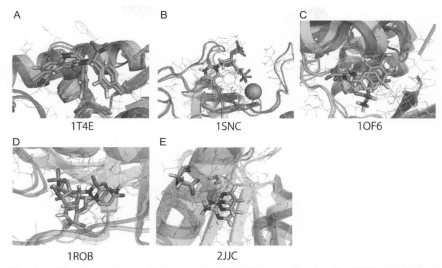

Figure 2.10 Comparisons of the predicted binding modes by the REUS simulations (green) with the experimental ligand binding modes from PDB (blue). (C) The REUS simulation predicted two different binding modes (green and pink), which were detected as equally stable ones on the free-energy landscape along the two principal component axes. The figures were created with PyMOL (De Lano, 2002). (For interpretation of the references to color in this figure legend, the reader is referred to the web version of this article.) *Reprinted from Kokubo et al. (2011) with kind permission of Wiley (2011).*

with the first local-minimum PMF, which has the lowest PMF inside the protein pocket. These collected structures were then analyzed by PCA.

Figure 2.10 shows the comparisons of the predicted binding modes (green) with the experimental ones (blue) from PDB. We see that the predicted binding modes are in excellent agreement with the experimental binding modes especially for (A) 1T4E and (B) 1SNC, while we obtained essentially the correct binding modes also for (C) 1OF6, (D) 1ROB, and (E) 2JJC, considering the fluctuations of both protein and ligand structures. Note that docking software GOLD made incorrect predictions for (B) 1SNC and (D) 1ROB and these two systems are categorized as "significant error" in the original paper by Jones et al. (1997).

4. CONCLUSIONS

In this chapter, we have introduced three powerful generalized-ensemble algorithms, namely, REM, MREM, and REUS, which can greatly enhance conformational sampling of biomolecular systems. We have presented the results of REMD simulations for ligand systems that are important for pharmaceutical design. We then presented the results of REUS for docking simulations of ligands to target proteins. We have obtained very promising results. We have succeeded in predicting the ligand binding structures in excellent agreement with the experimental data from PDB. Further investigation of the REUS-based docking method is now in progress. Future work will involve further improvements of the method. Two-dimensional replica exchange by the introduction of additional dimension to REUS may be effective for the enhancement of the ligand binding events.

ACKNOWLEDGMENTS

We are grateful to Dr. Masaki Tomimoto for his insight, discussions, and collegiality. The simulations and computations were performed on the TSUBAME Grid Cluster at the Global Scientific Information and Computing Center of Tokyo Institute of Technology, supported by the Ministry of Education, Culture, Sports, Science and Technology (MEXT) Open Advanced Research Facilities Initiative. Y. O. was supported, in part, by the collaboration funds from Takeda Pharmaceutical Co., Ltd. and Grants-in-Aid for Scientific Research (A) (No. 25247071), for Computational Materials Science Initiative, and for High Performance Computing Infrastructure from MEXT, Japan.

REFERENCES

Abagyan, R., & Argos, P. (1992). Optimal protocol and trajectory visualization for conformational searches of peptides and proteins. *Journal of Molecular Biology*, *225*, 519–532.

Amadei, A., Linssen, A. B. M., & Berendsen, H. J. C. (1993). Essential dynamics of proteins. *Proteins*, *17*, 412–425.

Bayly, C. I., Cieplak, P., Cornell, W., & Kollman, P. A. (1993). A well-behaved electrostatic potential based method using charge restraints for deriving atomic charges: The RESP model. *The Journal of Physical Chemistry*, *97*, 10269–10280.

Berrendsen, H. J. C., Postma, J. P. M., van Gunsteren, W. F., Dinola, A., & Haak, J. (1984). Molecular dynamics with coupling to an external bath. *The Journal of Chemical Physics*, *81*, 3684.

Case, D. A., Cheatham, T. E., Darden, T., Gohlke, H., Luo, R., Merz, K. M., et al. (2005). The AMBER biomolecular simulation programs. *Journal of Computational Chemistry*, *26*, 1668–1688.

Cieplak, P., Cornell, W., Bayly, C. I., & Kollman, P. A. (1995). Application of the multimolecule and multiconformational RESP methodology to biopolymers: Charge derivation for DNA, RNA and proteins. *Journal of Computational Chemistry*, *16*, 1357–1377.

Darden, T., York, D., & Pederson, L. (1993). Particle mesh Ewald: An $N \cdot \log(N)$ method for Ewald sums in large systems. *The Journal of Chemical Physics*, *98*, 10089–10092.

De Lano, W. L. (2002). *The PyMOL molecular graphics system*. Palo Alto, CA: De Lano Scientific.

Feig, M., Onufriev, A., Lee, M. S., Im, W., Case, D. A., & Brooks, C. L., III. (2004). Performance comparison of generalized Born and Poisson methods in the calculation of electrostatic solvation energies for protein structures. *Journal of Computational Chemistry*, *25*, 265–284.

Ferrenberg, A. M., & Swendsen, R. H. (1989). Optimized Monte Carlo data analysis. *Physical Review Letters*, *63*, 1195–1198.

Friesner, R. A., Murphy, R. B., Repasky, M. P., Frye, L. L., Greenwood, J. R., Halgren, T. A., et al. (2006). Extra precision glide: Docking and scoring incorporating a model of hydrophobic enclosure for protein-ligand complexes. *Journal of Medicinal Chemistry*, *49*, 6177–6196.

Fukunishi, F., Watanabe, O., & Takada, S. (2002). On the Hamiltonian replica exchange method for efficient sampling of biomolecular systems: Application to protein structure prediction. *The Journal of Chemical Physics*, *116*, 9058–9067.

Gallicchio, E., & Levy, R. (2011). Recent theoretical and computational advances for modeling protein-ligand binding affinities. *Advances in Protein Chemistry and Structural Biology*, *85*, 27–80.

Garcia, A. E. (1992). Large-amplitude nonlinear motions in proteins. *Physical Review Letters*, *68*, 2696–2699.

Gilson, M. K., & Zhou, H. X. (2007). Calculation of protein-ligand binding affinities. *Annual Review of Biophysics and Biomolecular Structure*, *36*, 21–42.

Grant, B. J., Gorfe, A. A., & McCammon, J. A. (2010). Large conformational changes in proteins: Signaling and other functions. *Current Opinion in Structural Biology*, *20*, 142–147.

Hansmann, U. H. E., & Okamoto, Y. (1999). New Monte Carlo algorithms for protein folding. *Current Opinion in Structural Biology*, *9*, 177–183.

Hornak, V., Abel, R., Okur, A., Strockbine, B., Roitberg, A., & Simmerling, C. (2006). Comparison of multiple Amber force fields and development of improved protein backbone parameters. *Proteins*, *65*, 712–725.

Hukushima, K., & Nemoto, K. (1996). Exchange Monte Carlo method and application to spin glass simulations. *Journal of the Physical Society of Japan*, *65*, 1604–1608.

Jakalian, A., Bush, B. L., Jack, D. B., & Bayly, C. I. (2000). Fast, efficient generation of high-quality atomic charges. AM1-BCC model: I. Method. *Journal of Computational Chemistry*, *21*, 132–146.

Jiang, W., Hodoscek, M., & Roux, B. (2009). Computation of absolute hydration and binding free energy with free energy perturbation distributed replica-exchange molecular dynamics. *Journal of Chemical Theory and Computation*, *5*, 2583–2588.

Jones, G., Willett, P., Glen, R. C., Leach, A. R., & Taylor, R. (1997). Development and validation of a genetic algorithm for flexible docking. *Journal of Molecular Biology*, *267*, 727–748.

Jorgensen, W. L., Chandrasekhar, J., Madura, J. D., Impey, R. W., & Klein, M. L. (1983). Comparison of simple potential functions for simulating liquid water. *The Journal of Chemical Physics*, *79*, 926–935.

Kirkpatrick, S., Gelatt, C. D., Jr., & Vecchi, M. P. (1983). Optimization by simulated annealing. *Science*, *220*, 671–680.

Kitao, A., Hirata, F., & Go, N. (1991). The effects of solvent on the conformation and the collective motions of protein: Normal mode analysis and molecular dynamics simulations of melittin in water and in vacuum. *Chemical Physics*, *158*, 447–472.

Kokubo, H., & Okamoto, Y. (2006). Replica-exchange methods and predictions of helix configurations of membrane proteins. *Molecular Simulation*, *32*, 791–801.

Kokubo, H., Tanaka, T., & Okamoto, Y. (2011). *Ab initio* prediction of protein-ligand binding structures by replica-exchange umbrella sampling simulations. *Journal of Computational Chemistry*, *32*, 2810–2821.

Kramer, B., Rarey, M., & Lengauer, T. (1999). Evaluation of the FLEXX incremental construction algorithm for protein-ligand docking. *Proteins*, *37*, 228–241.

Kumar, S., Bouzida, D., Swendsen, R. H., Kollman, P. A., & Rosenberg, J. M. (1992). The weighted histogram analysis method for free-energy calculations on biomolecules. 1. The method. *Journal of Computational Chemistry, 13*, 1011–1021.

Marinari, E., Parisi, G., & Ruiz-Lorenzo, J. J. (1997). Numerical simulations of spin glass systems. In A. P. Young (Ed.), *Spin glasses and random fields* (pp. 59–98). Singapore: World Scientific.

Meng, Y., Dashti, D. S., & Roitberg, A. E. (2011). Computing alchemical free energy differences with Hamiltonian replica exchange molecular dynamics (H-REMD) simulations. *Journal of Chemical Theory and Computation, 7*, 2721–2727.

Metropolis, N., Rosenbluth, A. W., Rosenbluth, M. N., Teller, A. H., & Teller, E. (1953). Equation of state calculations by fast computing machines. *The Journal of Chemical Physics, 21*, 1087–1092.

Mitsutake, A. (2009). Simulated-tempering replica-exchange method for the multidimensional version. *The Journal of Chemical Physics, 131*, 094105.

Mitsutake, A., & Okamoto, Y. (2009). Multidimensional generalized-ensemble algorithms for complex systems. *The Journal of Chemical Physics, 130*, 214105.

Mitsutake, A., Sugita, Y., & Okamoto, Y. (2001). Generalized-ensemble algorithms for molecular simulations of biopolymers. *Biopolymers, 60*, 96–123.

Mitsutake, A., Sugita, Y., & Okamoto, Y. (2003). Replica-exchange multicanonical and multicanonical replica-exchange Monte Carlo simulations of peptides. I. Formulation and benchmark test. *The Journal of Chemical Physics, 118*, 6664–6675.

Mori, Y., & Okamoto, Y. (2010a). Generalized-ensemble algorithms for the isobaric-isothermal ensemble. *Journal of the Physical Society of Japan, 79*, 074003.

Mori, Y., & Okamoto, Y. (2010b). Replica-exchange molecular dynamics simulations for various constant temperature algorithms. *Journal of the Physical Society of Japan, 79*, 074001.

Nosé, S. (1984a). A molecular dynamics method for simulations in the canonical ensemble. *Molecular Physics, 52*, 255–268.

Nosé, S. (1984b). A unified formulation of the constant temperature molecular dynamics methods. *The Journal of Chemical Physics, 81*, 511–519.

Okamoto, Y. (2004). Generalized-ensemble algorithms: Enhanced sampling techniques for Monte Carlo and molecular dynamics simulations. *Journal of Molecular Graphics and Modelling, 22*, 425–439.

Okamoto, Y., Tanaka, T., & Kokubo, H. (2010). Dependency of ligand free energy landscapes on charge parameters and solvent models. *Journal of Computer-Aided Molecular Design, 24*, 699–712.

Okumura, H., & Okamoto, Y. (2006). Multibaric-multithermal ensemble molecular dynamics simulations. *Journal of Computational Chemistry, 27*, 379–395.

Onufriev, A., Bashford, D., & Case, D. A. (2004). Exploring protein native states and large-scale conformational changes with a modified generalized born model. *Proteins, 55*, 383–394.

Ryckaert, J., Ciccotti, G., & Berendsen, H. J. C. (1977). Numerical integration of the Cartesian equations of motion of a system with constraints: Molecular dynamics of *n*-alkanes. *Journal of Computational Physics, 23*, 327–341.

Sugita, Y., Kitao, A., & Okamoto, Y. (2000). Multidimensional replica-exchange method for free-energy calculations. *The Journal of Chemical Physics, 113*, 6042–6051.

Sugita, Y., & Okamoto, Y. (1999). Replica-exchange molecular dynamics method for protein folding. *Chemical Physics Letters, 314*, 141–151.

Sugita, Y., & Okamoto, Y. (2002). Free-energy calculations in protein folding by generalized-ensemble algorithms. In T. Schlick & H. H. Gan (Eds.), *Lecture notes in computational science and engineering* (pp. 304–332). Berlin: Springer-Verlag e-print: cond-mat/0102296.

Teeter, M. M., & Case, D. A. (1990). Harmonic and quasiharmonic descriptions of crambin. *The Journal of Physical Chemistry, 94*, 8091–8097.

Torrie, G. M., & Valleau, J. P. (1997). Nonphysical sampling distributions in Monte Carlo free-energy estimation: Umbrella sampling. *Journal of Computational Physics, 23*, 187–199.

Wang, J., Wang, W., Caldwell, J. W., Kollman, P. A., & Case, D. A. (2006). Development and testing of a general AMBER force field. *Journal of Computational Chemistry, 25*, 1157–1174.

Wang, J., Wang, W., Kollman, P. A., & Case, D. A. (2006). Automatic atom type and bond type perception in molecular mechanical calculations. *Journal of Molecular Graphics and Modelling, 25*, 247–260.

CHAPTER THREE

Conformational Dynamics of Single Protein Molecules Studied by Direct Mechanical Manipulation

Pétur O. Heidarsson[*], Mohsin M. Naqvi[†], Punam Sonar[†], Immanuel Valpapuram[†], Ciro Cecconi[‡,1]

[*]Structural Biology and NMR Laboratory, Department of Biology, University of Copenhagen, Copenhagen N, Denmark
[†]Department of Physics, University of Modena and Reggio Emilia, Modena, Italy
[‡]CNR Institute of Nanoscience S3, University of Modena and Reggio Emilia, Modena, Italy
[1]Corresponding author: e-mail address: ciro.cecconi@gmail.com

Contents

Abstract

Advances in single-molecule manipulation techniques have recently enabled researchers to study a growing array of biological processes in unprecedented detail. Individual molecules can now be manipulated with subnanometer precision along a simple and well-defined reaction coordinate, the molecular end-to-end distance, and their

Advances in Protein Chemistry and Structural Biology, Volume 92
ISSN 1876-1623
http://dx.doi.org/10.1016/B978-0-12-411636-8.00003-1

conformational changes can be monitored in real time with ever-improving time res-olution. The behavior of biomolecules under tension continues to unravel at an accel-erated pace and often in combination with computational studies that reveal the atomistic details of the process under investigation. In this chapter, we explain the basic principles of force spectroscopy techniques, with a focus on optical tweezers, and describe some of the theoretical models used to analyze and interpret single-molecule manipulation data. We then highlight some recent and exciting results that have emerged from this research field on protein folding and protein–ligand interactions.

ABBREVIATIONS

AFM atomic force microscopy
CaM calmodulin
CFT Crooks' fluctuation theorem
HMM hidden Markov model
MD molecular dynamics
NMR nuclear magnetic resonance
WLC worm-like chain

1. INTRODUCTION

Sophisticated bulk methods have been developed to characterize the conformational changes of proteins as they carry out their biological func-tions. Traditional techniques such as circular dichroism, nuclear magnetic resonance (NMR) spectroscopy, and hydrogen/deuterium exchange mass spectrometry (Greenfield, 2007; Kern, Eisenmesser, & Wolf–Watz, 2005; Maity, Maity, Krishna, Mayne, & Englander, 2005) have been extensively used to study the structure and dynamics of proteins, both free in solution or bound to their molecular partners. These bulk studies have been very infor-mative but limited to the description of the overall properties of a large pop-ulation of proteins. This is because the output from these measurements is the ensemble average generated from a large, and often dephased, popula-tion of molecules, in which the time-dependent dynamics of the individual molecules, as well as rare but potentially important molecular events, are hidden. These technical limitations, which have restrained our ability to decipher the intricacies of many molecular processes, have recently been overcome with the advent of single-molecule methods. These novel exper-imental approaches enable us to follow the real-time trajectories of single molecules and describe the inherent heterogeneity of biological processes that are stochastic in nature.

Single-molecule detection methods are mainly based on fluorescence intensity or energy transfer efficiency between fluorophores, using either freely diffusing or surface-tethered molecules. Several excellent papers that review fluorescence-based single-molecule techniques have been published elsewhere (Borgia, Williams, & Clarke, 2008; Deniz, Mukhopadhyay, & Lemke, 2008; Joo, Balci, Ishitsuka, Buranachai, & Ha, 2008; Schuler & Eaton, 2008; Tinoco & Gonzalez, 2011). In this review, we focus exclusively on mechanical manipulation methods, specifically optical tweezers and atomic force microscopy (AFM) and illustrate the main experimental strategies used in these techniques. We then go on to describe some theoretical models employed in this field, before highlighting some pivotal single-molecule studies aimed at solving exciting biological problems. Due to space limitations and the fast expansion of the field, it is impossible for us to review all recently published studies and we refer the reader to other reviews for a more comprehensive overview of the field (Borgia et al., 2008; Bustamante, 2008; Deniz et al., 2008; Moffitt, Chemla, Smith, & Bustamante, 2008).

2. MECHANICAL MANIPULATION OF SINGLE PROTEIN MOLECULES

Just as chemicals and heat can be used to perturb the system under study, so can force be used for the same purpose. Force and the end-to-end distance of a protein molecule constitute the main variables during a mechanical manipulation experiment. Many methods have been developed to directly manipulate biomolecules. In this chapter, we briefly discuss the two most popular experimental methods, optical tweezers and AFM.

2.1. Optical tweezers

Ever since the pioneering work of Arthur Ashkin in the 1980s (Ashkin, Dziedzic, Bjorkholm, & Chu, 1986), optical tweezers have continued to evolve and improve to tackle ever more complex systems. An optical trap can be formed by focusing a collimated beam of light through a microscope objective with a high numerical aperture. In this way, small objects of high refractive index can be optically trapped and manipulated. For a detailed description of the principles behind optical trapping, which is beyond the scope of this chapter, we refer the reader to previous publications (Ashkin et al., 1986; Moffitt et al., 2008; Smith, Cui, & Bustamante, 2003). There are three main geometries that are nowadays used in the optical tweezers setups (Fig. 3.1).

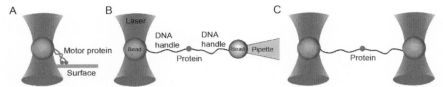

Figure 3.1 Optical tweezers experimental geometries. (A) A single-beam optical twee-zers setup where the molecule of interest is tethered between an optically trapped bead and a surface. The movements of the protein along the surface are revealed by the motions of the bead in the trap. (B) A single-beam optical tweezers experimental setup where the protein is attached to two polystyrene beads through two double-stranded DNA handles. One bead is held in the optical trap while the other is held at the end of a micropipette by suction. The micropipette can be mechanically moved relative to the trap, to induce the unfolding or refolding of a protein molecule. (C) A double-beam setup where the protein molecule is tethered through DNA handles between two opti-cally trapped beads. See text for details. (For color version of this figure, the reader is referred to the online version of this chapter.)

In all cases, the molecule under study is tethered between an optically trapped bead and: (i) a substrate, (ii) a second bead held at the end of micro-pipette by suction, or (iii) another bead held in a second optical trap. In all cases, the force applied to the molecule is modulated by varying the distance between the two tethering points. Different approaches can be taken to manipulate a molecule. In what follows, we briefly describe the experimental strategy used in: (i) constant-velocity (also force-ramp), (ii) constant-force (also force-clamp), (iii) passive-mode, and (iv) force-jump experiments. In constant-velocity experiments, the force is increased and relaxed at constant speed (nm/s), to obtain force versus extension cycles as shown in Fig. 3.2A. During stretching, the force is raised until the mol-ecule is observed to unfold. This event is marked by a sudden increase in the extension of the molecule, as it goes from its compact native state (N) to a more extended unfolded state (U). During relaxation, the molecule is typically observed to refold around at ∼5–10 pN, through a sharp transition that restores the original extension of the molecule. The changes in contour length of the protein associated with the unfolding and refolding events can be estimated by fitting the force–extension traces with the worm-like chain (WLC) model (Bustamante, Marko, Siggia, & Smith, 1994; Cecconi, Shank, Bustamante, & Marqusee, 2005; Liphardt, Onoa, Smith, Tinoco, & Bustamante, 2001). Constant-velocity experiments can provide information on both kinetics and thermodynamics of a protein folding reac-tion. Kinetic parameters, such as rate constants and position of the transition

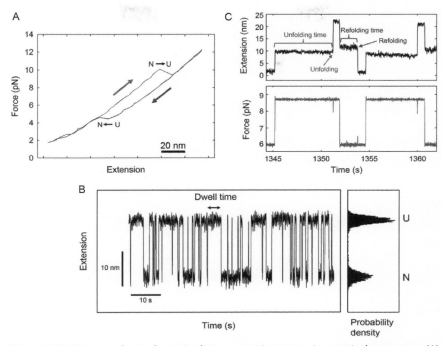

Figure 3.2 Basic mechanical manipulation experiments using optical tweezers. (A) Constant-velocity experiment showing a two-state unfolding/refolding event. The arrows indicate the pulling direction. (B) Constant-force experiment showing a molecule fluctuating between an unfolded (U) and folded (N) state. The dwell times of the unfolded and folded states contain both thermodynamic and kinetic information. (C) Force-jump experiment. The force is rapidly jumped between two different set values, to increase the probability of observing unfolding/refolding events, see text for details. *Adapted from Heidarsson et al. (2012) with permission.* (For color version of this figure, the reader is referred to the online version of this chapter.)

state along the reaction coordinate, can, for example, be estimated by analyzing force distributions. Thermodynamics information (e.g., unfolding free energy) can instead be recovered by analyzing irreversible work distributions (see below for details).

In constant-force experiments, the force applied to the molecule is kept constant through a feedback mechanism, while changes in the extension of the molecule are monitored over time (Fig. 3.2B). In these measurements, rate coefficients can be obtained directly from the lifetimes of the folded and unfolded states, and free energies can be calculated from the ratio of the kinetics coefficients. As the molecule unfolds or refolds, the force is kept constant by reducing or increasing the distance between the tethering surfaces, respectively. These movements, which are controlled by the feed-back mechanism, can, however, take place only at a certain rate and, because of this constant force, measurements can sometimes provide misleading results. This is the case, for example, when the response rate of the feedback mechanism is slower than the rate at which the molecule can fluctuate. In these cases, in fact, short-lived transitions are missed, leading to average dwell times biased toward larger values (Elms, Chodera, Bustamante, & Marqusee, 2012b). This potential instrumental artifact can be avoided, or at least reduced, by studying molecular fluctuations through passive-mode measurements. In this case, no feedback is employed. Instead, the distance between the tethering surface and the optical trap is kept constant, while force is allowed to increase or decrease as the molecule unfolds or refolds, respectively. In these experiments, the response time of the system is dictated by the corner frequency of the trapped bead, which is typically much higher than the response frequency of a feedback mechanism. Consequently, passive-mode measurements are better suited to study rapid conformational transitions. On the other hand, as no feedback mechanism is used, reliable passive-mode measurements require high mechanical and optical stability of the instrument itself.

Constant-force or passive-mode measurements can be effectively used to study molecular fluctuations at equilibrium only when, in a certain range of forces, the rate of both the forward and reverse reactions is high enough to allow the acquisition of a large number of events in a relatively short amount of time. This, however, is not always the case, as for some molecules unfolding and refolding occurs at quite different forces. In these instances, at any given force, either the unfolding or refolding rate is so low that the acquisition of a significant number of events would require very long recordings. To overcome this problem, force–jump experiments can be performed. In a force–jump experiment, the force is increased (jumped) or decreased (dropped) quickly to a preset force value and kept constant with a feedback mechanism until an unfolding or refolding event is observed (Fig. 3.2C). These experiments allow the direct measurement of rate constants in force ranges where the probability of observing either an unfolding or refolding event is high (Li, Collin, Smith, Bustamante, & Tinoco, 2006).

2.1.1 Alternative, novel, and hybrid optical tweezers instruments

Single-molecule manipulation instruments are advancing at a rapid pace. Considerable effort has been spent on designing optical tweezers that measure other single-molecule parameters such as rotation or torque in combination with force and extension (Neuman & Nagy, 2008). The ability to measure torque has significant importance because it addresses many aspects of cell and protein biology such as transcription, replication, recombination, and protein folding (De Vlaminck et al., 2010; Koster, Crut, Shuman, Bjornsti, & Dekker, 2010). The measurement of torque can be achieved with magnetic tweezers, a popular instrument because it is simple and cheap (De Vlaminck & Dekker, 2012). Spatial and temporal resolution of magnetic tweezers may in the near future advance to subnanometer precision, possibly with newly emerging camera technology that enhances the rate of data acquisition (De Vlaminck & Dekker, 2012).

The integration of dual beam optical tweezers with magnetic tweezers allows manipulation of a single molecule with nanometer precision. This hybrid instrument has been used to investigate a wide range of complex systems such as higher order chromatin interaction, DNA–DNA interaction mediated by proteins, measurement of intermolecular friction and localization, and binding strength analysis of DNA bound proteins (Noom, van den Broek, van Mameren, & Wuite, 2007). The combination of magnetic or optical tweezers and fluorescence microscopy is a powerful tool for DNA manipulation and has been used to study DNA supercoiling, the dynamics of diffusion, hopping of plectonemics in DNA, and the torque and twist DNA-breathing dynamics (De Vlaminck, Henighan, van Loenhout, Burnham, & Dekker, 2012; Sirinakis, Ren, Gao, Xi, & Zhang, 2012).

A recently designed Quad-trap optical tweezers instrument was used to study the condensation of bacterial chromosome DNA (Dame, Noom, & Wuite, 2006). The instrument was able to independently trap four polystyrene beads at the same time, allowing the simultaneous mechanical manipulation of two independent DNA molecules. With this method, the authors were able to detect the complex and dynamic interactions of DNA and histone-like nucleoid structuring protein (H-NS), which is involved in mediating DNA–DNA contact. This remarkable technical feature helped explain the mechanism and role of H-NS in chromosomal DNA condensation.

2.1.2 Sample preparation

A major issue in optical tweezers experiments is to find an efficient method to manipulate the molecule of interest. Biomolecules such as proteins and RNAs are typically too small to be directly manipulated with micrometer-sized optical tweezer beads, as the tethering surfaces would come so close to each other that they would interact. To avoid these unspecific and unwanted interactions, a method has been developed that relies on the use of two DNA molecular handles (Fig. 3.1B and C) (Cecconi, Shank, Dahlquist, Marqusee, & Bustamante, 2008). One end of each handle (~500–1000 bp DNA molecule) is attached covalently to the side chain of a cysteine residue. The other end is attached to a polystyrene bead through either biotin–streptavidin interactions or digoxigenin/antibody interactions. The handles act as spacers between the protein and the beads to avoid unwanted interactions between the tethering surfaces that would compromise the experiment. This optical tweezers manipulation method was employed for the first time in 2005 (Cecconi et al., 2005), and it is now used by a growing number of laboratories around the world (Gao, Sirinakis, & Zhang, 2011; Gebhardt, Bornschlogl, & Rief, 2010; Stigler & Rief, 2012; Xi, Gao, Sirinakis, Guo, & Zhang, 2012; Yu, Liu, et al., 2012). For details on how the DNA–protein coupling reaction is performed, we refer the reader to Cecconi, Shank, Marqusee, and Bustamante (2011).

2.2. Atomic force microscopy

AFM was initially developed for high-resolution imaging of surface contours of microscopic samples (Binnig, Quate, & Gerber, 1986) and it still primarily serves this function. Later, AFM also evolved into a versatile technique to manipulate single molecules and characterize their mechanical properties, in a mode of operation called "force spectroscopy" or "force measuring." AFM force spectroscopy has been used to study a number of biological systems, including binding of antibodies to their antigens (Raab et al., 1999), ligand to receptors (Florin, Moy, & Gaub, 1994), binding forces of complementary DNA strands (Lee, Chrisey, & Colton, 1994), conformational changes in biological polymers (Rief, Oesterhelt, Heymann, & Gaub, 1997), and to study the mechanical properties of a large variety of protein molecules (Bornschlogl & Rief, 2011; Garcia-Manyes, Dougan, Badilla, Brujic, & Fernandez, 2009; Ng, Randles, & Clarke, 2007; Rounsevell, Forman, & Clarke, 2004).

When AFM is used to study protein folding, the molecule of interest, typically a polyprotein made of a linear array of a globular domain, is tethered between a flat surface, usually made of gold, and a silicon nitrite AFM tip (Fig. 3.3A). One end of the polyprotein interacts with the gold surface through thiol groups of terminal cysteine residues, while the other end adheres to the tip unspecifically. The polyprotein is then stretched and

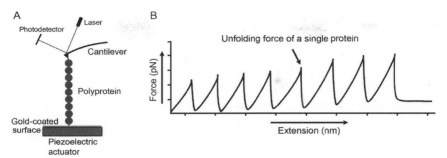

Figure 3.3 An AFM experiment. (A) In a typical AFM experiment, polyprotein constructs are picked up and unfolded by the cantilever tip. The position of the tip is determined by a photodetector from the deflection of a laser beam. (B) The sequential unfolding of single proteins within the polyprotein construct results in a characteristic sawtooth pattern in the force versus extension trace. (For color version of this figure, the reader is referred to the online version of this chapter.)

relaxed by modulating the distance between the tethering surfaces, while the force applied on the molecule is determined by measuring the deflection of the cantilever. As the tip is pulled away from the surface, the tension along the polymer increases until one globular domain stochastically unfolds. Upon unfolding, the contour length of the polymer suddenly increases, generating a sharp drop of the force on the cantilever. As the stretching of the polymer continues, all the other domains subsequently unfold, generating a series of sudden force drops that give rise to a peculiar sawtooth-like pattern in the force–extension curve, where each peak corresponds to the unfolding of one domain (Fig. 3.3B). The rising phase of each peak reflects the elastic properties of the stretched polymer, while the distance between peaks is proportional to the number of amino acids constituting the protein domain. As in optical tweezers experiments molecular handles are necessary to keep the two polystyrene beads away from each other, in AFM studies the need for polyproteins comes from the necessity of keeping a certain distance between the tip and the substrate. At short distances, in fact, tip–surface interactions could become the dominant features of a force trace. Different methods have been devised to generate a linear array of a globular domain, as described in Carrion-Vazquez et al. (1999), Cecconi et al. (2008), Dietz, Bertz, et al. (2006), Steward, Toca-Herrera, and Clarke (2002), and Yang et al. (2000). AFM force spectroscopy experiments on polyproteins can be performed at different pH and ionic strengths, as well as in the presence of other molecules to study the effect of binding partners on the energetics of the protein (Junker & Rief, 2009; Junker, Ziegler, & Rief, 2009).

AFM and optical tweezers are complementary techniques for the study of the mechanical properties of biomolecules. An AFM cantilever has a higher root–mean square force noise ($\sim$15 pN) than an optically trapped bead ($\sim$0.1 pN). As a consequence, optical tweezers is the technique of choice to study processes that take place at low forces, such as the refolding of a protein, that typically occur in a 5–10 pN force range. In addition, the shallower harmonic potential of an optical trap, compared to an AFM cantilever, makes laser tweezers a better technique to observe a molecule hopping between different molecular states. In fact, the steeper the harmonic potential, the larger is the kinetic barrier that the system molecule-force transducer must overcome to hop between a folded and unfolded state of the molecule. On the other hand, optical tweezers are usually unable to measure forces that are larger than 100 pN, while AFM can measure forces even in the nanoNewton range. It follows that AFM is the technique of choice to measure the rupture force of single covalent bonds ($\sim$2 nN) (Grandbois, Beyer, Rief, Clausen-Schaumann, & Gaub, 1999), the force at which polysaccharides switch to different conformations ($\sim$270 pN) (Marszalek, Oberhauser, Pang, & Fernandez, 1998), or to characterize the mechanical properties of proteins that unfold at high forces (Dietz, Berkemeier, Bertz, & Rief, 2006).

3. THEORETICAL MODELS OF SINGLE-MOLECULE FORCE SPECTROSCOPY

The mechanical manipulation of single molecules using optical tweezers and AFM can yield valuable information about the free-energy surface of a single-molecule reaction. In this section, using basic thermodynamic and kinetic principles, we explain the effect of mechanical force on the energy landscape of a molecule. Then, we discuss some theoretical models used to analyze and interpret single-molecule manipulation results.

3.1. Effect of force on the thermodynamics of a single-molecule reaction

The effect of force on a two–state reaction in which A is converted into B is depicted in Fig. 3.4. Along the mechanical reaction coordinate, which is an extension of the molecule, states A and B occupy free-energy minima separated by a distance Δx.

The free-energy difference between A and B at zero force is

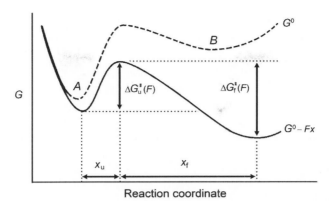

Figure 3.4 Energy landscape for a two-state unfolding reaction. The dotted-line curve is the free-energy surface at force = 0 and the solid-line curve at force = F. The force F tilts the energy landscape by a factor Fx, where x is the extension of the molecule.

$$\Delta G(F=0) = \Delta G^\circ + k_B T \ln\left(\frac{[B]}{[A]}\right) \tag{3.1}$$

where ΔG° is the standard state free energy, $k_B T$ is the thermal energy, and $[A]$ and $[B]$ give the probabilities of populating states A and B in single-molecule experiments. To a first approximation, when a force F is applied to a molecule, each point of its energy landscape is lowered by an amount equal to $F\Delta x$, where Δx is the distance between the point of interest and the native state. As a consequence, an applied force tilts the free-energy surface along the mechanical reaction coordinate (Bustamante, Chemla, Forde, & Izhaky, 2004), such that

$$\Delta G(F) = \Delta G^\circ - F(\Delta x) + k_B T \ln\left(\frac{[B]}{[A]}\right) \tag{3.2}$$

At equilibrium, $\Delta G = 0$ and

$$\Delta G^\circ = -k_B T \ln K_{eq}(F) + F\Delta x \tag{3.3}$$

Equation (3.3) holds true only if the position of A and B along the reaction coordinate are unaffected by force. For most reactions, however, this is not true. Let us consider, for example, the case of a protein that under tension transits from its native state to its unfolded state. In this case, the extension of the native state, and thus its position along the reaction coordinate, can to a good approximation be considered unchanged, but the extension of the unfolded state instead will significantly increase. This increment in the average end-to-end distance of the unfolded state corresponds to a change in the free energy of the system, as

$$\Delta G^{\circ} = -k_{\mathrm{B}} T \ln K_{\mathrm{eq}}(F) + F\Delta x - \Delta G_{\mathrm{stretch}}(F) \tag{3.4}$$

where $\Delta G_{\mathrm{stretch}}(F)$ is the free-energy change due to stretching of the unfolded state at force F. When a molecule is manipulated at close to equilibrium conditions, it unfolds and refolds through transitions that take place around a force ($F_{1/2}$) at which the molecule has equal probability of being either in its folded or unfolded state. At $F_{1/2}$, $K_{\mathrm{eq}} = 1$ and

$$\Delta G^{\circ} = F_{1/2}\Delta x - \Delta G_{\mathrm{stretch}}\left(F_{1/2}\right) \tag{3.5}$$

Under these experimental conditions, $F_{1/2}\Delta x$ can be calculated as the area under the unfolding/refolding rips observed in force versus extension curves. This free energy can then be compared with that measured in bulk measurements after subtraction of $\Delta G_{\mathrm{stretch}}(F_{1/2})$, which can to a good approximation be calculated as the area under the WLC force–extension curve integrated from zero to the extension of the unfolded molecule at $F_{1/2}$ (Cecconi et al., 2005; Liphardt et al., 2001).

3.2. Effect of force on the kinetics of a single-molecule reaction

As we know from transition state theory, the rates of unfolding (k_{u}^{0}) and refolding (k_{f}^{0}) of a molecule at zero force are

$$k_{\mathrm{u}}^{0} = A \exp\left(\frac{-\Delta G_{\mathrm{u}}}{k_{\mathrm{B}} T}\right) \tag{3.6}$$

$$k_{\mathrm{f}}^{0} = A \exp\left(\frac{-\Delta G_{\mathrm{f}}}{k_{\mathrm{B}} T}\right) \tag{3.7}$$

where A is the natural frequency of oscillation, ΔG_{u} and ΔG_{f} are the activation energies for unfolding and refolding, respectively. In the presence of force, the unfolding (ΔG_{u}) and refolding (ΔG_{f}) activation energies will be lowered by an amount equal to $F \cdot \Delta x_{\mathrm{u}}$ and $F \cdot \Delta x_{\mathrm{f}}$, respectively, where Δx_{u} and Δx_{f} are the distances to the transition state from the unfolded and folded states (Bell, 1978). It follows that the unfolding and refolding rates at force F are given by:

$$k_{\mathrm{u}}(F) = k_{\mathrm{u}}^{0} \exp\left(\frac{F\Delta x_{\mathrm{u}}}{k_{\mathrm{B}} T}\right) \tag{3.8}$$

$$k_f(F) = k_f^0 \exp\left(\frac{-F\Delta x_f}{k_B T}\right) \tag{3.9}$$

From the ratio of $k_u(F)$ and $k_f(F)$, the equilibrium constant can be calculated as:

$$K_{eq}(F) = K_{eq}^0 \exp\left(\frac{F\Delta x}{k_B T}\right) \tag{3.10}$$

where K_{eq}^0 is the equilibrium constant at zero force and $\Delta x = \Delta x_u + \Delta x_f$.

It is worth pointing out that in the above kinetics models, the position of the transition state along the reaction coordinate is considered to be independent from the applied force. This approximation is usually correct when we consider a very narrow range of forces. For more general cases, however, force-induced shifts in the position of the transition state must be accounted for. To this end, several improved models have been proposed and applied to different systems, as for example in Dudko, Graham, and Best (2011), Manosas, Collin, and Ritort (2006), and Schlierf, Berkemeier, and Rief (2007).

3.3. Extracting kinetic parameters from force distributions

In constant-velocity optical tweezers experiments (Fig. 3.2A), above 3–4 pN, the rate at which the force is applied on the molecule, loading rate r (dF/dt in units of pN/s), typically becomes constant (Cecconi et al., 2005; Liphardt et al., 2001). This allows us to analyze experimental data with analytical models, as explained below.

For a first-order reaction with negligible refolding rate, the time dependence of the probability that the molecule has not unfolded is (Tinoco & Bustamante, 2002):

$$\frac{dP_f(t)}{dt} = -k_u(t)P_f(t) \tag{3.11}$$

If force varies linearly with time t as $F = rt$, where r is the loading rate, then variable t can be changed to F in the above equation as

$$\frac{dP_f(F)}{dF} = -\frac{k_u(F)}{r}P_f(F) \tag{3.12}$$

Integrating the above equation from 0 to F and using Eq. (3.8), we get

$$\ln\{P_f(F)\} = \frac{k_u^0 k_B T}{rx_u}\left(1 - \exp\left(\frac{Fx_u}{k_B T}\right)\right) \tag{3.13}$$

From Eq. (3.13), the probability of unfolding as a function of force $(P_u(F))$ can be derived as

$$P_u(F) = 1 - P_f(F) = 1 - \exp\left(-\frac{k_u^0 k_B T}{(rx_u)(\exp(Fx_u/k_B T) - 1)}\right) \tag{3.14}$$

By differentiating $P_u(F)$, which is a sigmoidal function of force, we can calculate the probability density as

$$\frac{dP_u}{dF} = \frac{k_u^0}{r}\exp\left(\frac{Fx_u}{k_B T}\right) * \exp\left(-\frac{k_u^0 k_B T}{(rx_u)(\exp(Fx_u/k_B T) - 1)}\right) \tag{3.15}$$

Following similar steps, the probability density function for the refolding force can be calculated as

$$\frac{dP_f}{dF} = \frac{k_f^0}{r}\exp(-Fx_f/k_B T) * \frac{\exp\left(-\frac{k_f^0 k_B T}{rx_f}\left(\exp\left(-\frac{Fx_f}{k_B T}\right) - 1\right)\right)}{\exp\left(\left(-\frac{k_f^0 k_B T}{rx_f}\right) - 1\right)} \tag{3.16}$$

Normalized distributions of the unfolding and refolding forces of a molecule manipulated at constant loading rate can be fit to Eqs. (3.15) and (3.16) to estimate rate constants at zero force (k_u^0, k_f^0) and distances to the transition state (x_u, x_f) (Heidarsson et al., 2012; Schlierf, Li, & Fernandez, 2004).

Often, however, experimental force distributions are analyzed with a slightly different method. When $\exp(Fx_u/k_B T) > 10$, which is usually the case with biomolecules, Eq. (3.13) can be written as

$$\ln\{P_f(F)\} = -\frac{k_u^0 k_B T}{rx_u}\left(\exp\left(\frac{Fx_u}{k_B T}\right)\right) \tag{3.17}$$

which can then be linearized as

$$\ln\left\{r\ln\left(\frac{1}{P_f(F)}\right)\right\} = \ln\frac{k_u^0 k_B T}{x_u} + \left(\frac{x_u}{k_B T}\right)F \tag{3.18}$$

Through similar considerations, for the refolding process, we have

$$\ln\left\{-r\ln\left(\frac{1}{P_u(F)}\right)\right\} = \ln\frac{k_f^0 k_B T}{x_f} + \left(\frac{x_f}{k_B T}\right)F \tag{3.19}$$

Equations (3.18) and (3.19) are often used to fit $\ln[r \ln[1/N]]$ and $\ln[-r \ln [1/U]]$ versus force graphs, where N and U are the folded and unfolded fractions, respectively, which are calculated by integrating the histograms of the force distributions over the corresponding range of forces (Heidarsson et al., 2012; Li et al., 2006; Liphardt et al., 2001).

3.4. Extracting thermodynamic parameters from nonequilibrium measurements

When the rate at which force is applied on a molecule is faster than its slowest relaxation rate, the unfolding and refolding processes occur out of equilibrium. Under these experimental conditions, the free-energy change of the process is not equal to the work done on the molecule, and cannot be calculated as the area under the unfolding/refolding transitions. However, fluctuation theorems have been developed to extract equilibrium information from nonequilibrium measurements (Jarzynski, 2011). The first of these theorems to be successfully applied to single-molecule experimental data was presented by Jarzynski (1997). He derived an equality that relates the free-energy difference $\Delta G(z)$, separating states of a system at positions 0 and z along a reaction coordinate, to the work done to irreversibly switch the system between the two states:

$$\exp[-\beta \Delta G(z)] = \lim N_{N\_\infty} \langle \exp[-\beta w_i(z,r)] \rangle N \qquad (3.20)$$

where $\langle \rangle$ denotes averaging over N work trajectories, $w_i(z,r)$ represents the work of the ith of N trajectories, and r is the switching rate. The first application of this method in single-molecule force spectroscopy was reported by Liphardt, Dumont, Smith, Tinoco, and Bustamante (2002). They manipulated a P5ab RNA hairpin out of equilibrium with optical tweezers and applied Jarzynski's equality to the irreversible work trajectories to extract the unfolding free energy of the molecule. Although effective for near-equilibrium processes like that observed for the P5ab RNA hairpin, in far from equilibrium systems Jarzynski's equality is hampered by large statistical uncertainties due to the exponential averaging of low work values (Gore, Ritort, & Bustamante, 2003). These large uncertainties can be significantly reduced by using Crooks' fluctuation theorem (CFT). CFT relates the amount of work done on a molecule that unfolds and refolds out of equilibrium, with the free-energy change of the process (Crooks, 1999).

Let $P_U(W)$ and $P_R(W)$ denote the probability distributions of the work performed on a molecule that is pulled (U) and relaxed (R) an infinite number times. The CFT then predicts that

$$\frac{P_U(W)}{P_R(W)} = \exp\left(\frac{W - \Delta G}{k_B T}\right) \tag{3.21}$$

where ΔG is the free-energy change between the final and the initial states of the molecule, and thus equal to the reversible work associated with this process. The value of ΔG can be determined as the point of intersection of the unfolding and refolding work distributions, where $W = \Delta G$ (Fig. 3.5). When the overlapping region of the two distributions is small because the unfolding/refolding process occurs very far from equilibrium, Bennett's acceptance ratio method (Bennett, 1976) is often used to reduce the uncertainty in the estimation of ΔG. CFT has already been applied in several single-molecule manipulation studies, as in Collin et al. (2005), Gebhardt et al. (2010), and Shank, Cecconi, Dill, Marqusee, and Bustamante (2010).

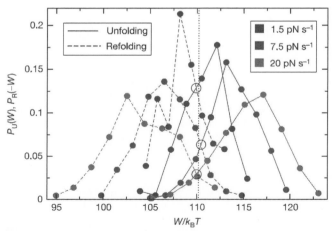

Figure 3.5 Work distributions $P_U(W)$ and $P_R(W)$ for RNA unfolding (solid curve) and refolding (dash curve) at different loading rates. The point of intersection of the two distributions gives $\Delta G = 110.3\ k_B T$. Adapted from Collin et al. (2005) with permission. (For color version of this figure, the reader is referred to the online version of this chapter.)

3.5. Extracting kinetics and thermodynamic parameters from equilibrium fluctuations

As discussed above, in constant-force and passive-mode measurements some molecules can be observed to fluctuate at equilibrium between different molecular conformations (Fig. 3.2B). Although different theoretical approaches could be used, hidden Markov models (HMMs) are often the methods of choice to extract kinetics and thermodynamic information from these experimental data (Alemany, Mossa, Junier, & Ritort, 2012; Elms, Chodera, Bustamante, & Marqusee, 2012a; Gao et al., 2011; Kaiser, Goldman, Chodera, Tinoco, & Bustamante, 2011; Stigler, Ziegler, Gieseke, Gebhardt, & Rief, 2011). HMMs are powerful statistical tools introduced by Baum and colleagues in the late 1960s and early 1970s (Baum, 1972; Baum & Petrie, 1966), and were later implemented for speech processing applications by Baker (1975). Figure 3.6 shows a general HMM, where the X_i represents the hidden state sequence, O_i the observation sequence, and A and B are the matrices of state transition and observation probabilities, respectively. HMM predicts the most likely state sequence X_i which has the maximum probability to give the observation sequence O_i by constructing a model using A and B matrices (Stamp & Le, 2005).

In the analysis of single-molecule force spectroscopy data, the HMM assumes that the observation sequence (force or extension trace) is generated by a Markov process in which the molecule makes history-independent transitions governed by a transition matrix among its different conformational states. At a given force, each state of a molecule can be defined by the distribution of extension or force values. The HMM analysis provides the transition probability matrix T that can be used to calculate the matrix for rate constants K using the relation $T=\exp(K\Delta t)$, where Δt is the data acquisition time (Chodera & Noe, 2010). Using this method, rate constants can be calculated at different forces (F) and plots of $\ln k$ versus F can be fit

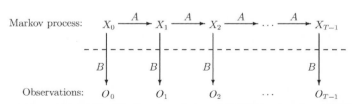

Figure 3.6 A general HMM where the hidden state sequence X_i is related to the observation sequence O_i by observation probability matrix B and transition probability matrix A. *With kind permission of Springer Science + Business Media (Stamp & Le, 2005).*

to linearized forms of Eqs. (3.8) and (3.9) to estimate the distances to the transition state and the rate constants at zero force. The ratio of the rate constants (K_{eq}) will provide information about the free energy of the reaction according to Eq. (3.4).

We would like to point out that the list of theoretical models presented above is by no means comprehensive. With the advent of new experimental strategies, novel theoretical models have also been developed and applied to different systems, as in Dudko et al. (2011), Dudko, Hummer, and Szabo (2008), Mossa, Huguet, and Ritort (2010), and Zoldak and Rief (2013).

4. BIOLOGICAL APPLICATIONS

4.1. Mechanical processes in the cell

A wide variety of cellular processes involves molecules that generate or are subjected to mechanical forces (Bustamante et al., 2004). Molecular motors such as kinesin convert chemical energy into mechanical work and movement, usually through ATP hydrolysis (Kolomeisky & Fisher, 2007). Mechanosensors go through subtle conformational changes in response to mechanical stimuli to initiate signaling cascades (Jaalouk & Lammerding, 2009). Proteins can also be unfolded by force, for example, by translocases, to target them for degradation or transport across membranes (King, Deshaies, Peters, & Kirschner, 1996; Maillard et al., 2011). Conversely, proteins perform work, and thus measurable force, through the compacting of their polypeptide chains during spontaneous folding. Below, we describe some recent studies on the conformational dynamics involved in protein folding and protein–ligand or protein–protein interactions.

4.2. Protein folding

Folding into a three-dimensional form is pivotal for the function and specificity of most proteins. The early classical experiments of Christian B. Anfinsen showed that a protein can spontaneously fold into thermodynamically stable states *in vitro* and that all the necessary information for the process is encoded in the amino acid sequence (Anfinsen, Haber, Sela, & White, 1961). Far from being a simple task, especially in the case of larger and of multidomain proteins, the mechanism of protein folding has been studied extensively over many decades. Yet, our understanding of this complex process is still incomplete. Proteins fold within a biologically relevant timescale of microseconds to seconds despite the astronomical amount of conformations available to them, a paradox so famously stated by Cyrus Levinthal (Levinthal, 1968; Zwanzig, Szabo, & Bagchi, 1992). The paradox, which highlights the fact that protein folding would be impossible with a random search through conformational space, ultimately through simplified models and pathways, led to the idea of folding funnels (Bryngelson, Onuchic, Socci, & Wolynes, 1995; Dill & Chan, 1997; Dill & MacCallum 2012; Leopold, Montal, & Onuchic, 1992; Oliveberg & Wolynes, 2005). In this view, the folding of protein molecules is described as diffusion of a statistical ensemble over a funnel-shaped energy landscape with possibly numerous parallel pathways (Fig. 3.7A) (Dill & Chan, 1997; Onuchic, Luthey-Schulten, & Wolynes, 1997).

The funnel shape provides an energetic bias toward folding into the native state and this can be described with two seemingly counteracting properties: a decrease in (1) configurational entropy (which is unfavorable) and (2) potential energy (which is favorable) (Karplus, 2011). The funneled energy landscape contains all the information to describe the path a protein can take en route to its native state. Theory predicts that a rugged energy landscape, with many small minima, will lead to slower folding

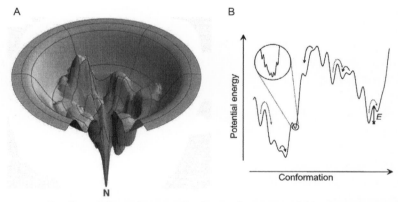

Figure 3.7 The funnel-shaped energy landscape for protein folding. (A) Proteins are currently thought to fold to their native state by diffusion over a multidimensional funnel-shaped energy landscape. (B) Two-dimensional scheme of the ruggedness of an energy landscape. Two main conformations are accessible with multiple smaller kinetic traps that are either short-lived (solid arrows) or long-lived (dotted arrows) conformations. E represents the height of the activation barrier for one such conformation. *(A) Adapted from Dill and Chan (1997) with permission. (B) Adapted from Milanesi et al. (2012) with permission.* (For color version of this figure, the reader is referred to the online version of this chapter.)

(Fig. 3.7B) and this was recently demonstrated experimentally with spectrin domains (Bryngelson & Wolynes, 1989; Wensley et al., 2010). The ensemble view suggests that due to the ruggedness of the energy landscape, a fraction of the molecules will fold slowly, due to trapping in energy minima, while other molecules will fold fast without populating intermediate states. This phenomenon is known as kinetic partitioning and explains the multiphasic kinetics observed in many systems (Thirumalai, Klimov, & Woodson, 1997). Kinetically trapped intermediate structures visited along a folding trajectory may be on–pathway, and thus productive toward the native state, or off–pathway misfolded states, which can form aggregates and lead to serious diseases (see section 4.2.3).

Traditional bulk techniques, such as NMR and protein engineering phi-value analysis (Matouschek, Kellis, Serrano, & Fersht, 1989), have unlocked a wealth of knowledge on the folding process but these methods are limited by an ensemble–averaged experimental output. Computational approaches to folding have also significantly enhanced our understanding but despite recent reports of millisecond–long simulations (Shaw et al., 2010) they have mostly been limited to probing a relatively short timescale using small, fast–folding proteins (Scheraga, Khalili, & Liwo, 2007). Single–molecule mechanical manipulation methods have added a new dimension to protein folding studies. The early mechanical manipulation instruments were only able to directly detect the unfolding process of single molecules, leaving the refolding process silent. The first of these experiments were performed simultaneously with AFM and optical tweezers on the giant muscle protein titin (Kellermayer, Smith, Granzier, & Bustamante, 1997; Rief, Gautel, Oesterhelt, Fernandez, & Gaub, 1997; Tskhovrebova, Trinick, Sleep, & Simmons, 1997). Quickly the field developed and the refolding of a single molecule into its native state was observed (Cecconi et al., 2005; Fernandez & Li, 2004). In the following years came a surge of exciting results, establishing single–molecule force spectroscopy, often in combination with molecular dynamics (MD) simulations, as a powerful tool to study protein folding.

4.2.1 Folding pathways and the energy landscape

Statistical ensemble views of protein folding predict that a rapid hydrophobic collapse into a subset of globular, unspecific structures, is the initial and necessary reduction in a polypeptide's available conformational space. This collapse has been difficult to study from the bulk viewpoint due to the averaging of pathways and ensembles (Agashe, Shastry, & Udgaonkar, 1995;

Dasgupta & Udgaonkar, 2010). AFM constant-force mode was used to study the early events of folding of individual ubiquitin molecules, induced by rapid force quenching (Garcia-Manyes et al., 2009). Garcia-Manyes et al. were able to directly observe a collapse into an ensemble of lower energy structures for ubiquitin prior to folding into the native state. Polyubiquitin molecules were first unfolded followed by quenching to low force, allowing the molecules to refold. Subsequently, the molecules were unfolded again to monitor whether the proteins had folded into their native state. The natively folded ubiquitin molecules unfolded in discrete steps but structures apparently collapsed during refolding, unfolded in various size steps and had less mechanical stability. By varying the refolding time, they were able to extract the kinetics of the refolding process and showed that the collapsed states folded in a fast phase followed by a slower phase involving a barrier-separated two-state mechanism.

The presence of multiple pathways is implied in the three-dimensional funnel-shaped energy landscape but experimental evidence for this phenomenon has remained limited (Radford, Dobson, & Evans, 1992; Wright, Lindorff-Larsen, Randles, & Clarke, 2003). This is in part due to differential populations of pathways, where some pathways may be rarely visited by a protein. He et al. studied a slipknot protein by combining AFM and course-grained Steered Molecular Dynamics (SMD) simulations (He, Genchev, Lu, & Li, 2012). They found that the protein could untie the slipknot and unfold through either a two-state or a three-state mechanism, suggesting parallel pathways. SMD simulations revealed that the intermediate state in the three-state pathway was formed only if the unfolding initiated from the C-terminus and key structural elements were found that prevent the formation of a tightened knot structure in both pathways. The results support the idea of kinetic partitioning and thus add to a growing number of evidence, suggesting it to be a general mechanism in protein folding.

Force acts on proteins locally in contrast to the global effect of temperature and chemicals. Furthermore, during mechanical manipulation, the molecule is tethered, restraining its conformational flexibility. Certainly the unfolded state in these experiments is stretched as opposed to a random coil for freely diffusing molecules. A question therefore arises on whether the folding pathways are the same when determined with chemical/thermal denaturants or with force. Originally, both experimental and computational studies suggested that force-induced unfolding differs from that induced by chemical/thermal denaturants (Best et al., 2003; Ng et al., 2005). Recently, however, evidence has emerged for similar pathways to exist. Simulations have suggested that at sufficiently low pulling speeds, where the molecule is able to sample a large part of its conformational space, mechanical unfolding pathways may start to resemble those observed in chemical denaturation studies (West, Olmsted, & Paci, 2006). A force-induced pathway-switch has indeed recently been observed experimentally for an SH3 domain in the low-force regime using optical tweezers (Jagannathan, Elms, Bustamante, & Marqusee, 2012). Here, two different pulling axes were tested and one of them revealed two distinct and parallel unfolding pathways, giving evidence of a multidimensional energy landscape. Using both constant-velocity experiments at different velocities and force-jump experiments, the authors determined the unfolding rates for both pulling axes. Although the absolute unfolding rates determined in bulk and using mechanical manipulation cannot generally be compared (due to contributions from the DNA handles and beads, however, a very recent study reports a data analysis method that accounts for these effects (Hinczewski, Gebhardt, Rief, & Thirumalai, 2013)), the effects of mutations on the rates could be compared in this study. The rates analysis indicated that the transition states in bulk and on the single-molecule level were populated to a similar extent, suggesting similar pathways.

Similar mechanical and chemical denaturation pathways have been suggested for acyl-CoA binding protein (ACBP) (Heidarsson et al., 2012). The structure and position along the reaction coordinate of the transition state of this small globular single-domain protein were investigated using a combination of optical tweezers and ratcheted MD simulations. Through both equilibrium and nonequilibrium optical tweezers experiments, Heidarsson et al. determined the "mechanical" Tanford β-value ($m\beta_T$), which is analogous to a Tanford β-value in bulk studies (Elms et al., 2012a; Tanford, 1970). This value, which takes values between 0 and 1, can be regarded as a measure of the nativeness of the transition state and the determined $m\beta_T$ value for ACBP was very similar to the β-value in bulk (Kragelund et al., 1996). An almost identical $m\beta_T$ value was also confirmed by ratcheted MD simulations, which were then used to estimate the structure of the transition state in atomic detail. The structure resembled the structure observed in bulk through phi-value analysis and NMR chemical shifts (Bruun, Iesmantavicius, Danielsson, & Poulsen, 2010; Kragelund et al., 1999). It may therefore be protein-specific whether mechanical and chemical/thermal transition states and unfolding pathways are similar or not.

Related to transition states is the transition path, which is the seemingly instantaneous transition across the transition state barrier and includes all the mechanistic details of how a process happens. Transition path times, the actual time it takes to cross the barrier (not to be confused with transition rates which describe the frequency of barrier crossing), are impossible to measure by bulk methods as this is entirely a single-molecule property. Yu et al. characterized the energy landscape of prion protein using optical tweezers (Yu, Gupta, et al., 2012). From constant-force experiments they reconstructed a detailed free-energy landscape and, using Kramers theory of barrier-limited diffusion, were able to directly determine the transition path times and folding rates. They found, in accordance with results from single-molecule fluorescence and MD results (Chung, McHale, Louis, & Eaton, 2012; Shaw et al., 2010), that transition path times are in the order of ~2 μs. Importantly, given the size of the prion protein, this study reiterated that transition path times are relatively insensitive to protein size, whereas folding rates can vary by many orders of magnitude.

4.2.2 Molecular response to force, secondary structure, and topology
A remarkable feature of mechanical manipulation techniques is the ability to probe folding using various reaction coordinates. Almost any pulling geometry can be chosen by changing the position of the residues that define the load application. This offers the unique opportunity to directly explore the anisotropy of the energy landscape. Carrion-Vazquez et al. were the first to exploit this and using AFM they found that the direction of force application can dramatically change the mechanical stability of ubiquitin molecules (Carrion-Vazquez et al., 2003). Similar energy landscape anisotropy has been demonstrated with the green fluorescent protein (GFP) (Dietz, Berkemeier, et al., 2006). Dietz et al. performed AFM experiments on GFP using five different pulling geometries by varying the placement of cysteine residues that serve as attachment points in the polyprotein constructs (Fig. 3.8A). The mechanical stability of GFP displayed large variability, depending on the pulling axis. From the resulting unfolding force distributions (Fig. 3.8B) and using Monte Carlo simulations, the height of the transition state barriers and the potential energy well width of the native state could be determined. Remarkably, whereas the unfolding rates were not significantly affected by pulling direction, the width of the potential energy well showed significant variation, indicating either brittle or compliant behavior, depending on the pulling geometry. Ultimately, the authors were able to describe the GFP structure in terms of different directional spring constants (Fig. 3.8C), highlighting the anisotropic nature of its energy landscape. These and similar studies demonstrate how directional control can

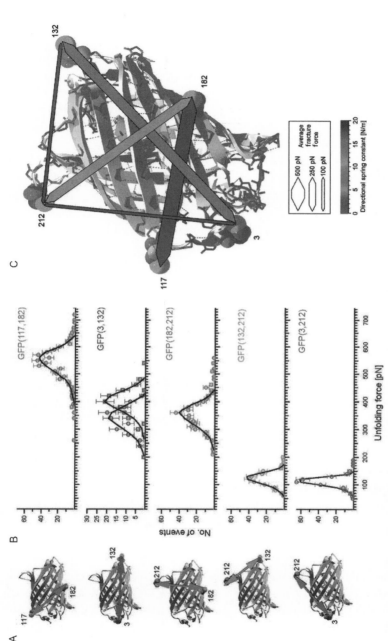

Figure 3.8 The anisotropy of GFP energy landscape probed with AFM. (A) Five different pulling geometries of GFP were studied by generating polyproteins with various attachment points through cysteine mutations. The numbers correspond to the cysteine residues and the arrows show the pulling axis. (B) Unfolding force distributions of the different variants. The solid lines represent results of a Monte Carlo simulation to reproduce the experimental data. (C) Mechanical stability (indicated by the width of arrows) and directional spring constants of GFP (indicated by the color of arrows). *Adapted from Dietz, Berkemeier, et al. (2006) with permission.* (For color version of this figure, the reader is referred to the online version of this chapter.)

allow exploration of regions of the energy landscape that are inaccessible to conventional methods.

The determinants of anisotropic mechanical response are likely to be encoded in the structural architecture of proteins, involving features such as secondary structure, topology, and long-range contacts. It has been demonstrated that the mechanical stability of proteins is highly dependent on secondary structure. Proteins consisting mainly of β-sheets are better able to resist mechanical denaturation than α-helical proteins, and this has been attributed to the extensive network of hydrogen bonding in β-sheet proteins in contrast to mainly hydrophobic interactions between helices (Brockwell et al., 2005; Schlierf & Rief, 2005). Secondary structure also correlates with the distance from the native state to the transition state (x_u), where helical proteins have larger distances indicating softer, more compliant structures (Li, 2007).

Topology, the arrangement of secondary elements along the sequence, has been shown to be important for cooperative folding during mechanical manipulation experiments. In T4 lysozyme (T4L), the A–helix is at the amino-terminal end but is part of the C-terminal structural domain. Shank et al. generated a circular permutant of T4L, effectively placing the A–helix at the C-terminal, and performed constant-velocity experiments using optical tweezers (Shank et al., 2010). Even though the experiments provided only nonequilibrium information, equilibrium properties of the system could be revealed using CFT (see section 3.4). In traditional ensemble studies, T4L unfolds in a single cooperative transition and this was also observed for the wild-type protein in constant-velocity experiments. However, the circular permutant displayed a three-state mechanism in which the N-domain now unfolded before the C-domain, indicating a decoupling of the two domains, which reduced cooperativity.

The mechanical properties of partially folded intermediate states are particularly interesting. Molten-globule intermediate states have significant native-like secondary structure but lack the characteristic stable tertiary interactions of the native state. Elms et al. studied the molten-globular state of apomyoglobin using optical tweezers (Elms et al., 2012a). They found that compared to native states, the molten globule is highly deformable (compliant), rendering its unfolding rate more sensitive to force, and speculated that this may be a general feature of molten globules. Surprisingly, the native state of ACBP has been shown to be even more deformable than the molten globule of apomyoglobin, a feature that may be important for its function as a lipid transporter (Heidarsson et al., 2012). The possibility that some intermediate states and even some proteins are more sensitive to force than others implies also that nature may have evolved this feature into proteins so that cells use less energy to unfold proteins that are targeted for degradation or translocation.

4.2.3 Intermediate states, misfolded states, and beyond

The role of intermediate states in protein folding has been debated and whether they are productive to the native state has remained an open question. Cecconi et al. used optical tweezers to characterize the refolding of ribonuclease H (RNase H) (Cecconi et al., 2005). RNase H has been suggested to refold through a molten–globular intermediate state but it was unclear whether this state was on- or off-pathway (Raschke, Kho, & Marqusee, 1999). By using constant-force experiments, they observed that within a narrow range of forces the molecule "hopped" between the unfolded state and the intermediate state. In some cases, the hopping came to an abrupt stop followed by a further compaction into the native state. This compaction occurred in the vast majority of cases directly from the intermediate state, providing direct evidence that it was on–pathway to the native state.

Predicted intermediate states have also been confirmed using force spectroscopy. Soluble N-ethylmaleimide-sensitive factor attachment protein receptor (SNARE) proteins mediate membrane fusion and are particularly important in vesicle fusion for neurotransmitter release (Ramakrishnan, Drescher, & Drescher, 2012). Different SNARE proteins, attached to the vesicle membrane and the plasma membrane, are thought to assemble into a parallel four-helix bundle and it has been suggested that the bundle then zippers toward the membrane, producing sufficient force to enable fusion. The zippering action has, however, not been supported by direct evidence and the assembly intermediates have eluded detection. Using optical tweezers and some clever protein engineering, Gao et al. showed that a neuronal SNARE complex zippers in three distinct stages (Gao et al., 2012). By applying forces in the same range as occurs during fusion, they were able to stabilize a half-zippered intermediate. The proposed zippering mechanism spawned from this study significantly impacts the study of neurotransmitter release.

Protein folding mechanisms have not evolved to perfection and can sometimes lead to incorrectly folded states. Misfolded proteins have attracted significant attention due to their well–established link to many severe diseases such as Alzheimer's, Parkinson's, and Creutzfeldt–Jakobs' (Dobson,

2003). As the prime cause of the latter, the infectious form of the prion protein is one of the most widely studied misfolding systems. Yu et al. studied the prion protein using optical tweezers and found that folding occurred in a two-state mechanism, contrary to previous reports. They observed that besides the native state, three distinct short-lived misfolded states were also populated (Fig. 3.9A) (Yu, Liu, et al., 2012). These states were only accessible from the unfolded state, and could thus be classified as off-pathway, and remarkably, they were more frequently accessed than the native state under tension. A mutant that had higher aggregation propensity in bulk, populated two of the three misfolded states more frequently. The results of this study challenge the assumption that an on-pathway intermediate is responsible for aggregation and suggest that prion misfolding is mediated from the unfolded state, not the native state (Fig. 3.9B).

Interestingly, proteins not directly linked to misfolding diseases have been shown to populate misfolded states. The ubiquitously expressed protein calmodulin (CaM) transduces changing levels in intracellular Ca^{2+} concentrations and is one of the most extensively studied calcium binding proteins. Despite extensive information on CaM structural states and complex formations from traditional methods, single-molecule methods have been able to impressively mine new features of its folding mechanism. Besides characterizing in detail the folding network of four distinct on-pathway states, Stigler et al. observed two additional misfolded states that were the result of incorrect pairing of EF-hands (Stigler et al., 2011). This misfolding slowed the overall folding kinetics of otherwise fast-folding individual domains.

Nonnative interactions are not necessarily unwanted as elegantly demonstrated by Forman et al. (2009). The mechanosensory polycystin-1 PKD domain has been shown to be mechanically stronger than its native state predicts and MD simulations had suggested that this is due to a rearrangement leading to nonnative hydrogen bonds that resist unfolding. Using genetic engineering, the authors produced mutant proteins that were designed to prevent formation of these nonnative bonds. They subsequently showed through AFM experiments and MD simulations that, despite having a negligible effect on the native state stability, the mutations caused dramatic mechanical destabilization. Thus, formation of nonnative interactions during the mechanical unfolding of the PKD domain seems crucial for mechanical stability and, by inference, mechanosensory function.

Functional nonnative interactions have also been suggested for coiled coils, molecules that have various roles in the cell often connected with

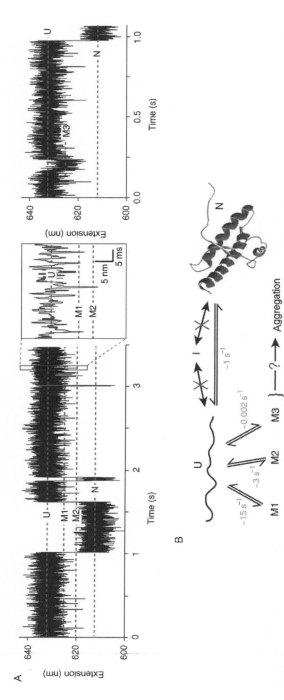

Figure 3.9 Misfolded states in single-molecule experiments. (A) Extension versus time traces of the prion protein. Left panel: the unfolded state frequently transitions into two distinct and short-lived misfolded states, M1 and M2. The inset shows a zoom into the region defined by the blue square. Right panel: rarely, a third and longer-lived misfolded state was populated. (B) An illustration of the folding/misfolding scheme of the prion protein along with the transition rates to the different misfolded states under tension. *Adapted from Yu, Liu, et al. (2012) with permission.* (For interpretation of the references to color in this figure legend, the reader is referred to the online version of this chapter.)

sensing or generating force. The coiled coils have characteristic *abcdefg* heptad repeats that engage in the dimer interface and native coiled coil dimer formation depends on the correct pairing of the *a* and *d* positions. Sometimes, alternative pairings are formed when one helix is shifted, usually by one heptad repeat, and coiled coils have been suggested to undergo helix sliding to populate these so-called staggered helices. The nonnative conformations may be important for biological function but have eluded detailed characterization due to their instability. Using high-resolution optical tweezers, Xi et al. studied two very stable coiled coils: a variant of the GCN4 leucine zipper (pIL) and a heterodimeric coiled coil (pER) (Xi et al., 2012). The authors were able to directly observe partially folded states where the difference in the number of folded residues suggested these states to correspond to one helix shifted by one to three heptad repeats or, in other words, staggered helices. These states were populated less than 2% of the time with lifetimes decreasing inversely with the number of shifted heptad repeats. Usually, the partially folded states would only be accessed from the unfolded state (misfolding) but occasionally (<10%) the native state would transit directly into these states, which the authors interpreted as evidence for helix sliding.

Protein folding studies have greatly benefited from the advent of single-molecule manipulation studies from which new and exciting information continues to emerge. Kinetic partitioning seems to emerge as a general feature of protein folding kinetics and the frequent observation of misfolded states, even in small proteins thought to be efficient folders, challenges the idea of a highly evolved folding process. However, the traditional focus of protein folding studies needs to broaden from mostly single-domain proteins to more complex systems, as only a handful of larger proteins and double-domain systems has been explored on the single-molecule level. The current development of novel and hybrid instruments promises the ability to tackle ever more complex proteins and determine the details of multidomain protein folding.

4.3. Protein–ligand and protein–protein interactions

Proteins interact with a multitude of binding partners *in vivo* and these interactions naturally affect their structure and energetics. Molecular chaperones interact with proteins to improve the fidelity of their folding process, sometimes by rescuing misfolded states (Hartl, Bracher, & Hayer-Hartl, 2011). The binding of various ions can activate proteins by inducing conformational change, such as in the case of calcium sensors, to initiate signaling pathways (Chazin, 2011). Conformational dynamics and flexibility are the hallmark of protein interactions (Teilum, Olsen, & Kragelund, 2011) and single-molecule force spectroscopy is now being used to probe these events in significant detail.

Mechanical stability is largely governed by specific noncovalent interactions in the protein and ligand binding can induce conformational changes in the protein structure that lead to changes in mechanical stability. Alterations in mechanical stability upon ligand binding can therefore serve as an intrinsic reporter to identify the functional state of protein at the single-molecule level (Cao, Balamurali, Sharma, & Li, 2007). The effect of ligand binding on the mechanical stability of a protein is not easy to predict. Using AFM, only moderate increases in mechanical stability were observed upon ligand binding to mouse dihydrofolate reductase (DHFR) (Junker, Hell, Schlierf, Neupert, & Rief, 2005) and to Im9 (Hann et al., 2007) while strong enhancement of mechanical stability occurs when protein G interacts with an Fc fragment from IgG (Cao et al., 2007). Stabilization of enzymatically inactive conformations by inhibitors might have significant importance in biomedical applications and drug design. For instance, the role of DHFR in cell proliferation and growth has rendered this enzyme a target for anticancer drug therapy (Li et al., 2000). Through AFM experiments Ainavarapu et al. showed that binding of the cancer chemotherapeutic agent methotrexate (MTX) to DHFR increases the mechanical stability of the enzyme (Ainavarapu, Li, Badilla, & Fernandez, 2005). The correlation between this result and the decrease in the degradation rate of DHFR inside the cell supports the idea that force-induced denaturation is necessary for translocation and degradation of this protein. Also, through similar experiments, Junker et al., have shown that simultaneous binding of MTX and the cofactor NADHP to DHFR increases the lifetime of one intermediate state of the protein, suggesting that MTX and NADHP could inhibit DHFR by trapping it in an enzymatically inactive intermediate structure (Junker et al., 2005).

Several proteins, like CaM, are activated by calcium ion binding, which allows them to interact with a vast array of binding partners. Using optical tweezers, Stigler and Rief studied CaM folding under conditions approaching physiological Ca^{2+} concentrations (Stigler & Rief, 2012). Variations in Ca^{2+} concentrations led to dramatic changes in the folding/unfolding kinetics of CaM. At high Ca^{2+} concentrations, CaM folding was characterized by a complex network of on- and off-pathway intermediate states (Fig. 3.10A). At low Ca^{2+} concentrations, the folding network remained the same but the

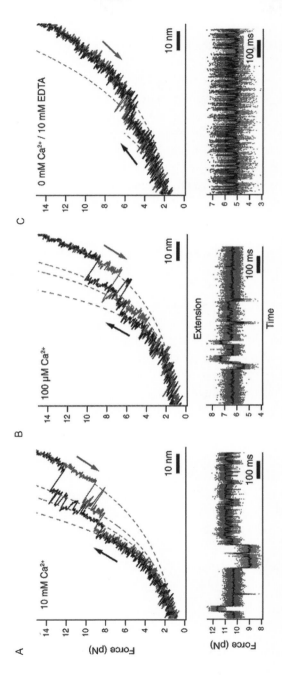

Figure 3.10 Calcium binding modulates the mechanical folding behavior of CaM. Force versus extension cycles of CaM with varying Ca^{2+} concentration. (A) 10 mM Ca^{2+}, (B) 100 μm Ca^{2+}, and (C) 0 mM Ca^{2+}. The lower panels show force versus time traces, where the two beads tethering the molecule are kept at a constant distance. At high Ca^{2+} concentrations, CaM populates four intermediate structures between the unfolded and folded states, each color-coded. At lower Ca^{2+} concentrations, the same folding pattern is observed but at low forces. Under apo-conditions (0 mM Ca^{2+}), the kinetic pattern changes drastically and CaM fluctuates between only two states. Dotted lines in upper panels represent WLC fits to the data. Arrows indicate pulling direction. *Adapted from Stigler and Rief (2012) with permission.* (For color version of this figure, the reader is referred to the online version of this chapter.)

mechanical stability of the protein drops and the unfolding and refolding rate constants changed significantly (Fig. 3.10B). In the absence of Ca^{2+}, the CaM folding network was essentially reduced to a two–state mechanism (Fig. 3.10C).

Chaperones interact directly with polypeptide chains to increase the efficiency of folding. Some chaperones can prevent misfolding and aggregation, or facilitate protein translocation across membranes or for degradation (Maillard et al., 2011). However, the mechanism by which folding pathways are affected by chaperones is poorly understood. Bechtluft and coworkers used optical tweezers in combination with MD simulations to study the effect of the chaperone SecB on the folding pathway of maltose binding protein (MBP) (Bechtluft et al., 2007). They observed that, in the absence of SecB, MBP populated a molten–globule–like compacted state before folding into its native structure. Upon addition, SecB binds to the molten–globule state, stabilizing it and preventing the formation of stable native tertiary contacts of MBP. The effect of SecB on the structure of MBP might explain the mechanism by which this chaperone facilitates translocation of this protein through cellular membranes, as the absence of stable tertiary contacts likely facilitates the passage of the protein through the translocation machinery. This work demonstrated how effective optical tweezers approaches can be to elucidate the effect of chaperones on protein folding landscapes.

Kim et al. developed an elegant optical tweezers method to perform repeated measurements of the binding/release kinetics between receptor and ligand (Kim, Zhang, Zhang, & Springer, 2010). Using this novel method, they were able to investigate the interaction between the A1 domain of von Willebrand factor (VWF) and the leucine-rich repeat (LRR) domain of glycoprotein Ib α subunit (GPIbα) by tethering the two molecules together via a flexible linker (Fig. 3.11A). Force experiments at different pulling rates and constant-force showed that the A1–GPIbα interaction exists in two states (Fig. 3.11B and C), which the authors referred to as a flex–bond. One state was observed at low force whereas the other was mechanically more stable and had a much longer lifetime. The kinetics of the bond formation (Fig. 3.11D) help to explain how platelets bound to VWF are able to resist force to plug arterioles and how increased flow activates platelet plug formation.

5. FUTURE PERSPECTIVES

Single-molecule methods are making an impact in many aspects of cellular and molecular biology and several new and exciting developments are on the horizon. Intrinsically disordered proteins (IDPs) are a recently

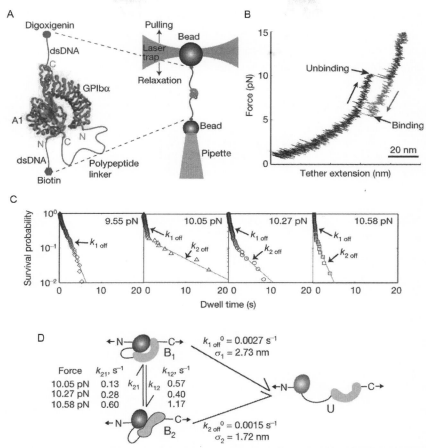

Figure 3.11 Novel approach to studying receptor–ligand binding/unbinding kinetics. (A) The experimental setup. Left: the A1 domain–GPIbα LRR domain construct. Right: the optical tweezers setup. (B) Force versus extension trace for one cycle of stretching (black) and relaxing (red). (C) Constant-force experiments were performed to determine the survival probability of the bound state as a function of time at different force values. (D) A model of the A1–GPIbα flex-bond and the associated rate constants. *Adapted from Kim et al. (2010) with permission.* (For interpretation of the references to color in this figure legend, the reader is referred to the online version of this chapter.)

recognized group of proteins where the central paradigm of a folded functional protein is challenged (Tompa, 2011). These proteins have seemingly no defined structure, allowing significant functional plasticity but their properties are just beginning to be studied at the single-molecule level. So far, IDPs have exclusively been studied with single-molecule detection techniques but mechanical manipulation should offer an attractive approach to tackle IDPs as well. Optical tweezers have already successfully manipulated objects in living cells (Oddershede, 2012). The dream is that near-future instrumental and methodological advances will allow us to directly observe life unfold by watching individual molecules *in vivo*, in real time.

ACKNOWLEDGMENTS

P. O. H. acknowledges the Carlsberg Foundation for financial support. C. C. gratefully acknowledges financial support from Fondazione Cassa di Risparmio di Modena, the EU through a Marie Curie International Reintegration Grant (No. 44952), the Italian MIUR (Grant No. 17DPXLNBEK), and partial support from Italian MIUR FIRB RBPR05JH2P "ITALNANONET." Dr. Birthe B. Kragelund is thanked for critical reading of parts of this manuscript.

REFERENCES

Agashe, V. R., Shastry, M. C., & Udgaonkar, J. B. (1995). Initial hydrophobic collapse in the folding of barstar. *Nature, 377*(6551), 754–757.

Ainavarapu, S. R., Li, L., Badilla, C. L., & Fernandez, J. M. (2005). Ligand binding modulates the mechanical stability of dihydrofolate reductase. *Biophysical Journal, 89*(5), 3337–3344.

Alemany, A., Mossa, A., Junier, I., & Ritort, F. (2012). Experimental free-energy measurements of kinetic molecular states using fluctuation theorems. *Nature Physics, 8*, 688–694.

Anfinsen, C. B., Haber, E., Sela, M., & White, F. H., Jr. (1961). The kinetics of formation of native ribonuclease during oxidation of the reduced polypeptide chain. *Proceedings of the National Academy of Sciences of the United States of America, 47*, 1309–1314.

Ashkin, A., Dziedzic, J. M., Bjorkholm, J. E., & Chu, S. (1986). Observation of a single-beam gradient force optical trap for dielectric particles. *Optics Letters, 11*(5), 288–290.

Baker, J. K. (1975). The dragon system—An overview. *IEEE Transactions on Acoustics, Speech, & Signal Processing, 23*, 24–29.

Baum, L. E. (1972). An inequality and associated maximization technique in statistical estimation for probabilistic functions of Markov processes. *Inequalities, 3*, 1–8.

Baum, L. E., & Petrie, T. (1966). Statistical inference for probabilistic functions of finite state Markov chains. *Annals of Mathematical Statistics, 37*, 1554–1563.

Bechtluft, P., van Leeuwen, R. G. H., Tyreman, M., Tomkiewicz, D., Nouwen, N., Tepper, H. L., et al. (2007). Direct observation of chaperone-induced changes in a protein folding pathway. *Science, 318*(5855), 1458–1461.

Bell, G. I. (1978). Models for the specific adhesion of cells to cells. *Science, 200*(4342), 618–627.

Bennett, C. H. (1976). Efficient estimation of free energy differences from Monte Carlo data. *Journal of Computational Physics, 22*, 245–268.

Best, R. B., Fowler, S. B., Herrera, J. L., Steward, A., Paci, E., & Clarke, J. (2003). Mechanical unfolding of a titin Ig domain: Structure of transition state revealed by combining atomic force microscopy, protein engineering and molecular dynamics simulations. *Journal of Molecular Biology, 330*(4), 867–877.

Binnig, G., Quate, C. F., & Gerber, C. (1986). Atomic force microscope. *Physical Review Letters, 56*(9), 930–933.

Borgia, A., Williams, P. M., & Clarke, J. (2008). Single-molecule studies of protein folding. *Annual Review of Biochemistry, 77*, 101–125.

Bornschlogl, T., & Rief, M. (2011). Single-molecule protein unfolding and refolding using atomic force microscopy. *Methods in Molecular Biology, 783*, 233–250.

Brockwell, D. J., Beddard, G. S., Paci, E., West, D. K., Olmsted, P. D., Smith, D. A., et al. (2005). Mechanically unfolding the small, topologically simple protein L. *Biophysical Journal, 89*(1), 506–519.

Bruun, S. W., Iesmantavicius, V., Danielsson, J., & Poulsen, F. M. (2010). Cooperative formation of native-like tertiary contacts in the ensemble of unfolded states of a four-helix protein. *Proceedings of the National Academy of Sciences of the United States of America, 107*(30), 13306–13311.

Bryngelson, J. D., Onuchic, J. N., Socci, N. D., & Wolynes, P. G. (1995). Funnels, pathways, and the energy landscape of protein folding: A synthesis. *Proteins, 21*(3), 167–195.

Bryngelson, J. D., & Wolynes, P. G. (1989). Intermediates and barrier crossing in a random energy-model (with applications to protein folding). *The Journal of Physical Chemistry, 93*(19), 6902–6915.

Bustamante, C. (2008). In singulo biochemistry: When less is more. *Annual Review of Biochemistry, 77*, 45–50.

Bustamante, C., Chemla, Y. R., Forde, N. R., & Izhaky, D. (2004). Mechanical processes in biochemistry. *Annual Review of Biochemistry, 73*, 705–748.

Bustamante, C., Marko, J. F., Siggia, E. D., & Smith, S. (1994). Entropic elasticity of lambdaphage DNA. *Science, 265*(5178), 1599–1600.

Cao, Y., Balamurali, M. M., Sharma, D., & Li, H. (2007). A functional single-molecule binding assay via force spectroscopy. *Proceedings of the National Academy of Sciences of the United States of America, 104*(40), 15677–15681.

Carrion-Vazquez, M., Li, H., Lu, H., Marszalek, P. E., Oberhauser, A. F., & Fernandez, J. M. (2003). The mechanical stability of ubiquitin is linkage dependent. *Nature Structural Biology, 10*(9), 738–743.

Carrion-Vazquez, M., Oberhauser, A. F., Fowler, S. B., Marszalek, P. E., Broedel, S. E., Clarke, J., et al. (1999). Mechanical and chemical unfolding of a single protein: A comparison. *Proceedings of the National Academy of Sciences of the United States of America, 96*(7), 3694–3699.

Cecconi, C., Shank, E. A., Bustamante, C., & Marqusee, S. (2005). Direct observation of the three-state folding of a single protein molecule. *Science, 309*(5743), 2057–2060.

Cecconi, C., Shank, E. A., Dahlquist, F. W., Marqusee, S., & Bustamante, C. (2008). Protein-DNA chimeras for single molecule mechanical folding studies with the optical tweezers. *European Biophysics Journal, 37*(6), 729–738.

Cecconi, C., Shank, E. A., Marqusee, S., & Bustamante, C. (2011). DNA molecular handles for single-molecule protein-folding studies by optical tweezers. *Methods in Molecular Biology, 749*, 255–271.

Chazin, W. J. (2011). Relating form and function of EF-hand calcium binding proteins. *Accounts of Chemical Research, 44*(3), 171–179.

Chodera, J. D., & Noe, F. (2010). Probability distributions of molecular observables computed from Markov models. II. Uncertainties in observables and their time-evolution. *The Journal of Chemical Physics, 133*(10), 105102.

Chung, H. S., McHale, K., Louis, J. M., & Eaton, W. A. (2012). Single-molecule fluorescence experiments determine protein folding transition path times. *Science, 335*(6071), 981–984.

Collin, D., Ritort, F., Jarzynski, C., Smith, S. B., Tinoco, I., Jr., & Bustamante, C. (2005). Verification of the Crooks fluctuation theorem and recovery of RNA folding free energies. *Nature, 437*(7056), 231–234.

Crooks, G. E. (1999). Entropy production fluctuation theorem and the nonequilibrium work relation for free energy differences. *Physical Review E: Statistical Physics, Plasmas, Fluids, and Related Interdisciplinary Topics, 60*(3), 2721–2726.

Dame, R. T., Noom, M. C., & Wuite, G. J. (2006). Bacterial chromatin organization by H-NS protein unravelled using dual DNA manipulation. *Nature*, *444*(7117), 387–390.

Dasgupta, A., & Udgaonkar, J. B. (2010). Evidence for initial non-specific polypeptide chain collapse during the refolding of the SH3 domain of PI3 kinase. *Journal of Molecular Biology*, *403*(3), 430–445.

Deniz, A. A., Mukhopadhyay, S., & Lemke, E. A. (2008). Single-molecule biophysics: At the interface of biology, physics and chemistry. *Journal of the Royal Society Interface*, *5*(18), 15–45.

De Vlaminck, I., & Dekker, C. (2012). Recent advances in magnetic tweezers. *Annual Review of Biophysics*, *41*, 453–472.

De Vlaminck, I., Henighan, T., van Loenhout, M. T., Burnham, D. R., & Dekker, C. (2012). Magnetic forces and DNA mechanics in multiplexed magnetic tweezers. *PLoS One*, *7*(8), e41432.

De Vlaminck, I., Vidic, I., van Loenhout, M. T., Kanaar, R., Lebbink, J. H., & Dekker, C. (2010). Torsional regulation of hRPA-induced unwinding of double-stranded DNA. *Nucleic Acids Research*, *38*(12), 4133–4142.

Dietz, H., Berkemeier, F., Bertz, M., & Rief, M. (2006). Anisotropic deformation response of single protein molecules. *Proceedings of the National Academy of Sciences of the United States of America*, *103*(34), 12724–12728.

Dietz, H., Bertz, M., Schlierf, M., Berkemeier, F., Bornschlogl, T., Junker, J. P., et al. (2006). Cysteine engineering of polyproteins for single-molecule force spectroscopy. *Nature Protocols*, *1*(1), 80–84.

Dill, K. A., & Chan, H. S. (1997). From Levinthal to pathways to funnels. *Nature Structural Biology*, *4*(1), 10–19.

Dill, K. A., & MacCallum, J. L. (2012). The protein-folding problem, 50 years on. *Science*, *338*(6110), 1042–1046.

Dobson, C. M. (2003). Protein folding and misfolding. *Nature*, *426*(6968), 884–890.

Dudko, O. K., Graham, T. G., & Best, R. B. (2011). Locating the barrier for folding of single molecules under an external force. *Physical Review Letters*, *107*(20), 208301.

Dudko, O. K., Hummer, G., & Szabo, A. (2008). Theory, analysis, and interpretation of single-molecule force spectroscopy experiments. *Proceedings of the National Academy of Sciences of the United States of America*, *105*(41), 15755–15760.

Elms, P. J., Chodera, J. D., Bustamante, C., & Marqusee, S. (2012a). The molten globule state is unusually deformable under mechanical force. *Proceedings of the National Academy of Sciences of the United States of America*, *109*(10), 3796–3801.

Elms, P. J., Chodera, J. D., Bustamante, C. J., & Marqusee, S. (2012b). Limitations of constant-force-feedback experiments. *Biophysical Journal*, *103*(7), 1490–1499.

Fernandez, J. M., & Li, H. (2004). Force-clamp spectroscopy monitors the folding trajectory of a single protein. *Science*, *303*(5664), 1674–1678.

Florin, E. L., Moy, V. T., & Gaub, H. E. (1994). Adhesion forces between individual ligand-receptor pairs. *Science*, *264*(5157), 415–417.

Forman, J. R., Yew, Z. T., Qamar, S., Sandford, R. N., Paci, E., & Clarke, J. (2009). Non-native interactions are critical for mechanical strength in PKD domains. *Structure*, *17*(12), 1582–1590.

Gao, Y., Sirinakis, G., & Zhang, Y. (2011). Highly anisotropic stability and folding kinetics of a single coiled coil protein under mechanical tension. *Journal of the American Chemical Society*, *133*(32), 12749–12757.

Gao, Y., Zorman, S., Gundersen, G., Xi, Z., Ma, L., Sirinakis, G., et al. (2012). Single reconstituted neuronal SNARE complexes zipper in three distinct stages. *Science*, *337*(6100), 1340–1343.

Garcia-Manyes, S., Dougan, L., Badilla, C. L., Brujic, J., & Fernandez, J. M. (2009). Direct observation of an ensemble of stable collapsed states in the mechanical folding of ubiquitin. *Proceedings of the National Academy of Sciences of the United States of America, 106*(26), 10534–10539.

Gebhardt, J. C., Bornschlogl, T., & Rief, M. (2010). Full distance-resolved folding energy landscape of one single protein molecule. *Proceedings of the National Academy of Sciences of the United States of America, 107*(5), 2013–2018.

Gore, J., Ritort, F., & Bustamante, C. (2003). Bias and error in estimates of equilibrium free-energy differences from nonequilibrium measurements. *Proceedings of the National Academy of Sciences of the United States of America, 100*(22), 12564–12569.

Grandbois, M., Beyer, M., Rief, M., Clausen-Schaumann, H., & Gaub, H. E. (1999). How strong is a covalent bond? *Science, 283*(5408), 1727–1730.

Greenfield, N. J. (2007). Using circular dichroism spectra to estimate protein secondary structure. *Nature Protocols, 1*(6), 2876–2890.

Hann, E., Kirkpatrick, N., Kleanthous, C., Smith, D. A., Radford, S. E., & Brockwell, D. J. (2007). The effect of protein complexation on the mechanical stability of Im9. *Biophysical Journal, 92*(9), L79–L81.

Hartl, F. U., Bracher, A., & Hayer-Hartl, M. (2011). Molecular chaperones in protein folding and proteostasis. *Nature, 475*(7356), 324–332.

He, C., Genchev, G. Z., Lu, H., & Li, H. (2012). Mechanically untying a protein slipknot: Multiple pathways revealed by force spectroscopy and steered molecular dynamics simulations. *Journal of the American Chemical Society, 134*(25), 10428–10435.

Heidarsson, P. O., Valpapuram, I., Camilloni, C., Imparato, A., Tiana, G., Poulsen, F. M., et al. (2012). A highly compliant protein native state with a spontaneous-like mechanical unfolding pathway. *Journal of the American Chemical Society, 134*(41), 17068–17075.

Hinczewski, M., Gebhardt, J. C., Rief, M., & Thirumalai, D. (2013). From mechanical folding trajectories to intrinsic energy landscapes of biopolymers. *Proceedings of the National Academy of Sciences of the United States of America, 110*, 4500–4505.

Jaalouk, D. E., & Lammerding, J. (2009). Mechanotransduction gone awry. *Nature Reviews. Molecular Cell Biology, 10*(1), 63–73.

Jagannathan, B., Elms, P. J., Bustamante, C., & Marqusee, S. (2012). Direct observation of a force-induced switch in the anisotropic mechanical unfolding pathway of a protein. *Proceedings of the National Academy of Sciences of the United States of America, 109*(44), 17820–17825.

Jarzynski, C. (1997). Nonequilibrium equality for free energy differences. *Physical Review Letters, 78*, 2690–2693.

Jarzynski, C. (2011). Equalities and inequalities: Irreversibility and the second law of thermodynamics at the nanoscale. *Annual Reviews of Condensed Matter Physics, 2*, 329–351.

Joo, C., Balci, H., Ishitsuka, Y., Buranachai, C., & Ha, T. (2008). Advances in single-molecule fluorescence methods for molecular biology. *Annual Review of Biochemistry, 77*, 51–76.

Junker, J. P., Hell, K., Schlierf, M., Neupert, W., & Rief, M. (2005). Influence of substrate binding on the mechanical stability of mouse dihydrofolate reductase. *Biophysical Journal, 89*(5), L46–L48.

Junker, J. P., & Rief, M. (2009). Single-molecule force spectroscopy distinguishes target binding modes of calmodulin. *Proceedings of the National Academy of Sciences of the United States of America, 106*(34), 14361–14366.

Junker, J. P., Ziegler, F., & Rief, M. (2009). Ligand-dependent equilibrium fluctuations of single calmodulin molecules. *Science, 323*(5914), 633–637.

Kaiser, C. M., Goldman, D. H., Chodera, J. D., Tinoco, I., Jr., & Bustamante, C. (2011). The ribosome modulates nascent protein folding. *Science, 334*(6063), 1723–1727.

Karplus, M. (2011). Behind the folding funnel diagram. *Nature Chemical Biology, 7*(7), 401–404.

Kellermayer, M. S., Smith, S. B., Granzier, H. L., & Bustamante, C. (1997). Folding-unfolding transitions in single titin molecules characterized with laser tweezers. *Science*, *276*(5315), 1112–1116.

Kern, D., Eisenmesser, E. Z., & Wolf-Watz, M. (2005). Enzyme dynamics during catalysis measured by NMR spectroscopy. *Methods in Enzymology*, *394*, 507–524.

Kim, J., Zhang, C. Z., Zhang, X., & Springer, T. A. (2010). A mechanically stabilized receptor-ligand flex-bond important in the vasculature. *Nature*, *466*(7309), 992–995.

King, R. W., Deshaies, R. J., Peters, J. M., & Kirschner, M. W. (1996). How proteolysis drives the cell cycle. *Science*, *274*(5293), 1652–1659.

Kolomeisky, A. B., & Fisher, M. E. (2007). Molecular motors: A theorist's perspective. *Annual Review of Physical Chemistry*, *58*, 675–695.

Koster, D. A., Crut, A., Shuman, S., Bjornsti, M. A., & Dekker, N. H. (2010). Cellular strategies for regulating DNA supercoiling: A single-molecule perspective. *Cell*, *142*(4), 519–530.

Kragelund, B. B., Hojrup, P., Jensen, M. S., Schjerling, C. K., Juul, E., Knudsen, J., et al. (1996). Fast and one-step folding of closely and distantly related homologous proteins of a four-helix bundle family. *Journal of Molecular Biology*, *256*(1), 187–200.

Kragelund, B. B., Osmark, P., Neergaard, T. B., Schiodt, J., Kristiansen, K., Knudsen, J., et al. (1999). The formation of a native-like structure containing eight conserved hydrophobic residues is rate limiting in two-state protein folding of ACBP. *Nature Structural Biology*, *6*(6), 594–601.

Lee, G. U., Chrisey, L. A., & Colton, R. J. (1994). Direct measurement of the forces between complementary strands of DNA. *Science*, *266*(5186), 771–773.

Leopold, P. E., Montal, M., & Onuchic, J. N. (1992). Protein folding funnels: A kinetic approach to the sequence-structure relationship. *Proceedings of the National Academy of Sciences of the United States of America*, *89*(18), 8721–8725.

Levintha, C. (1968). Are there pathways for protein folding. *Journal de Chimie Physique et de Physico-Chimie Biologique*, *65*(1), 44–45.

Li, M. S. (2007). Secondary structure, mechanical stability, and location of transition state of proteins. *Biophysical Journal*, *93*(8), 2644–2654.

Li, P. T., Collin, D., Smith, S. B., Bustamante, C., & Tinoco, I., Jr. (2006). Probing the mechanical folding kinetics of TAR RNA by hopping, force-jump, and force-ramp methods. *Biophysical Journal*, *90*(1), 250–260.

Li, R., Sirawaraporn, R., Chitnumsub, P., Sirawaraporn, W., Wooden, J., Athappilly, F., et al. (2000). Three-dimensional structure of *M. tuberculosis* dihydrofolate reductase reveals opportunities for the design of novel tuberculosis drugs. *Journal of Molecular Biology*, *295*(2), 307–323.

Liphardt, J., Dumont, S., Smith, S. B., Tinoco, I., Jr., & Bustamante, C. (2002). Equilibrium information from nonequilibrium measurements in an experimental test of Jarzynski's equality. *Science*, *296*(5574), 1832–1835.

Liphardt, J., Onoa, B., Smith, S. B., Tinoco, I., Jr., & Bustamante, C. (2001). Reversible unfolding of single RNA molecules by mechanical force. *Science*, *292*(5517), 733–737.

Maillard, R. A., Chistol, G., Sen, M., Righini, M., Tan, J., Kaiser, C. M., et al. (2011). ClpX(P) generates mechanical force to unfold and translocate its protein substrates. *Cell*, *145*(3), 459–469.

Maity, H., Maity, M., Krishna, M. M., Mayne, L., & Englander, S. W. (2005). Protein folding: The stepwise assembly of foldon units. *Proceedings of the National Academy of Sciences of the United States of America*, *102*(13), 4741–4746.

Manosas, M., Collin, D., & Ritort, F. (2006). Force-dependent fragility in RNA hairpins. *Physical Review Letters*, *96*(21), 218301.

Marszalek, P. E., Oberhauser, A. F., Pang, Y. P., & Fernandez, J. M. (1998). Polysaccharide elasticity governed by chair-boat transitions of the glucopyranose ring. *Nature*, *396*(6712), 661–664.

Matouschek, A., Kellis, J. T., Jr., Serrano, L., & Fersht, A. R. (1989). Mapping the transition state and pathway of protein folding by protein engineering. *Nature*, *340*(6229), 122–126.

Milanesi, L., Waltho, J. P., Hunter, C. A., Shaw, D. J., Beddard, G. S., Reid, G. D., et al. (2012). Measurement of energy landscape roughness of folded and unfolded proteins. *Proceedings of the National Academy of Sciences of the United States of America*, *109*(48), 19563–19568.

Moffitt, J. R., Chemla, Y. R., Smith, S. B., & Bustamante, C. (2008). Recent advances in optical tweezers. *Annual Review of Biochemistry*, 77, 205–228.

Mossa, A., Huguet, J. M., & Ritort, F. (2010). Investigating the thermodynamics of small biosystems with optical tweezers. *Physica E: Low-dimensional Systems and Nanostructures*, *42*(3), 666–671.

Neuman, K. C., & Nagy, A. (2008). Single-molecule force spectroscopy: Optical tweezers, magnetic tweezers and atomic force microscopy. *Nature Methods*, *5*(6), 491–505.

Ng, S. P., Randles, L. G., & Clarke, J. (2007). Single molecule studies of protein folding using atomic force microscopy. *Methods in Molecular Biology*, *350*, 139–167.

Ng, S. P., Rounsevell, R. W., Steward, A., Geierhaas, C. D., Williams, P. M., Paci, E., et al. (2005). Mechanical unfolding of TNfn3: The unfolding pathway of a fnIII domain probed by protein engineering, AFM and MD simulation. *Journal of Molecular Biology*, *350*(4), 776–789.

Noom, M. C., van den Broek, B., van Mameren, J., & Wuite, G. J. (2007). Visualizing single DNA-bound proteins using DNA as a scanning probe. *Nature Methods*, *4*(12), 1031–1036.

Oddershede, L. B. (2012). Force probing of individual molecules inside the living cell is now a reality. *Nature Chemical Biology*, *8*(11), 879–886.

Oliveberg, M., & Wolynes, P. G. (2005). The experimental survey of protein-folding energy landscapes. *Quarterly Reviews of Biophysics*, *38*(3), 245–288.

Onuchic, J. N., Luthey-Schulten, Z., & Wolynes, P. G. (1997). Theory of protein folding: The energy landscape perspective. *Annual Review of Physical Chemistry*, *48*, 545–600.

Raab, A., Han, W., Badt, D., Smith-Gill, S. J., Lindsay, S. M., Schindler, H., et al. (1999). Antibody recognition imaging by force microscopy. *Nature Biotechnology*, *17*(9), 901–905.

Radford, S. E., Dobson, C. M., & Evans, P. A. (1992). The folding of hen lysozyme involves partially structured intermediates and multiple pathways. *Nature*, *358*(6384), 302–307.

Ramakrishnan, N. A., Drescher, M. J., & Drescher, D. G. (2012). The SNARE complex in neuronal and sensory cells. *Molecular and Cellular Neurosciences*, *50*(1), 58–69.

Raschke, T. M., Kho, J., & Marqusee, S. (1999). Confirmation of the hierarchical folding of RNase H: A protein engineering study. *Nature Structural Biology*, *6*(9), 825–831.

Rief, M., Gautel, M., Oesterhelt, F., Fernandez, J. M., & Gaub, H. E. (1997). Reversible unfolding of individual titin immunoglobulin domains by AFM. *Science*, *276*(5315), 1109–1112.

Rief, M., Oesterhelt, F., Heymann, B., & Gaub, H. E. (1997). Single molecule force spectroscopy on polysaccharides by atomic force microscopy. *Science*, *275*(5304), 1295–1297.

Rounsevell, R., Forman, J. R., & Clarke, J. (2004). Atomic force microscopy: Mechanical unfolding of proteins. *Methods*, *34*(1), 100–111.

Scheraga, H. A., Khalili, M., & Liwo, A. (2007). Protein-folding dynamics: Overview of molecular simulation techniques. *Annual Review of Physical Chemistry*, *58*, 57–83.

Schlierf, M., Berkemeier, F., & Rief, M. (2007). Direct observation of active protein folding using lock-in force spectroscopy. *Biophysical Journal*, *93*(11), 3989–3998.

Schlierf, M., Li, H., & Fernandez, J. M. (2004). The unfolding kinetics of ubiquitin captured with single-molecule force-clamp techniques. *Proceedings of the National Academy of Sciences of the United States of America*, *101*(19), 7299–7304.

Schlierf, M., & Rief, M. (2005). Temperature softening of a protein in single-molecule experiments. *Journal of Molecular Biology*, *354*(2), 497–503.

Schuler, B., & Eaton, W. A. (2008). Protein folding studied by single-molecule FRET. *Current Opinion in Structural Biology*, *18*(1), 16–26.

Shank, E. A., Cecconi, C., Dill, J. W., Marqusee, S., & Bustamante, C. (2010). The folding cooperativity of a protein is controlled by its chain topology. *Nature*, *465*(7298), 637–640.

Shaw, D. E., Maragakis, P., Lindorff-Larsen, K., Piana, S., Dror, R. O., Eastwood, M. P., et al. (2010). Atomic-level characterization of the structural dynamics of proteins. *Science*, *330*(6002), 341–346.

Sirinakis, G., Ren, Y., Gao, Y., Xi, Z., & Zhang, Y. (2012). Combined versatile high-resolution optical tweezers and single-molecule fluorescence microscopy. *The Review of Scientific Instruments*, *83*(9), 093708.

Smith, S. B., Cui, Y., & Bustamante, C. (2003). Optical-trap force transducer that operates by direct measurement of light momentum. *Methods in Enzymology*, *361*, 134–162.

Stamp, M., & Le, E. (2005). Hamptonese and hidden Markov models. In W. Dayawansa, A. Lindquist, & Y. Zhou (Eds.), *New directions and applications in control theory*; Vol. 321, (pp. 367–378). Berlin, Heidelberg: Springer.

Steward, A., Toca-Herrera, J. L., & Clarke, J. (2002). Versatile cloning system for construction of multimeric proteins for use in atomic force microscopy. *Protein Science*, *11*(9), 2179–2183.

Stigler, J., & Rief, M. (2012). Calcium-dependent folding of single calmodulin molecules. *Proceedings of the National Academy of Sciences of the United States of America*, *109*, 17814–17819.

Stigler, J., Ziegler, F., Gieseke, A., Gebhardt, J. C., & Rief, M. (2011). The complex folding network of single calmodulin molecules. *Science*, *334*(6055), 512–516.

Tanford, C. (1970). Protein denaturation. C. Theoretical models for the mechanism of denaturation. *Advances in Protein Chemistry*, *24*, 1–95.

Teilum, K., Olsen, J. G., & Kragelund, B. B. (2011). Protein stability, flexibility and function. *Biochimica et Biophysica Acta*, *1814*(8), 969–976.

Thirumalai, D., Klimov, D. K., & Woodson, S. A. (1997). Kinetic partitioning mechanism as a unifying theme in the folding of biomolecules. *Theoretical Chemistry Accounts*, *96*(1), 14–22.

Tinoco, I., Jr., & Bustamante, C. (2002). The effect of force on thermodynamics and kinetics of single molecule reactions. *Biophysical Chemistry*, *101–102*, 513–533.

Tinoco, I., Jr., & Gonzalez, R. L., Jr. (2011). Biological mechanisms, one molecule at a time. *Genes & Development*, *25*(12), 1205–1231.

Tompa, P. (2011). Unstructural biology coming of age. *Current Opinion in Structural Biology*, *21*(3), 419–425.

Tskhovrebova, L., Trinick, J., Sleep, J. A., & Simmons, R. M. (1997). Elasticity and unfolding of single molecules of the giant muscle protein titin. *Nature*, *387*(6630), 308–312.

Wensley, B. G., Batey, S., Bone, F. A., Chan, Z. M., Tumelty, N. R., Steward, A., et al. (2010). Experimental evidence for a frustrated energy landscape in a three-helix-bundle protein family. *Nature*, *463*(7281), 685–688.

West, D. K., Olmsted, P. D., & Paci, E. (2006). Mechanical unfolding revisited through a simple but realistic model. *The Journal of Chemical Physics*, *124*(15), 154909.

Wright, C. F., Lindorff-Larsen, K., Randles, L. G., & Clarke, J. (2003). Parallel protein-unfolding pathways revealed and mapped. *Nature Structural Biology, 10*(8), 658–662.

Xi, Z., Gao, Y., Sirinakis, G., Guo, H., & Zhang, Y. (2012). Single-molecule observation of helix staggering, sliding, and coiled coil misfolding. *Proceedings of the National Academy of Sciences of the United States of America, 109*(15), 5711–5716.

Yang, G., Cecconi, C., Baase, W. A., Vetter, I. R., Breyer, W. A., Haack, J. A., et al. (2000). Solid-state synthesis and mechanical unfolding of polymers of T4 lysozyme. *Proceedings of the National Academy of Sciences of the United States of America, 97*(1), 139–144.

Yu, H., Gupta, A. N., Liu, X., Neupane, K., Brigley, A. M., Sosova, I., et al. (2012). Energy landscape analysis of native folding of the prion protein yields the diffusion constant, transition path time, and rates. *Proceedings of the National Academy of Sciences of the United States of America, 109*(36), 14452–14457.

Yu, H., Liu, X., Neupane, K., Gupta, A. N., Brigley, A. M., Solanki, A., et al. (2012). Direct observation of multiple misfolding pathways in a single prion protein molecule. *Proceedings of the National Academy of Sciences of the United States of America, 109*, 5283–5288.

Zoldak, G., & Rief, M. (2013). Force as a single molecule probe of multidimensional protein energy landscapes. *Current Opinion in Structural Biology, 23*(1), 48–57.

Zwanzig, R., Szabo, A., & Bagchi, B. (1992). Levinthal's paradox. *Proceedings of the National Academy of Sciences of the United States of America, 89*(1), 20–22.

CHAPTER FOUR

Generation of Ligand Specificity and Modes of Oligomerization in β-Prism I Fold Lectins

Thyageshwar Chandran, Alok Sharma, Mamannamana Vijayan[1]
Molecular Biophysics Unit, Indian Institute of Science, Bangalore, India
[1]Corresponding author: e-mail address: mv@mbu.iisc.ernet.in

Contents

Abstract

β-Prism I fold lectins constitute one of the five widely occurring structural classes of plant lectins. Each single domain subunit is made up of three Greek key motifs arranged in a threefold symmetric fashion. The threefold symmetry is not reflected in the sequence except in the case of the lectin from banana, a monocot, which carries two sugar-binding sites instead of the one in other lectins of known three-dimensional structure, all from dicots. This is believed to be a consequence of the different

Advances in Protein Chemistry and Structural Biology, Volume 92
ISSN 1876-1623
http://dx.doi.org/10.1016/B978-0-12-411636-8.00004-3
135

evolutionary paths followed by the lectin in monocots and dicots. The galactose-specific lectins among them have two chains produced by posttranslational proteolysis and contain three aromatic residues at the binding site. The extended binding sites of galactose- and mannose-specific lectins have been thoroughly characterized. Ligand binding at the sites involves both conformational selection and induced fit. Molecular plasticity of some of the lectins in the family has been characterized. The plasticity appears to be such as to promote variability in quaternary association which could be dimeric, tetrameric, or octameric. Structural and evolutionary reasons for the variability have been explored, and the relation of oligomerization to ligand binding and conformational selection investigated.

1. INTRODUCTION

Lectins, commonly described as multivalent carbohydrate-binding proteins, were first identified in plants and their best known property was the ability to agglutinate blood. Therefore, they used to be called phytohemagglutinins. Subsequently, they were shown to exist in all kingdoms of life and viruses (Chandra et al., 2006; Hamblin & Kent, 1973). In addition to biochemical and biological studies, the three-dimensional structures of lectins from very different sources have been determined primarily using X-ray crystallography (http://www.cermav.cnrs.fr/lectines). Lectin molecules come in a variety of shapes and sizes and they assume widely different folds. The one property shared by all of them is the ability to selectively bind different carbohydrates and cause biological effects by doing so. Lectins assumed considerable importance with the realization that most of the recognitive processes on the cell surface are mediated by specific protein–carbohydrate interactions (Brandley & Schnaar, 1986; Frazier & Glaser, 1979; Sharon & Lis, 2004). They are known to be involved in a plethora of biological processes such as symbiosis (Bohlool & Schmidt, 1974; De Hoff, Brill, & Hirsch, 2009), cell–cell interactions (Crocker, 2002), and innate immunity (Epstein, Eichbaum, Sheriff, & Ezekowitz, 1996; Fujita, 2002; Turner, 1996). Some of them are believed to be antifungal (Broekaert, Van Parijs, Leyns, Joos, & Peumans, 1989; Ngai & Ng, 2007), antiretroviral (Balzarini et al., 1991; Wang, Cole, Hong, Waring, & Lehrer, 2003), and antiproliferative (Zhang, Sun, Wang, & Ng, 2009).

Those from plants and animals constitute the most extensively studied groups of lectins (Drickamer & Taylor, 1993; Rini & Lobsanov, 1999; Rudiger & Gabius, 2001; Vijayan & Chandra, 1999). In particular, the functional roles of animal lectins have been elucidated in considerable detail (Drickamer & Taylor, 1993; Taylor & Drickamer, 2007). Among the bacterial lectins, toxins have received considerable attention (Merritt et al., 1998; Swaminathan & Eswaramoorthy, 2000; Zhang, Scott, et al., 1995; Zhang, Westbrook, et al., 1995). Viral hemagglutinins, particularly those from influenza virus, have been studied thoroughly and their biological function is established in some detail (Skehel & Wiley, 2000; Wilson, Indiveri, Quaranta, Pellegrino, & Ferrone, 1981). Much less extensively studied are fungal, yeast, and algal lectins.

2. SETTING THE STAGE: ISSUES ADDRESSED

Although lectins were first identified in plants, their biological role is not yet clearly understood. Some of them are known to be involved in nodulation on the roots of leguminous plants (Bohlool & Schmidt, 1974). They are believed to be part of the defense mechanism of plants (Chrispeels & Raikhel, 1991). However, historically and in terms of the elucidation of the structural basis of protein–carbohydrate association, plant lectins have been very important. Concanavalin A (ConA) from the legume *Canavalia ensiformis* was the first lectin to be structure analyzed using X-ray crystallography (Becker, Reeke, Wang, Cunningham, & Edelman, 1975; Hardman, Wood, Schiffer, Edmundson, & Ainsworth, 1972). Since then, the structure and interactions of ConA have been extensively investigated, and it has remained an important model for protein–sugar interactions. ConA is mannose/glucose specific and is a tetramer made up of two dimers. Dimerization in the structure involves joining together of two six-stranded β-sheets, one from each monomer, to form an extended 12-stranded β-sheet in the dimer. Arrangements similar to that found in this "canonical" dimer have since been observed in many protein structures. Different modes of dimerization were subsequently observed in lectin IV from *Griffonia simplicifolia* (Delbaere et al., 1993), which binds complex carbohydrates, and the galactose-specific *Erythrina corallodendron* lectin (EcorL) (Shaanan, Lis, & Sharon, 1991). The departure from the canonical dimerization in these two lectins was then attributed to steric clashes involving covalently linked sugar. The subsequent structure determination of peanut lectin, a nonglycosylated galactose-specific tetrameric legume lectin which exhibits an unusual quaternary association, demonstrated that noncanonical modes of dimerization are not necessarily caused by interactions involving covalently linked sugar but are likely to have resulted from factors intrinsic to the protein (Banerjee et al., 1994). Further work, particularly that involving winged bean lectins and recombinant EcorL (Kulkarni et al., 2004; Manoj, Srinivas, Surolia, Vijayan, & Suguna, 2000; Prabu et al., 1998), showed that legume lectins constitute a family of proteins in which small alterations in essentially the same tertiary structure lead to large differences in quaternary association, possibly as a means of generating different types of multivalency (Chandra, Prabu, Suguna, & Vijayan, 2001; Prabu, Suguna, & Vijayan, 1999). Other studies further contributed to the elaboration of the variability in the quaternary association of legume lectins (Bouckaert, Hamelryck, Wyns, & Loris, 1999; Sinha, Gupta, Vijayan, & Surolia, 2007).

Most of the legume lectins with known three-dimensional structure are either mannose (Man)/glucose (Glc) specific or galactose (Gal)/ N-acetylgalactosamine (GalNAc) specific at the primary binding site. The primary binding site is substantially preformed. Yet, the affinities of a given lectin for a monomeric sugar can exhibit variability. For instance, Gal/GalNAc-specific lectins show considerable variability in the relative affinities for Gal and GalNAc. Substantial variability occurs in the specificity for disaccharides and higher oligomers. For example, peanut lectin is specific for the tumor-associated T-antigenic disaccharide Gal-β-1,3-GalNAc. It binds to this disaccharide with 20-fold higher affinity than that for the structurally similar lactose (Banerjee et al., 1996). The higher affinity for the T-antigenic disaccharide in this case is caused by an additional water bridge (Ravishankar, Ravindran, Suguna, Surolia, & Vijayan, 1997). Thus, peanut lectin uses water bridge as a strategy for generation of ligand specificity. In another instance, variation of loop length has been used as a strategy for generating specificity (Manoj et al., 2000). Among other things, there is also an instance where difference in the internal structure of related blood group substances leads to difference in specificity (Kulkarni, Katiyar, Surolia, Vijayan, & Suguna, 2007).

Those from leguminous plants are undoubtedly the most thoroughly studied family of lectins. Chronologically, the second structural family of plant lectins to be identified was those with hevein domain, through the X-ray analysis of wheat germ agglutinin (WGA) (Wright, 1977). Since then, the interactions of sugars with WGA and other lectins of the family have been explored. However, studies on this family of lectins have been far less extensive than those on legume lectins. Lectins with β-trefoil fold were the third family of plant lectins to be characterized through the structure determination of ricin (Rutenber et al., 1991). In most known cases involving them, a lectin chain with two trefoil domains is covalently linked through a disulfide bridge to a catalytic domain, to give rise to a type II ribosome-inactivating protein (RIP). Each domain carries one carbohydrate-binding site which helps the molecule to interact with the cell surface. The catalytic domain cleaves a ribosomal RNA. Thus, type II RIP is a two–chain single subunit protein. There are instances where two such subunits are linked together to form a four–chain molecule, in a manner reminiscent to dimerization (Chandran, Sharma, & Vijayan, 2010; Sweeney et al., 1998). Primary carbohydrate-binding sites on the two lectin domains have been well characterized, but the changes in the binding sites in inactive RIP homologues such as those found in edible plants are an interesting issue. Snowdrop lectin is the first to be structure analyzed from a family referred to as monocot lectins or β-prism II fold lectins (Hester, Kaku, Goldstein, & Wright, 1995). Structures of many lectins from this family are now known. They also exhibit variability in oligomerization. In fact, oligomerization has been suggested to be a strategy for generating carbohydrate specificity in these lectins (Chandra et al., 1999). The fifth major structural family of plant lectins is made up of β-prism I fold lectins and forms the subject matter of this review.

In short, unlike animal and bacterial lectins, the number of folds exhibited by plant lectins is few in number. Five distinct folds account for almost all the plant lectins of known three–dimensional structure. However, substantial structural diversity is generated through variability in oligomerization. Thus, mode of oligomerization is an interesting issue when dealing with plant lectins. Another interesting issue is the limited variability in the essentially preformed primary binding sites. Yet, another issue is the strategies for generation of specificity for different types of monomeric sugars, for disaccharides, and for higher oligomers. In the rest of this chapter, these issues are critically examined in relation to β-prism I fold lectins using results obtained from crystallographic, modeling, and molecular dynamics (MD) studies.

3. β-PRISM I FOLD LECTINS: THE BASICS

β-Prism I fold lectins were first identified in this laboratory through the crystal structure analysis of jacalin, one of the two lectins in jackfruit seeds (Sankaranarayanan et al., 1996). Subsequently, the structure of the second lectin, artocarpin, was also determined (Pratap et al., 2002). The lectins are homologues; both form tetramers. However, each subunit of jacalin is made up of two chains of unequal length, produced by posttranslational proteolysis, while artocarpin has a single-chain, 149–amino acid–long subunit. Jacalin is glycosylated while artocarpin is not. Perhaps more significantly, jacalin is galactose specific while artocarpin is mannose specific at the primary binding site. Thus, the structure of jacalin, initially determined as a complex with methyl-α-galactose, and that of artocarpin, determined in the apo form and in complex with methyl-α-mannose, form a framework for examining galactose as well as mannose-specific β-prism I fold lectins.

Each subunit of jacalin and artocarpin is essentially made up of three Greek keys (Fig. 4.1). However, one Greek key is broken in jacalin on account of posttranslational proteolysis. Nevertheless, the 20–amino acid–long shorter fragment (β-chain) forms an integral part of three-dimensional structure of the subunit. The corresponding stretch in artocarpin forms its N-terminal region. This region (and the short peptide in jacalin) forms an

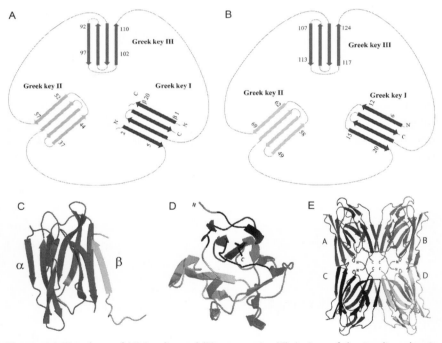

Figure 4.1 Topology of (A) jacalin and (B) artocarpin. (C) A view of the jacalin subunit perpendicular to the approximate threefold axis. (D) A view of the artocarpin subunit down the approximate threefold axis. The three Greek keys are colored differently. (E) The tetrameric molecule of artocarpin. The four subunits are colored differently. (For color version of this figure, the reader is referred to the online version of this chapter.)

outer strand of the Greek key involving the C–terminal stretch of the longer polypeptide chain (α–chain). The three-dimensional structure of the subunit exhibits an approximate threefold symmetry. However, this symmetry is not reflected in the amino acid sequence. It was, however, hypothesized that each subunit could have resulted from the successive gene duplication and fusion of a primitive, approximately 40 amino acid long, carbohydrate-binding module with a Greek key topology (Sankaranarayanan et al., 1996).

Each subunit of jacalin and artocarpin carries one carbohydrate-binding site. The primary binding sites in the two lectins have similar structures but, as shown in Fig. 4.2, there are crucial differences. Gly 1 in jacalin, corresponding to Gly 15 in artocarpin, is a positively charged terminal amino group of the longer chain generated by posttranslational modification. Its location in jacalin is such as to produce an interaction with O4 in the bound galactose. Mannose has a different configuration at C4 and therefore O4 has a different orientation in mannose. Thus, it would appear that the posttranslational proteolysis that generates the additional N-terminal amino group is designed for specific interaction with galactose. Furthermore, Tyr 78 in jacalin stacks against the B face of the bound galactose. The corresponding residue is a threonine in artocarpin. Also, Tyr 122 and Trp 123, two aromatic residues, are part of the binding site in jacalin. The corresponding residues in artocarpin are Leu and Asp, respectively. Thus, the presence of a terminal amino group generated by posttranslational modification and three aromatic residues is the characteristic features of galactose-specific jacalin. These features are observed in other homologous galactose-binding lectins as well.

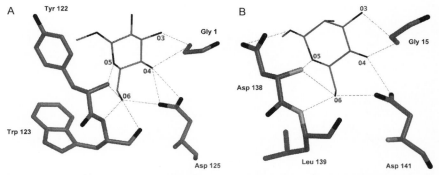

Figure 4.2 Hydrogen-bonded interactions in (A) jacalin with methyl-α-Gal and (B) artocarpin with methyl-α-Man. (For color version of this figure, the reader is referred to the online version of this chapter.)

While the general architecture of the binding site is similar, these features are absent in artocarpin and other homologous single-chain, single domain mannose-binding lectins.

4. AVAILABLE STRUCTURES AND RELATIONSHIPS AMONG THEM

Structures of nine multimeric lectins with each subunit made up of a single β-prism I fold domain and several of their complexes have been determined using X-ray crystallography. These are listed in Table 4.1. Alignment of the sequences of these lectins is shown in Fig. 4.3. All of them are homologous to one another with varying degrees of relatedness. The sequence identities among them are listed in Table 4.2. As can be seen from Table 4.2, sequence identities correlate well with relatedness in three-dimensional structure as evidenced by the root mean square deviations (RMSDs) in C^{α} positions on pairwise superposition.

Among the nine, jacalin, *Artocarpus hirsuta* lectin (AHL) (Rao, Suresh, Katre, Gaikwad, & Khan, 2004), and *Maclura pomifera* agglutinin (MPA) (Lee et al., 1998) have been demonstrated to be galactose specific with structural characteristics mentioned earlier. All of them form tetramers in a similar manner. Moringa M (Rabijns et al., 2005), artocarpin, banana lectin (Banlec) (Meagher, Winter, Ezell, Goldstein, & Stuckey, 2005; Singh

Table 4.1 Single domain β-prism I fold plant lectins of known three-dimensional structure

Name	Oligomeric state	Number of native structure	Number of sugar complexes
Jacalin	Tetramer	1	7
Artocarpus hirsuta lectin (AHL)	Tetramer	0	2
Maclura pomifera agglutinin (MPA)	Tetramer	1	3
Moringa M	Tetramer	1	1
Artocarpin	Tetramer	2	3
Banlec	Dimer	1	6
Heltuba	Octamer	1	2
Calsepa	Dimer	1	0
Ipomoelin	Tetramer	1	3

```
Jacalin      ----------NEQSGKSQTVIVGSWGAKVS---GKAFDDGAFTGIREINLSYNKETAIGDFQVVY 52
AHL          ----------DENSGKSQTVIVGPWGAKVS---GKAFDDGAFTGIREINLSYNKETAIGDFQVVY 52
MPA          ---------GRNGKSQSIIVGPWGDRVTN-GVTFDDGAYTGIREINFEYNSETAIGGLRVTY 52
MoringM      AGTSTNTQTTGTSQTVEVGLWGGPGG---NAWDDGSYTGIREINLSHG--DAIGAFSVIY 55
Artocarpin   -----------ASQTITVGPWGGPGG---NGWDDGSYTGIRQIELSYK--EAIGSFSVIY 44
Banlec       --------MNG---AIKVGAWGGNGGS-AFDMGPAYRIISVKIFSGDVVD-----AVDVTF 44
Heltuba      --------MAASDIAVQAGPWGGNGGKRWLQTAHGGKITSIIIKGGTCIF-----SIQFVY 48
Calsepa      --------MAVPMDTISGPWGNNGGN-FWSFRPVNKINQIVISGGGGN----NPIALTF 47
Ipomoelin    ------MALQLAAHSDARSGPVGSNGGQ-FWSFRPVRPLNKIVLSFSGSPDQTLNLISITF 54
                               *   *                :            . . :
Jacalin      DLN----GSPYVGQNHKSFITGFTPVKISLDFPSEYIMEVSGYTG------NVSGYVVVRSLT 105
AHL          DLN----GSPYVGQNHKSFITGFTPVKISLDFPSEYIMEVSGYTG------NVSGYVVVRSLT 105
MPA          DLN----GMPFVAEDHKSFITGFKPVKISLEFPSEYIVEVSGYVG------KVEGYTVIRSLT 105
MoringM      DLN----GQPFTGPTHPGNEPSFKTVKITLDFPNEFLVSVSGYTGVLPRLATGKDVIRSLT 112
Artocarpin   DLN----GEPFSGPKHTSKLP-YKNVKIELRFPDEFLESVSGYTAPFSALATPTPVVRSLT 100
Banlec       TYY------GKTETR-HFGGSSGTPHEIVLQEGEYLVGMKGEFG----NYHGVVVVGKLG 93
Heltuba      KDK-----DNIEYHSGKFGVLGDKAETITFAEDEDITAISGTFG----AYYHMTVVTSLT 99
Calsepa      SSTKADGSKDTITVGGGGPDSITGTEMVNIGTDEYLTGISGTFG-----IYLDNNVLRSIT 103
Ipomoelin    SSN----PTDIITVGGVGPEPLTYTETVNI--DGDIIEISGMIA----NYKGYNVIRSIK 104
                          .    :   :.*  .              *:  .:
Jacalin      FKTNK-KTYGPYGVTSGTPFNLPIENG-LIVGFKGSIGYWLDYFSMYLSL- 153
AHL          FKTNK-KTYGPYGVTSGTPFNLPIENG-LIVGFKGSIGYWLDYFSMYLSL- 153
MPA          FKTNK-QTYGPYGVTNGTPFSLPIENG-LIVGFKGSIGYWLDYFSIYLSL- 153
MoringM      FKTNK-KTYGPYGKEEGTPFSLPIENG-LIVGFKGRSGFVVDAIGVHLSL- 160
Artocarpin   FKTNKGRTFGPYGDEEGTYFNLPIENG-LIVGFKGRTGDLLDAIGVHMAL- 149
Banlec       FSTNK-KSYGPFGNTGGTPFSLPIAAG-KISGFFGRGGDFIDAIGVYLEP- 141
Heltuba      FQTNK-KVYGPFGTVASSSFSLPLTKG-KFAGFFGNSGDVLDSIGGVVVP- 147
Calsepa      FTTNL-KAHGPYGQKVGTPFSSANVVGNEIVGFLGRSGYYVDAIGTYNRHK 153
Ipomoelin    FTTNK-KEYGPYGANAGTPFNIKIPDGNKIVGFFGNSGWYVDAIGAYYTAK 154
             * **   : .**:*   .: *.       *   : **  *   *   :* :.
```

Figure 4.3 Multiple sequence alignment of the nine β-prism I fold lectins. (For color version of this figure, the reader is referred to the online version of this chapter.)

et al., 2005), and heltuba (Bourne et al., 1999) have been demonstrated to be mannose specific through, apart from solution studies, X-ray analysis of complexes with mannose or/and their derivatives. They have single-chain subunits with no aromatic residues in the binding site. Moringa M and artocarpin form tetramers. Banlec is dimeric, whereas heltuba is octameric in its crystal structure. The tetrameric lectins are highly homologous among themselves. Banlec and heltuba are only moderately homologous to each other. The homology of each of them with Moringa M and artocarpin is also moderate.

Calsepa (Bourne et al., 2004) and ipomoelin (Chang, Liu, Hsu, Jeng, & Cheng, 2012), whose structure has been determined recently, present interesting cases. They exhibit a sequence identity of about 50% between them. Their sequence identity with the galactose- and mannose-specific lectins mentioned earlier ranges between 22% and 35%. Unlike in the galactose-specific lectins and as in the mannose-specific lectins, ipomoelin has a single-chain subunit with no posttranslational cleavage. However, like in galactose-specific lectins, the crucial aromatic residues are present at the binding site. Mannose as well as galactose specificities of the lectin have been

Table 4.2 Sequence identities among β-prism I fold lectins

	Jacalin (1WS4)	Artocarpus hirsuta lectin (1TOQ)	Maclura pomifera agglutinin (1JOT)	Moringa M (1XXQ)	Artocarpin (1J4S)	Banlec (1X1V)	Heltuba (1C3K)	Calsepa (1OUW)	Ipomoelin (3R50)
Jacalin (1WS4)									
Artocarpus hirsuta lectin (1TOQ)	95.4 (0.3)								
Maclura pomifera agglutinin (1JOT)	75.3 (0.3)	74.0 (0.3)							
Moringa M (1XXQ)	54.9 (0.6)	54.9 (0.6)	43.7 (0.6)						
Artocarpin (1J4S)	53.8 (0.8)	44.2 (0.8)	42.4 (0.8)	64.8 (0.5)					
Banlec (1X1V)	36.1 (1.3)	36.8 (1.2)	30.4 (1.2)	36.3 (1.1)	33.3 (1.2)				
Heltuba (1C3K)	30 (1.5)	30.4 (1.5)	27.3 (1.5)	29.9 (1.3)	33.1 (1.3)	36.7 (1.1)			
Calsepa (1OUW)	28.3 (1.9)	28.9 (1.9)	21.9 (1.9)	33.5 (1.7)	31.7 (1.7)	34.4 (1.3)	33.8 (1.4)		
Ipomoelin (3R50)	30.6 (2.0)	30.9 (2.0)	28.2 (2.0)	28.2 (2.0)	25.9 (1.9)	32.7 (1.5)	31.5 (1.4)	50.6 (0.9)	

RMSDs in C^{α} positions are given in parenthesis.

demonstrated using crystallography and thermodynamic measurements. In fact, ipomoelin is the only β-prism I fold lectin in which dual specificity has been clearly demonstrated in crystals as well as in solution. There was an earlier report on the dual specificity of jacalin (Bourne et al., 2002). However, it was demonstrated that the affinity of mannose to jacalin is 20 times weaker than that of galactose (Jeyaprakash et al., 2005). Specificity ceases to be meaningful when affinity is extremely weak. Like ipomoelin, calsepa also has a single-chain subunit. Among the three aromatic residues in the binding site of the galactose-specific lectins, the tyrosine that stacks the B face of the sugar is missing in calsepa. The other two residues in the galactose-binding site are a tyrosine and a tryptophan. The tyrosine is retained in calsepa, but the tryptophan is changed to tyrosine. Crystal structures of calsepa complexed with galactose, mannose, or their derivatives are not available. However, solution studies indicate the lectin to be mannose specific. Thus, it would appear that features necessary for galactose binding and mannose binding are evenly matched in ipomoelin while those necessary for mannose binding have an edge in calsepa. In any case, the two lectins together form a good system for delineating the finer structural features that discriminate between galactose and mannose.

Among the nine lectins considered, five, namely, jacalin, AHL, MPA, Moringa M, and artocarpin are from the *Moraceae* family. Sequence identity between every pair among them is higher than 40% and the RMSD between pairs is well below 1 Å. Calsepa and ipomoelin belong to the *Convolvulaceae* family. Despite the subtle difference in carbohydrate specificity exhibited by them, the sequence identity and the RMSD in C^α positions between them are high at 50% and low at 0.9 Å, respectively. The four lectins in the group with unambiguously established mannose (glucose) specificity belong to three different taxonomic families. Pairwise sequence identity among those belonging to different families is always lower than 40% and the RMSD in C^α positions is always higher than 1 Å. They also exhibit different modes of quaternary association. Thus, structural features are essentially determined by lineage and are comparatively unaffected by ligand specificity.

In addition to the nine lectins considered here, the structures of two more involving the β-prism I fold are available. One of them is parkia lectin from a primitive leguminous plant (Gallego del Sol, Nagano, Cavada, & Calvete, 2005). This appears to be the only legume lectin which exhibits a fold other than the classical legume lectin fold of the type found in ConA and its homologues. In this lectin, the β-prism I domains are organized in a very different way from those in other β-prism I fold lectins. Three domains occurring consecutively in a single polypeptide chain form a subunit in parkia lectin. Two such subunits come together to form the dimeric lectin. The structure of griffithsin, a lectin from an algal source, has also been reported (Ziolkowska et al., 2006). The structure involves domain–swapped β–prism I folds. These two lectins thus do not lend themselves to comparison with the nine considered here.

5. PROTEIN–CARBOHYDRATE INTERACTIONS

5.1. Interactions at the extended binding site of jacalin and related lectins

Jacalin is the most thoroughly studied Gal/GalNAc-specific β–prism I fold lectin. The structure and interactions of AHL and MPA are very similar to those of jacalin. The crystal structures of seven complexes of jacalin with galactose, its derivatives and disaccharides involving it, have been determined (Jeyaprakash et al., 2002, 2003; Sankaranarayanan et al., 1996). A composite view of the binding sites of these complexes is given in Fig. 4.4. The site in all of them is made of the free amino group of the α–chain and loops 46–52, 76–82, and 122–125 and involves three subsites. The side chains of Phe 47, Tyr 78, and Asp 125, the main chain nitrogen and oxygen atoms of Tyr 122 and Trp 123, and the free amino terminus

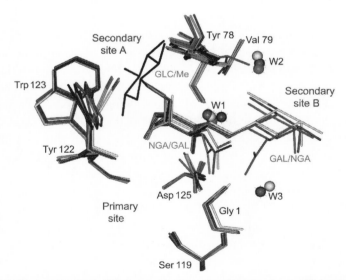

Figure 4.4 Composite view of the sugar-binding site of jacalin along with bound sugars. NGA stands for GalNAc. *Reproduced from Jeyaprakash et al. (2003).* (For color version of this figure, the reader is referred to the online version of this chapter.)

of the α-chain constitute the primary binding site. Secondary binding site A involves the side chains of three aromatic residues, namely, Tyr 78, Tyr 122, and Trp 123. The backbone nitrogen and oxygen atoms of Val 79, Ser 119 OG, Asp 125 OD1, and the carboxyl terminal region of the shorter β-chain make up secondary binding site B. The locations of different regions of the ligands in their respective complexes are indicated in Table 4.3. The anomeric oxygen of Gal located in the primary binding site points toward secondary binding site A. α-Substituents at this position, whether it is a methyl group or a sugar residue, as in Gal-α-1,6-Glc, interact with secondary binding site A. β-Substitutions, however, lead to unacceptable steric contacts with the protein. Therefore, β-substituted disaccharides bind to the lectin with the reducing sugar at the primary site and the nonreducing sugar at the secondary binding site B.

There are slight differences in the occurrence of protein–sugar hydrogen bonds among the four subunits in each complex and among the complexes. However, the consensus observed in the complexes involving Gal and GalNAc-β-1,3-Gal-α-O-Me, shown in Fig. 4.5, represents all the interactions at the primary binding site and most of the interactions at the secondary sites. Only Gal-α-1,6-Glc has direct hydrogen–bonded interactions at secondary binding site A. However, this secondary site provides evidence for interactions involving the π electrons of the side chain of Tyr 122. In cases where the anomeric oxygen points toward this site, as in the complexes with Gal and T-antigen, α and β anomers are seen. The —OH group of the β-anomer in such cases is involved in an OH...π hydrogen bond with Tyr 122 (Fig. 4.5A). When there is an α-methyl substitution, as in

Table 4.3 Modes of sugar binding to jacalin in complexes of Gal and carbohydrates involving it

Complex	Primary site	Secondary site A	Secondary site B
Gal	Gal	–	–
Me-α-Gal	Gal	Me	–
Me-α-GalNAc	GalNAc	Me	–
Galβ1–3GalNAc	GalNAc	–	Gal
Galβ1–3GalNAc-α-O-Me	GalNAc	Me	Gal
GalNAcβ1–3Gal-α-O-Me	Gal	Me	GalNAc
Gal1α–6Glc	Gal	Glc	–

Reproduced from Jeyaprakash et al. (2003).

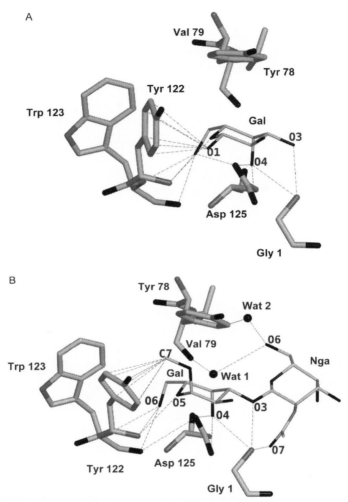

Figure 4.5 Interactions of (A) β-Gal and (B) methyl-T-antigen with jacalin. (For color version of this figure, the reader is referred to the online version of this chapter.)

GalNAc-β-1,3-Gal-α-O-Me, the methyl group is involved instead in a CH...π hydrogen bond with the same residue (Fig. 4.5B). This, of course, involves a movement of the tyrosyl residue from its position in complexes with no α-substitution. The methyl group and the Glc residue in Gal-α-1,6-Glc nestle in the hydrophobic groove generated by the three aromatic rings. This also contributes to the movement of Tyr 122 and Tyr 78. This is an example of crystal structures themselves providing information on the internal mobility of the protein molecule. The Glc residue of Gal-α-1,6-Glc also forms hydrogen bonds with the hydroxyl groups of Tyr 122 and Tyr 78.

In the four disaccharide complexes of jacalin analyzed crystallographically, all except Gal-α-1,6-Glc involve a β-linkage between the residues.
The second residue in them is located at secondary binding site B. The interactions of this residue with the lectin are wholly made up of water bridges.
These bridges, on the one hand, involve O4 or O6 or both. On the other
hand, the main chain nitrogen and oxygen of Val 79, 119 Ser OG, and 125
Asp OD1 are involved in one or more water bridges. Interestingly, the side
chain of Asp 125 is part of the primary binding site as well. Likewise, the
side chain of Tyr 78 forms part of the primary binding site as well as secondary binding site A. Thus, the extended binding site of jacalin is a contiguous
region of the molecule with most of the interactions occurring at the central
patch which constitute the primary binding site.

As mentioned earlier, Me-α-Man and Me-α-Glc also weakly interact
with jacalin. The crystal structures of the corresponding complexes, prepared under very high molar concentrations of the sugars, are also available
(Bourne et al., 2002; Jeyaprakash et al., 2005). The sugars bind to the lectin
in a similar manner. The interactions of Me-α-Glc with jacalin in the crystal
structure of the complex are shown in Fig. 4.6. They are to an extent similar
to those in Me-α-Gal. A critical difference is the absence of an interaction
between Gly 1, generated through posttranslational proteolysis, and O4.
This is not surprising as O4 has a different location in Man/Glc compared
to that in Gal. This further strengthens the observation that the generation of

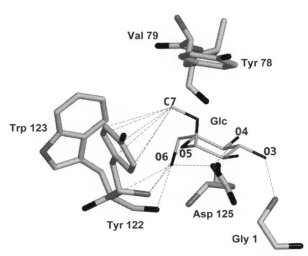

Figure 4.6 Interactions of methyl-α-Glc with jacalin. (For color version of this figure, the
reader is referred to the online version of this chapter.)

the free amino group is a strategy for creating specificity for Gal. On account of the different orientation of O4, O4...125 Asp OD2 hydrogen bond also disappears in Me-α-Man/Glc complexes. It is interesting that the methyl group α-linked to Glc/Man has the same kind of interactions with the Tyr 122 and Tyr 78 as the methyl group α-linked to Gal has. Both the Me-α-Gal and the Me-α-Man bind to jacalin 27 times more strongly than Gal and Man, respectively, do. Gal binds to jacalin 20 times more strongly than Man does. These and other results indicate that the binding of Me-α-Man and Me-α-Glc to jacalin is substantially due to interactions involving the methyl group.

Detailed MD simulations of jacalin and several of its complexes were carried out (Sharma, Sekar, & Vijayan, 2009). As illustrated in Fig. 4.7, the results provide a dynamic picture of the carbohydrate-binding site and the sugars located at the site. The broad features of the affinity of the sugars derived from simulations provide a dynamic rationale for the higher specificity of the lectin for galactose compared to that for mannose or glucose. These results, as indeed the X-ray results, are consistent with the results of thermodynamic measurements. The correlation between X-ray and MD

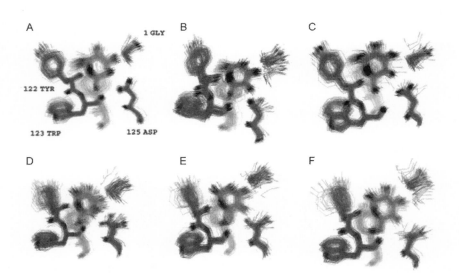

Figure 4.7 Scatter of carbohydrate-binding residues in (A) methyl-α-Gal, (B) methyl-α-Man, (C) methyl-α-Glc, (D) β-Gal, (E) β-Man, and (F) β-Glc. *Reproduced from Sharma et al. (2009).* (For color version of this figure, the reader is referred to the online version of this chapter.)

results are good, although the MD results indicate some possible additional features of jacalin–sugar interactions.

Location of water molecules in the combining site of unliganded lectins at positions corresponding to sugar oxygen atoms in lectin–carbohydrate complexes has been commented on earlier. Results of MD simulations provide further evidence for this phenomenon. To explore this phenomenon further, the occupancies of water oxygens in simulations involving uncomplexed jacalin were carefully examined. Interestingly, water oxygens with reasonable occupancies are seen at all the locations of sugar oxygens in the complexes. In the complexes, the sugar oxygen which is involved in the maximum number of hydrogen bonds is O6, followed by O4 of Gal. Strikingly, the occupancy of the water oxygen is the highest at the location corresponding to O6, followed by that corresponding to O4. Thus, water oxygens in the binding site tend to mimic the role of sugar oxygens in the complexes in terms of locations and hydrogen bonds. That could be the reason why the preformed binding site, primarily made up of loops, remains structurally stable.

5.2. Lectin–sugar interactions in artocarpin and heltuba

Despite the low sequence identity between them, the primary binding site of artocarpin and heltuba, both mannose specific, is remarkably similar (Fig. 4.8). In both the cases, two loops, which have similar lengths in the two lectins (14–16 and 137–141 in artocarpin numbering), constitute the primary binding sites. The situation is similar in Moringa M as well. Protein–sugar interactions at the monosaccharide level are similar in the two lectins. However, they exhibit different affinities for oligosaccharides. This is best illustrated by comparing the complexes of artocarpin and heltuba with Man-α-1,3-Man-α-1,6-Man (Bourne et al., 2002; Jeyaprakash, Srivastav, Surolia, & Vijayan, 2004). As shown in Fig. 4.9A, the two loops which bear the primary binding site have the same length in the two lectins. The third loop (88–96), which decisively influences the binding, is longer in artocarpin than in heltuba. All the sugar residues are ordered in the complex of artocarpin with the trisaccharide. Only the first two residues are ordered in the corresponding complex of heltuba. A superposition of the binding sites along with the ligands in the two complexes is shown in Fig. 4.9B. The location, orientation, and the interactions of the first sugar residue at the primary binding site are nearly the same in the two complexes. This is not true about the second residue. The orientations in the two are

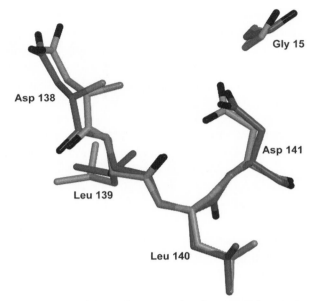

Figure 4.8 Superposition of the carbohydrate-binding sites of artocarpin (green) and heltuba (magenta). (For interpretation of the references to color in this figure legend, the reader is referred to the online version of this chapter.)

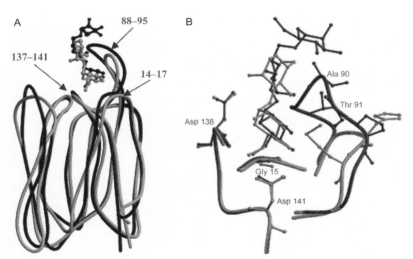

Figure 4.9 (A) Structural superposition of heltuba (cyan) on artocarpin (magenta) with bound mannotriose. (B) Sugar-binding sites along with ligands in the two lectins. *Reproduced from Jeyaprakash et al. (2004).* (For interpretation of the references to color in this figure legend, the reader is referred to the online version of this chapter.)

clearly different. The residue makes four hydrogen bonds with the lectin in the artocarpin complex. The corresponding number is 1 in the heltuba complex. The third residue in the artocarpin complex is involved in several van der Waals interactions with the 88–96 loop. The third residue in the heltuba complex can be modeled in several ways. However, none of the models indicate interactions with the lectin molecule. The comparison of the two complexes thus not only provides a structural rationale for the known higher affinity of artocarpin for the trisaccharide compared to that of heltuba, but also shows how change in loop length can be used as a strategy for generating higher affinity.

5.3. Banlec: A monocot lectin with two primary binding sites on each subunit

Mannose-specific dimeric Banlec presents an interesting case where each subunit carries two binding sites, one on Greek key I and the other on Greek key II (Fig. 4.10), unlike other β-prism I fold plant lectins of known structure, all of which carry only one binding site per subunit (Meagher et al., 2005; Singh et al., 2005). The two sites have similar geometrics and each

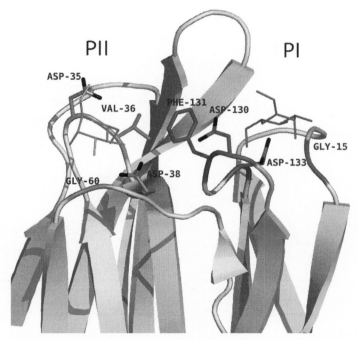

Figure 4.10 The primary binding sites PI and PII with bound sugars in Banlec. (For color version of this figure, the reader is referred to the online version of this chapter.)

of them is similar to the primary binding site in artocarpin and heltuba. The difference in the number of binding sites in lectin from banana, a monocot, and the other well-characterized β-prism I fold lectins, all from dicots, appears to have evolutionary implications. It had been suggested earlier that β-prism I fold lectins could have arisen through gene duplication and fusion of a primitive carbohydrate-binding motif. Each Greek key thus could correspond to this primitive motif. As can be seen from Fig. 4.11, there is hardly any significant sequence similarity among the three keys in the lectins from dicots. Though small, the sequence similarity among the keys is higher in banana lectin. In particular, the sugar-binding motif is preserved in Greek key II as well. Thus, it would appear that β-prism I fold lectins arose before dicots and monocots diverged. However, they underwent different evolutionary paths in the two groups. The evolutionary progression has been such as to obliterate similarities among the three keys in the lectins from dicots. Some similarity, particularly in relation to binding sites, is retained in those from monocots. Natural selection also might have helped in the retention of more sugar-binding sites in monocot lectins. Dicots are protected by a cork cambium layer. In the absence of such a layer in them, monocots have to strengthen other defense mechanisms such as that involving lectins. Larger number of binding sites could lead to stronger binding to carbohydrates and thus strengthen the defense mechanism involving lectins.

In the β-prism I fold lectins from dicots, Greek key I carries the primary binding site and Greek key III carries the secondary binding site which determines specificity at the level of disaccharides and higher oligomers. In Banlec, Greek keys I and II carry one primary binding site each. The orientations of sugars at these sites are such that they have a common secondary

```
ban_1  120  KISGFFGRGG-----DFIDAIGVYLE-----KVGAWGGNGGSAFDMGPA--Y---    24
ban_2   25  RIISVKIFSG-----DVVDAVDVTFTYYGKTETRHFGGSGGTPHEIV-LQ-EGEY   72
ban_3   73  -LVGMKGEFGNYHGVVVVGKLGFSTNKK---SYGPFGNTGGTPFSLP-IAAG---  119

art_1  128  LIVGFKGRTG--------DLLDAIGIHM--S--TV-GSWGGP-G-G-NGWDEGSY-T---  24
art_2   25  -GIRQIELSYKE-------AIGSFSVIYDLNGDPFS-GPKHTSKLPYKNVKIELKF-PDEF  75
art_3   76  -LESVSGYTGPFSALATPTPVVRSLTFKT--NKGRTFGPYGDE-E-G-TYFNLP-IENG--  127

hel_1  126  KFAGFFGNSG-----DVLDSIGGVVV---DIAVQAGPWGGNGGKRWLQTAHGG    28
hel_2   29  KITSIIIKGG-----TCIFSIQFVYKDKDNIEYHSGKFGVLGDKAETITFAED   76
hel_3   79  -ITAISGTFGAYYHMTVVTSLTFQTN-----KKVYGPFGTVASSSFSLPLTK-  124
```

Figure 4.11 Alignment of Greek keys I, II, and III in Banlec, artocarpin, and heltuba. The sugar-binding residues are boxed. (For color version of this figure, the reader is referred to the online version of this chapter.)

binding site on Greek III. Modeling indicates that this juxtaposition of the two primary binding sites and the common secondary binding site to be the basis of the affinity of Banlec for branched oligosaccharides (Singh et al., 2005).

5.4. Ipomoelin and its dual specificity

Crystal structures of the complexes of tetrameric ipomoelin with Me-α-Man, Me-α-Glc, and Me-α-Gal have been reported very recently (Chang et al., 2012). The binding site of the lectin bears close similarity with those of the mannose-specific artocarpin as well as the galactose-specific jacalin. However, unlike the other two lectins, ipomoelin binds the two sugars (and their methyl derivatives) with comparable affinities. The lectin uses the same site to bind Me-α-Man (and Me-α-Glc) and Me-α-Gal. This is achieved primarily through a lateral shift and a rotation of the ligands with respect to each other (Fig. 4.12). A small shift of the ligand is observed when the binding sites of artocarpin and ipomoelin with bound Me-α-Man are superposed. The shift is larger in the corresponding superposition of the primary binding sites of jacalin and ipomoelin bound to Me-α-Gal. Although

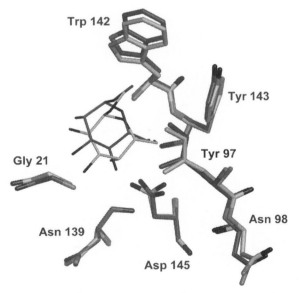

Figure 4.12 Superposition of the binding sites of ipomoelin complexes of methyl-α-Gal (magenta) and methyl-α-Man (green). (For interpretation of the references to color in this figure legend, the reader is referred to the online version of this chapter.)

the crystallographic and thermodynamic results reported so far are interesting, the structural basis for the dual specificity of the lectin merits further exploration.

5.5. Influence of the anomeric nature of the glycosidic linkage on the orientation of the bound sugar

One set of results which brought into focus the influence of the anomeric nature of the glycosidic linkage on the occupancy of the primary binding site involved the structures of the T-antigenic disaccharide Gal-β-1,3-GalNAc with different lectins. For instance, Gal occupies the primary site in peanut lectin while GalNAc does in jacalin. Subsequently, the structures of two more complexes of jacalin with β-1,3 linked disaccharides were determined. The reducing end occupies the primary binding site in these two as well. It was clear that β-1,3 linked disaccharides with the nonreducing end at the primary site would result in severe steric clashes of the second sugar residue with secondary binding site A.

The issue came into sharper focus when dealing with the disaccharide complexes of Banlec. The complexes analyzed using X-ray crystallography include those with Man-α-1,3-Man (Sharma & Vijayan, 2011) and Glc-β-1,3-Glc (laminaribose) (Meagher et al., 2005). In the mannobiose complex, the nonreducing end occupies the primary binding site while the reducing end does in the laminaribose complex (Fig. 4.13). The only difference between Man and Glc is in the orientation of O2 on account of the different configurations at C2. O2 in both structures points to the solution and it is not

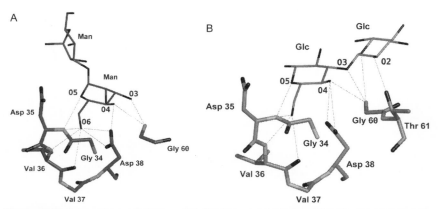

Figure 4.13 Interactions of Banlec with (A) Man-α-1,3-Man and (B) Glc-β-1,3-Glc at site PII. (For color version of this figure, the reader is referred to the online version of this chapter.)

involved in any interaction with the lectin molecule. Therefore, the orientation of O2 is unlikely to have any effect on the binding of the two disaccharides. To further explore the reason for the difference in the binding modes of the two disaccharides, all possible protein–sugar contacts in all possible conformations of α and β-1,3-linked disaccharides were examined using simple modeling. The contact distances clearly indicate that there is a definite preference for the nonreducing end to be at the primary binding site when the linkage is α while the location of the reducing end at the site is possible. When the linkage is β, location of the nonreducing end at the site is highly unfavorable, whereas that of the reducing end at the primary binding site is possible. Thus, the differential behavior of the two disaccharides in terms of the location of the residue at the primary binding site follows a sound structural rationale. The different orientations the disaccharide moiety assumes in complexes with lectins depending on the anomeric nature of the glycosidic linkage could be important in determining the affinity of lectins to glycoproteins and other glycoconjugates.

5.6. Conformational selection and induced fit

Conformational selection, which is often projected as an alternative to the conventional induced fit mechanism, is a topic of considerable current interest (Boehr & Wright, 2008; Kumar, Ma, Tsai, Sinha, & Nussinov, 2000; Tsai, Ma, & Nussinov, 1999). In the context of ligand binding, such selection would involve the utilization of those conformations that are most complementary to the ligand from among the conformational ensembles around the rugged bottom of the protein tunnel. The results of MD simulations carried out on jacalin (Sharma et al., 2009) and Banlec (Sharma & Vijayan, 2011) and their sugar complexes are of considerable interest in this context.

Sugar-binding sites of lectins are substantially preformed. However, some variation in the geometry of the binding site does occur as can be seen from the population distribution in terms of RMSDs of the atoms of the binding site residues from the initial structure, during the simulation of the uncomplexed jacalin molecule (Fig. 4.14A). The distribution involves a major peak and a minor peak corresponding to two distinct structural states. Only the state corresponding to the major peak is utilized on complexation, illustrating a clear case of conformational selection. A nuanced picture emerges when details are examined. For instance, the distribution of the torsion angle χ^2 of 122 Tyr, which describes the orientation of the aromatic ring in the side chain, in uncomplexed jacalin and jacalin

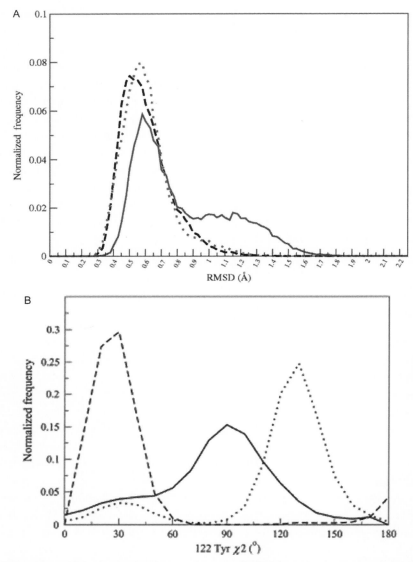

Figure 4.14 (A) Population distribution in terms of RMSD from the starting model in the structures of unliganded jacalin (blue solid line) and complexes involving α-anomeric sugars (red dashed line) and those involving β-anomeric sugars (green dotted line). (B) The population distribution of χ^2 in simulations involving the free lectin (solid line), in complexes with α-anomeric sugars (dashed line) and in those with β-anomeric sugars (dotted line). *Adapted from fig. 8 of Sharma et al. (2009).* (For interpretation of the references to color in this figure legend, the reader is referred to the online version of this chapter.)

complexed with two different kinds of sugar is shown in Fig. 4.14B. As expected, the distribution of χ^2 has a broad peak at around 90° in the free lectin. When the unsubstituted anomeric oxygen points toward the secondary binding site A, as in the complexes involving galactose and T-antigen, the β-anomer is stabilized through O—H...π interactions with the aromatic ring of 122 Tyr. When it is methyl substituted, as in methyl-α-Gal and methyl-T-antigen, the side chain of the tyrosyl residue moves and forms a C—H...π hydrogen with the methyl group. In the former case, the distribution of χ^2 has a major peak at around 135° and a very minor peak around 30°. In the latter case, there is only one single peak around 30°. In both cases, induced fit is obviously taking place.

A combination of conformational selection and induced fit is evident when dealing with the free and the bound ligands as well. For instance, the distributions of the torsion angles that define conformation of the methyl-T-antigen and T-antigen (Fig. 4.15A) in the simulations of the free molecules and in those of their complexes with jacalin are shown in Fig. 4.15B. Clearly, conformational selection operates in the case of methyl-T-antigen. A combination of conformational selection and induced fit operates in the case of T-antigen.

Results of the MD simulation on banana lectin are in conformity with the conclusion mentioned above. The population distribution in terms of RMSDs of atoms constituting the binding site is given in Fig. 4.16A. Similar distributions corresponding to the four binding sites in the dimer in the complexes with Man-α-1,3-Man and Glc-β-1,3-Glc are shown in Fig. 4.16B and C, respectively. The situation in the latter corresponds to conformational selection. That in the complex involving Man-α-1,3-Man presents a picture involving conformational selection and induced fit. The distribution of the ϕ and ψ angles in simulations involving the free disaccharides and those in the simulations of the respective complexes are presented in Fig. 4.17. They again present a situation involving a combination of conformational selection and induced fit.

6. PLASTICITY OF THE MOLECULE: A CASE STUDY PRIMARILY INVOLVING JACALIN

Jacalin is the most thoroughly studied β-prism I fold lectin. The structures of 12 different crystals of jacalin, grown under different conditions, belonging to different space groups and complexed with a variety of ligands, are available. The 42 crystallographically independent subunits belonging to

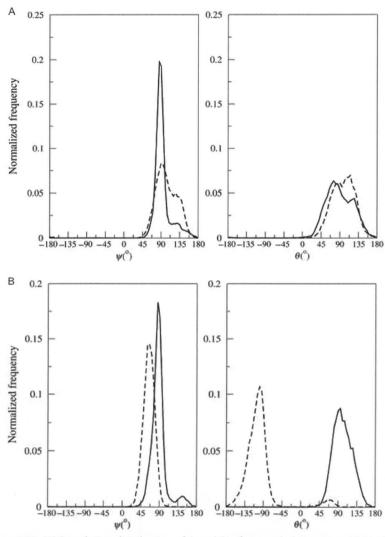

Figure 4.15 (A) Population distribution of ψ and θ in free methyl-T-antigen (dotted line) and in its complex with jacalin (solid line). (B) The corresponding distribution in T-antigen. *Reproduced from Sharma et al. (2009).*

them have been used to delineate the relatively rigid and relatively flexible regions of the molecule using program ESCET (Fig. 4.18A) (Schneider, 2004). MD simulations on 11 complexed subunits, each involving a unique ligand, and four uncomplexed subunits with slightly different structures have been reported (Sharma et al., 2009). Also reported are simulations on two

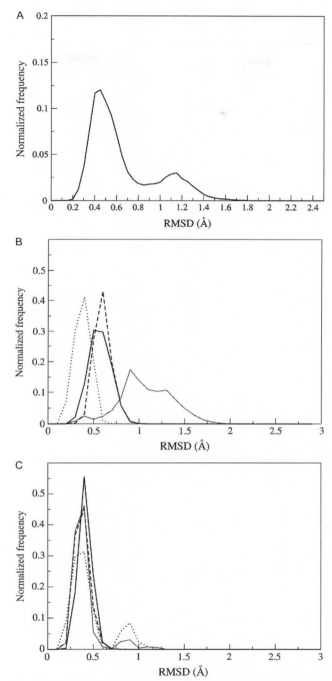

Figure 4.16 (A) Population distribution in terms of all-atom RMSD of carbohydrate-binding residues from those in the X-ray structure of lectin-methyl-α-Man complex of Banlec in the simulation of the native lectin. Corresponding distributions in simulations of complexes involving (B) Man-α-1,3-Man and (C) Glc-β-1,3-Glc. Thick lines: chain A, site PI; thin line: chain A, site PII; dashed line: chain b, site PI; dotted line: chain B, site P2. *Reproduced from Sharma and Vijayan (2011).*

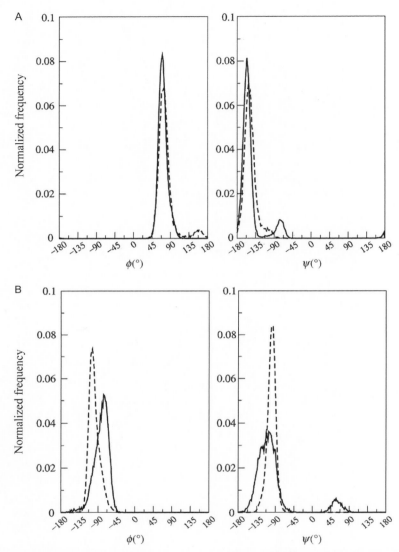

Figure 4.17 Population distribution of φ and ψ in the simulations involving (A) Man-α-1,3-Man and (B) Glc-β-1,3-Glc. Dashed and solid lines represent distribution in the free and Banlec-bound forms, respectively. *Reproduced from Sharma and Vijayan (2011).*

whole tetramers derived from two different crystal structures. The results of these simulations complement the information obtained from crystallo-graphic studies.

The variation of root mean square fluctuations (RMSFs) along the polypeptide chain, averaged over all simulations involving a single subunit,

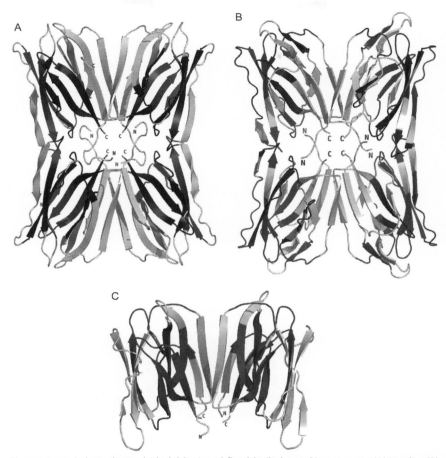

Figure 4.18 Relatively rigid (dark blue) and flexible (light pink) regions in (A) jacalin, (B) artocarpin, and (C) Banlec. (For interpretation of the references to color in this figure legend, the reader is referred to the online version of this chapter.)

is shown in Fig. 4.19. Also shown in the same figure for comparison are RMSFs averaged over eight subunits in simulations involving tetrameric molecules. These RMSFs exhibit correlation with the average B-factors (displacement parameters) derived from X-ray studies, thus pointing to the reliability of the results of simulations. The variation in RMSFs obtained from simulations on independent subunits and that derived from simulation of tetramers follows the same trend except in a few specific stretches. These stretches understandably include regions involved in oligomerization, namely, the N- and C-terminal stretches, the β-chain, and the 102–117 stretch.

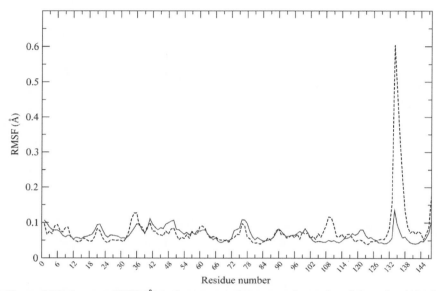

Figure 4.19 Average RMSF (Å) in the tetramer (green continuous) and the subunit (red dotted) in jacalin. Residues 1–133 correspond to the α-chain and 134–148 to the β-chain. *Adapted from fig. 4 of Sharma et al. (2009).* (For interpretation of the references to color in this figure legend, the reader is referred to the online version of this chapter.)

Average B-factors and RMSFs provide an indication of the variations along the polypeptide chain of the displacements from equilibrium conformation or vibrations of atoms in the molecule in an equilibrium conformation. The delineation of relatively rigid and flexible regions using ESCET, on the other hand, is concerned with variations in the equilibrium conformations themselves as a function of environmental effects such as crystallization conditions, crystal packing, ligand binding, etc. The two are related, but need not be the same. On an average, regular secondary structures such as strands in β-sheets tend to be relatively rigid with low B-factors and RMSFs, while loops are likely to be relatively flexible with higher B-factors and RMSFs.

In each subunit of the tetramer, a contiguous rigid segment is flanked by flexible regions, one large and the other small (Fig. 4.18A). Two dimers, AB and CD, make up a tetramer. The major flexible regions of the subunits come together to form a central large flexible region of the dimer. The inter-dimer interface in the tetramer has rigid and flexible elements. The flexible elements are the C-terminus of the α-chain, the N-terminus of the β-chain, both from Greek key I, and a loop connecting Greek keys I and III. These elements connect the flexible interfacial regions of the two dimers to form an elongated flexible region running through the length of the tetramer. This elongated flexible region is flanked by two rigid slices. As indicated earlier and discussed in greater detail later, β-prism I fold lectins exhibit considerable variability in quaternary association. Flexibility of the interfacial regions is perhaps related to this variability. Interestingly, calculations indicate that the observed pattern of rigidity and flexibility could have been caused in part by exposure to solution. Each subunit is by design nearly equally divided between rigid and flexible regions. However, accessible area of the flexible regions is high at 4976 Å^2 compared to an accessible area of 2883 Å^2 for the rigid region. This asymmetry is retained at different levels of oligomerization. The corresponding values for each subunit in the dimer are 4112 and 2423 Å^2, respectively. The values are 3589 and 2423 Å^2, respectively, in the tetramer. The ratio between the accessible surface area of the flexible regions and that of the rigid region decreases on dimerization as the interface in the dimer is made up of flexible regions. The ratio in the tetramer remains nearly the same as that in the dimer, as the dimer–dimer interface is made up of rigid as well as flexible regions.

In addition to hydrophobic forces, hydrogen bonds and water bridges contribute to the stability of protein structures. In jacalin, as indeed in most protein structures, main chain–main chain hydrogen bonds are more numerous than those involving side chains. By and large, main chain–main chain hydrogen bonds identified in X-ray structures occur with reasonable occupancy in MD simulations. Not surprisingly, the correlation between X-ray and MD results are weaker when dealing with hydrogen bonds involving side chains. Also, while main chain–main chain hydrogen bonds dominate interactions among strands, those involving side chains have an important role in stabilizing loop regions.

Among the water molecules that surround and interact with the protein, a few occur at relatively fixed positions. They play an important role in interactions that stabilize protein structure. Using the jacalin structures determined at 2 Å resolution or better and employing methods developed earlier in this laboratory, 10 such invariant water molecules could be identified. Their locations are shown in Fig. 4.20. Each of them is seen to have an important structural role by way of interconnecting different regions of the molecule. In particular, a cluster of five water molecules occurs at one end of the prismatic subunit. Four of these five are completely buried. They are involved in interactions with residues which are invariant or nearly invariant in all β-prism I fold lectins of known three-dimensional structure. This

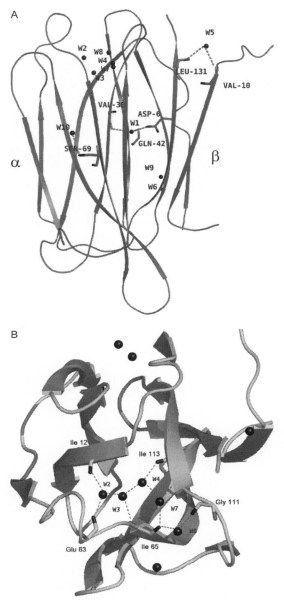

Figure 4.20 (A) Location of invariant water molecules in the jacalin subunit. (B) A cluster of water molecules at the bottom of the subunit. *Adapted from fig. 7 of Sharma et al. (2009).* (For color version of this figure, the reader is referred to the online version of this chapter.)

feature could therefore occur in different lectins belonging to the family. Not unexpectedly, water molecules with good occupancy occur in simulations at most of the locations of the invariant water molecules identified from crystal structures. Interestingly, the level of occupancy at these locations is related to accessibility. For instance, average occupancy of the buried sites is 0.65 while that of sites with accessibility varying between 1 and 5 Å^2 is 0.43.

Several crystal structures involving artocarpin and Banlec are also available, although structural studies on them are not as extensive as on jacalin. However, the number of available crystallographically independent subunits of artocarpin (18) and Banlec (14) are large enough for delineation of rigid and flexible regions using ESCET. The regions so delineated are illustrated in Fig. 4.18B and C, respectively. In both cases, each subunit has, as in jacalin, a central rigid region flanked by flexible regions; the intersubunit interface in the dimer is again made up of flexible regions. Artocarpin, like jacalin, is a tetramer, and the distribution of the rigid and flexible regions in the tetrameric molecule is similar to that in jacalin. Therefore, jacalin, artocarpin, Banlec, and, by inference, other β-prism I fold lectins appear to have some common features in their plasticity.

7. VARIABILITY IN QUATERNARY ASSOCIATION
7.1. Natural and modeled oligomers

Except ipomoelin, a recently characterized dual-specific lectin, all β-prism I fold lectins of known three-dimensional structure are either galactose specific or mannose specific. Among them, all galactose-specific lectins are tetrameric. Of the mannose-specific lectins, Banlec and calsepa are dimeric, artocarpin is tetrameric, and heltuba is octameric (Fig. 4.21). The association of protomers in the dimer of Banlec, designated as AB, is entirely different from that in calsepa, described as A/B/. Grossly, the AB dimer has a parallel arrangement of subunits, whereas the A/B/ dimer has an antiparallel arrangement. Both are, as expected, twofold symmetric. Among other regions, N- and C-terminal stretches are involved in the interfaces of the AB dimer as well as the A/B/dimer. Tetrameric artocarpin is a dimer of two AB-type dimers (AB and CD) with 222 symmetry. Octameric heltuba with 422 symmetry is made up of four AB dimers. Ipomoelin is in fact a dimer of two A'B' dimers, but oligomerization in it is yet to be explored.

A strategy adopted in the exploration of the variability in quaternary association is construction of models involving subunits of one protein with

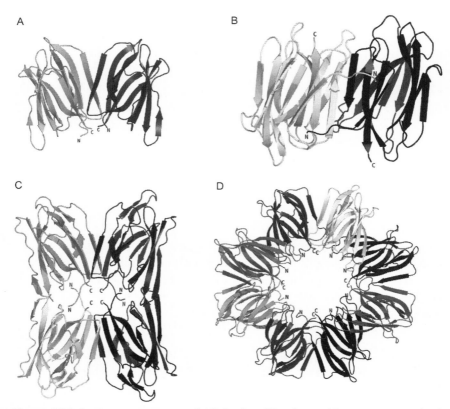

Figure 4.21 Quaternary structure of (A) Banlec, (B) calsepa, (C) artocarpin, and (D) heltuba. Subunits are colored differently. (For color version of this figure, the reader is referred to the online version of this chapter.)

the quaternary structure of another (Sharma & Vijayan, 2011). For example, Banlec on artocarpin was constructed by superposing the Banlec subunit on each of the subunits of artocarpin. In this manner, models of Banlec on calsepa, on artocarpin, and on heltuba; calsepa on Banlec, on artocarpin, and on heltuba; artocarpin on heltuba; and heltuba on artocarpin were constructed. Banlec on artocarpin was not viable on account of steric clashes involving loop 45–50 in Banlec on dimerization of the AB dimer. The 45–62 stretch including this loop in Banlec has a different conformation from that observed in all other β-prism I fold lectins (Singh et al., 2005). Calsepa on artocarpin was also not viable on account of the steric clashes involving the N-terminal stretch of calsepa. The constructed models, except the two mentioned above, and the native structures were

energy minimized. MD simulations were carried out on these energy minimized structures and also on the individual subunits of the four lectins
(Sharma & Vijayan, 2011).

7.2. Subunits and natural oligomers

Results of simulations on individual subunits and the respective oligomers
provide information on the intrinsic features of the subunit conformation
and the changes brought about on account of oligomerization. The RMSDs
in C^α positions among the most probable structures of the subunits of the
four lectins are less than 2 Å on pairwise superposition. Some regions, particularly the N- and the C-terminal stretches and loops, exhibit larger deviations. By and large, the regions which exhibit large RMSFs in individual
simulations are the same as those which exhibit deviations among subunits
belonging to different lectins, indicating a correlation between variations
among structures and mobility in each of them.

Changes in oligomerization in all cases involve movement of the N- and
the C-terminal stretches. The most striking movement on oligomerization
occurs in the N-terminal stretch of calsepa. This stretch is close to the bulk
of the structure in each subunit. On dimerization, the stretch in one subunit
moves across and remains close to the bulk of the other subunit (Fig. 4.22).
The swapping of the N-terminal stretch presumably lends additional stability
to the dimer. In all cases, oligomerization necessitates some readjustments in
the subunits. The effect of the readjustment is mainly felt in loops and the
terminal stretches. The magnitude of readjustments appears to be related to
the number of interfaces in the oligomer. The results of MD simulations lead
to a picture of the β-prism I fold subunit with a comparatively rigid core
surrounded by flexible regions that vary among the members of the family
and also undergo changes during oligomerization. Interestingly, the internal
symmetries of the natural oligomers are maintained, even though no symmetry restraints were employed during simulation. On the whole, depar-
tures from the expected rotation angles remain within $\pm 10°$, thereby
indicating the robustness of the observed quaternary structures.

7.3. Comparison involving modeled oligomers

Banlec and calsepa represent two distinctly different modes of dimerization.
In terms of different parameters such as interaction energy, surface area buried, number of hydrogen bonds, and shape complementary, the natural
Banlec dimer is only marginally better than Banlec on calsepa. On the contrary, the native structure of calsepa is emphatically superior to calsepa on

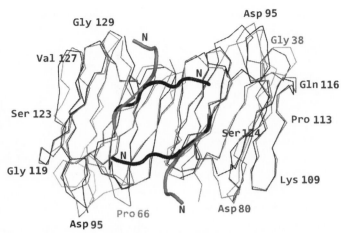

Figure 4.22 Superposition of structures corresponding to peaks in the population distribution of monomer (yellow) and dimer (magenta) simulations of calsepa. First 10 N-terminus residues are shown in thick lines. *Adapted from Sharma and Vijayan (2011).* (For interpretation of the references to color in this figure legend, the reader is referred to the online version of this chapter.)

Banlec. A close examination indicates that the difference in the mode of dimeric association between Banlec and calsepa is caused by critical changes in amino acids in the 110–119 stretch (Banlec numbering) which is involved in dimerization. This stretch is substantially conserved not only in the mannose-specific lectins Banlec, artocarpin, Moringa M, and heltuba but also in the two-chain galactose-specific lectins jacalin, MPA, and AHL. Substantial departure from the consensus sequences in them is observed in calsepa. In particular, a hydrophobic triplet –L-P-I/L- is replaced by –S-A-N- in calsepa. This substitution appears to be responsible for a shift in the course of polypeptide chain illustrated in Fig. 4.23. This shift adversely affects a few hydrogen bonds in addition to reducing the nonpolar surface area buried on dimerization.

As indicated earlier, Banlec on artocarpin and calsepa on artocarpin could not be constructed on account of severe steric clashes. However, heltuba on artocarpin remained intact with reasonable interaction parameters during MD simulations indicating that heltuba could form artocarpin-like tetramer in addition to the octamer observed in the crystal structure. This is in consonance with the observation of a tetrameric species of heltuba in solution (Van Damme, Barre, Mazard, Verhaert, & Horman, 1999). Energy-minimized models of Banlec on heltuba, artocarpin on heltuba, and calsepa on heltuba could be constructed. The first two octameric models fell apart

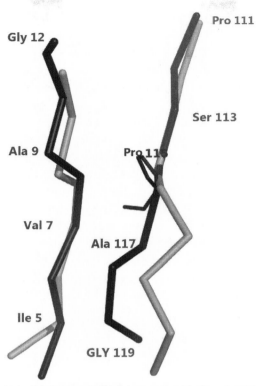

Figure 4.23 The 110–120 segment which interacts with the other subunit in the dimer in Banlec (blue) and calsepa on Banlec (yellow). Numbers correspond to Banlec. *Adapted from Sharma and Vijayan (2011).* (For interpretation of the references to color in this figure legend, the reader is referred to the online version of this chapter.)

during MD simulations. The third ended up with poor interaction parameters during simulations. These results appear to indicate that only heltuba can form robust octamer.

7.4. Evolutionary implications: Conformational selection

β-Prism I fold lectins with known three-dimensional structure bind to carbohydrates with a consensus sequence. G. . .GXXXD. Banlec is the only one among them which has two such motifs and hence two carbohydrate-binding sites. The distal glycine in the second binding site (P II) is Gly 60 which belongs to the 45–62 stretch. The main chain torsion angles of this residue are appropriate only for glycine. An L-amino acid occurs at this location in the other lectins under consideration and the second binding site is then no longer viable. This difference is also a contributory factor to the altered conformation of the 45–62 stretch which appears to be responsible for the prevention of the formation of an artocarpin-type tetramer by Banlec. Thus, we have here an interesting example of a relation among evolution, oligomerization, and ligand binding.

MD simulations on Banlec and Banlec on calsepa provide another insight on the relation between mode of oligomerization and ligand binding. The population distribution of the atoms in the binding site P II of banana lectin has a major peak and a minor peak. The structure corresponding to the major peak is adopted in Banlec–sugar complexes, in an instance of conformational selection (Fig. 4.24). The site in the structure corresponding to the minor peak is wider and appears to be inappropriate for the tight binding of sugar. The population distribution of Banlec on calsepa has a single peak at the location of the minor peak in the distribution of native Banlec. Thus, it would appear that P II would be comparatively ineffective if Banlec were to adopt the mode of dimerization found in calsepa. This is a good example of oligomerization-dependent conformational selection.

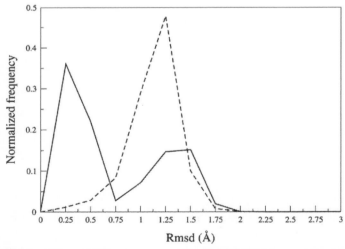

Figure 4.24 Population distribution in terms of RMSD (A) of all atoms, from the relevant crystal structure, of carbohydrate-binding residues of site PII in Banlec (solid line) and Banlec on calsepa (dashed line) simulations. *Reproduced from Sharma and Vijayan (2011).*

8. CONCLUSIONS

β-Prism I fold was identified as a lectin fold in 1996 through the X-ray analysis of tetrameric, galactose-specific jacalin, one of the two lectins in jackfruit seeds. The structure of the second lectin from the seeds, again tetrameric with β-prism I fold, but with mannose specificity, was also subsequently determined. Jacalin defined the basic characteristics of galactose binding in this class of lectins. On the other hand, artocarpin and heltuba, the structures of which were analyzed in the meantime, helped in delineating the structural features important for mannose specificity. The jacalin subunit has two chains, a longer α-chain and a shorter β-chain, generated by posttranslational proteolysis. The new amino group generated by proteolysis is important for recognition of galactose through its interaction with O4. The galactose-binding site is also characterized by the presence of three aromatic residues, the side chain of one of which stacks against the B face of the sugar. On the other hand, the mannose-specific artocarpin and heltuba have single-chain subunits and the primary binding site is devoid of aromatic residues. Further structural studies on jacalin and artocarpin and a few other β-prism I fold lectins and MD simulations on a couple of them have resulted in a thorough characterization of the extended binding sites in this class of lectins and in unraveling the strategies employed for generating ligand specificity.

At the time when β-prism I fold was identified as a lectin fold, it was suggested that the lectin could have arisen from successive gene duplication, fusion, and divergent evolution of a primitive sugar-binding Greek key motif (Sankaranarayanan et al., 1996). Subsequently, it was further suggested that monocot or β-prism II fold lectins could also have arisen in a similar fashion, but with a different assembly of Greek keys, thus establishing an evolutionary relationship between the two classes of lectins with threefold symmetric subunits (Sharma, Chandran, Singh, & Vijayan, 2007). Be as it may, the crystal structure of Banlec brought to light evidence in support of the hypothesis, mentioned above, on the evolutionary history of β-prism I fold lectins. The occurrence of the second binding site in Banlec, from a monocot, and the detectable, though weak, sequence similarity among the three Greek keys indicate that β-prism I fold lectins originated before dicots and monocots diverged and that the lectins followed somewhat different evolutionary paths in the two classes of plant lectins.

An interesting issue pertaining to lectin–sugar interactions is the effect of the anomeric nature of the glycosidic linkage in deciding whether the reducing end or the nonreducing end occurs at the primary binding site. This issue was addressed crystallographically and through modeling. The results indicate that the influence is exerted through steric factors and nonbonded interactions. Another issue of general interest pertaining to ligand binding is the relative importance of conformational selection and induced fit. Detailed MD simulations on jacalin, Banlec, and their sugar complexes suggest that both these mechanisms operate simultaneously. The two are complementary and not mutually exclusive.

The availability of several copies of the molecule in the different crystal structures of jacalin and, to a lesser extent, of artocarpin and Banlec permits the delineation of the rigid and the flexible regions of the molecules. Results of MD simulations supplement the information obtained from crystal structures. It appears that, on the whole, the flexibility of a region is related to its solvent accessibility. MD simulations point to the importance of main chain–main chain hydrogen bonds in maintaining the integrity of the core structure and that of hydrogen bonds involving side chains in stabilizing loops. Crystallographic studies and MD simulations also help in identifying invariant water molecules and in elucidating their role in holding the structure together.

The delineation of the rigid and flexible regions appears to indicate the plasticity of intersubunit interfaces in β-prism I fold lectins. That should be a factor that promotes the variability of quaternary association in them. In particular, mannose-binding lectins Banlec and calsepa are dimeric in different ways, artocarpin and Moringa M are tetrameric, and heltuba is octameric. This variability has been explored using crystallography, modeling, and MD simulations. The variability could be explained in terms of structural and evolutionary factors. Among other things, the investigations provide a rationale for the observation of tetrameric heltuba in solution and insights into relations among evolution, oligomerization, and ligand binding.

Structural and related studies of β-prism I fold lectins have been important not only in terms of their implication to glycobiology but also in relation to general issues concerned with the structure and function of proteins. In particular, they have provided valuable insights into the plasticity and internal dynamics of proteins, variability in the quaternary association of proteins with essentially the same tertiary structure, strategies for generating ligand specificity, and the relation between structural features and evolutionary factors.

ACKNOWLEDGMENTS

The work is supported by the Department of Science & Technology, Government of India. M. V. is the Albert Einstein Professor of the Indian National Science Academy.

REFERENCES

Balzarini, J., Schols, D., Neyts, J., Van Damme, E., Peumans, W., & De Clercq, E. (1991). Alpha-(1-3)- and alpha-(1-6)-D-mannose-specific plant lectins are markedly inhibitory to human immunodeficiency virus and cytomegalovirus infections *in vitro. Antimicrobial Agents and Chemotherapy, 35*, 410–416.

Banerjee, R., Das, K., Ravishankar, R., Suguna, K., Surolia, A., & Vijayan, M. (1996). Conformation, protein-carbohydrate interactions and a novel subunit association in the refined structure of peanut lectin-lactose complex. *Journal of Molecular Biology, 259*, 281–296.

Banerjee, R., Mande, S. C., Ganesh, V., Das, K., Dhanaraj, V., Mahanta, S. K., et al. (1994). Crystal structure of peanut lectin, a protein with an unusual quaternary structure. *Proceedings of the National Academy of Sciences of the United States of America, 91*, 227–231.

Becker, J. W., Reeke, G. N., Jr., Wang, J. L., Cunningham, B. A., & Edelman, G. M. (1975). The covalent and three-dimensional structure of concanavalin A. III. Structure of the monomer and its interactions with metals and saccharides. *The Journal of Biological Chemistry, 250*, 1513–1524.

Boehr, D. D., & Wright, P. E. (2008). Biochemistry. How do proteins interact? *Science, 320*, 1429–1430.

Bohlool, B. B., & Schmidt, E. L. (1974). Lectins: A possible basis for specificity in the Rhizobium—Legume root nodule symbiosis. *Science, 185*, 269–271.

Bouckaert, J., Hamelryck, T., Wyns, L., & Loris, R. (1999). Novel structures of plant lectins and their complexes with carbohydrates. *Current Opinion in Structural Biology, 9*, 572–577.

Bourne, Y., Astoul, C. H., Zamboni, V., Peumans, W. J., Menu-Bouaouiche, L., Van Damme, E. J., et al. (2002). Structural basis for the unusual carbohydrate-binding specificity of jacalin towards galactose and mannose. *The Biochemical Journal, 364*, 173–180.

Bourne, Y., Roig-Zamboni, V., Barre, A., Peumans, W. J., Astoul, C. H., Van Damme, E. J., et al. (2004). The crystal structure of the Calystegia sepium agglutinin reveals a novel quaternary arrangement of lectin subunits with a beta-prism fold. *The Journal of Biological Chemistry, 279*, 527–533.

Bourne, Y., Zamboni, V., Barre, A., Peumans, W. J., Van Damme, E. J., & Rouge, P. (1999). Helianthus tuberosus lectin reveals a widespread scaffold for mannose-binding lectins. *Structure, 7*, 1473–1482.

Brandley, B. K., & Schnaar, R. L. (1986). Cell-surface carbohydrates in cell recognition and response. *Journal of Leukocyte Biology, 40*, 97–111.

Broekaert, W. F., Van Parijs, J., Leyns, F., Joos, H., & Peumans, W. J. (1989). A chitin-binding lectin from stinging nettle rhizomes with antifungal properties. *Science, 245*, 1100–1102.

Chandra, N. R., Kumar, N., Jeyakani, J., Singh, D. D., Gowda, S. B., & Prathima, M. N. (2006). Lectindb: A plant lectin database. *Glycobiology, 16*, 938–946.

Chandra, N. R., Prabu, M. M., Suguna, K., & Vijayan, M. (2001). Structural similarity and functional diversity in proteins containing the legume lectin fold. *Protein Engineering, 14*, 857–866.

Chandra, N. R., Ramachandraiah, G., Bachhawat, K., Dam, T. K., Surolia, A., & Vijayan, M. (1999). Crystal structure of a dimeric mannose-specific agglutinin from garlic: Quaternary association and carbohydrate specificity. *Journal of Molecular Biology, 285*, 1157–1168.

Chandran, T., Sharma, A., & Vijayan, M. (2010). Crystallization and preliminary X-ray studies of a galactose-specific lectin from the seeds of bitter gourd (Momordica charantia). *Acta Crystallographica. Section F, Structural Biology and Crystallization Communications, 66*, 1037–1040.

Chang, W. C., Liu, K. L., Hsu, F. C., Jeng, S. T., & Cheng, Y. S. (2012). Ipomoelin, a jacalin-related lectin with a compact tetrameric association and versatile carbohydrate binding properties regulated by its N terminus. *PLoS One, 7*, e40618.

Chrispeels, M. J., & Raikhel, N. V. (1991). Lectins, lectin genes, and their role in plant defense. *The Plant Cell, 3*, 1–9.

Crocker, P. R. (2002). Siglecs: Sialic-acid-binding immunoglobulin-like lectins in cell-cell interactions and signalling. *Current Opinion in Structural Biology, 12*, 609–615.

De Hoff, P. L., Brill, L. M., & Hirsch, A. M. (2009). Plant lectins: The ties that bind in root symbiosis and plant defense. *Molecular Genetics and Genomics, 282*, 1–15.

Delbaere, L. T., Vandonselaar, M., Prasad, L., Quail, J. W., Wilson, K. S., & Dauter, Z. (1993). Structures of the lectin IV of Griffonia simplicifolia and its complex with the Lewis b human blood group determinant at 2.0 A resolution. *Journal of Molecular Biology, 230*, 950–965.

Drickamer, K., & Taylor, M. E. (1993). Biology of animal lectins. *Annual Review of Cell Biology, 9*, 237–264.

Epstein, J., Eichbaum, Q., Sheriff, S., & Ezekowitz, R. A. (1996). The collectins in innate immunity. *Current Opinion in Immunology, 8*, 29–35.

Frazier, W., & Glaser, L. (1979). Surface components and cell recognition. *Annual Review of Biochemistry, 48*, 491–523.

Fujita, T. (2002). Evolution of the lectin-complement pathway and its role in innate immunity. *Nature Reviews. Immunology, 2*, 346–353.

Gallego del Sol, F., Nagano, C., Cavada, B. S., & Calvete, J. J. (2005). The first crystal structure of a Mimosoideae lectin reveals a novel quaternary arrangement of a widespread domain. *Journal of Molecular Biology, 353*, 574–583.

Hamblin, J., & Kent, S. P. (1973). Possible role of phytohaemagglutinin in Phaseolus vulgaris L. *Nature: New Biology, 245*, 28–30.

Hardman, K. D., Wood, M. K., Schiffer, M., Edmundson, A. B., & Ainsworth, C. F. (1972). X-ray crystallographic studies of concanavalin A. *Cold Spring Harbor Symposia on Quantitative Biology, 36*, 271–276.

Hester, G., Kaku, H., Goldstein, I. J., & Wright, C. S. (1995). Structure of mannose-specific snowdrop (Galanthus nivalis) lectin is representative of a new plant lectin family. *Nature Structural Biology, 2*, 472–479.

Jeyaprakash, A. A., Geetha Rani, P., Banuprakash Reddy, G., Banumathi, S., Betzel, C., Sekar, K., et al. (2002). Crystal structure of the jacalin-T-antigen complex and a comparative study of lectin-T-antigen complexes. *Journal of Molecular Biology, 321*, 637–645.

Jeyaprakash, A. A., Jayashree, G., Mahanta, S. K., Swaminathan, C. P., Sekar, K., Surolia, A., et al. (2005). Structural basis for the energetics of jacalin-sugar interactions: Promiscuity versus specificity. *Journal of Molecular Biology, 347*, 181–188.

Jeyaprakash, A. A., Katiyar, S., Swaminathan, C. P., Sekar, K., Surolia, A., & Vijayan, M. (2003). Structural basis of the carbohydrate specificities of jacalin: An X-ray and modeling study. *Journal of Molecular Biology, 332*, 217–228.

Jeyaprakash, A. A., Srivastav, A., Surolia, A., & Vijayan, M. (2004). Structural basis for the carbohydrate specificities of artocarpin: Variation in the length of a loop as a strategy for generating ligand specificity. *Journal of Molecular Biology, 338*, 757–770.

Kulkarni, K. A., Katiyar, S., Surolia, A., Vijayan, M., & Suguna, K. (2007). Generation of blood group specificity: New insights from structural studies on the complexes of A- and B-reactive saccharides with basic winged bean agglutinin. *Proteins, 68*, 762–769.

Kulkarni, K. A., Srivastava, A., Mitra, N., Sharon, N., Surolia, A., Vijayan, M., et al. (2004). Effect of glycosylation on the structure of Erythrina corallodendron lectin. *Proteins, 56,* 821–827.

Kumar, S., Ma, B., Tsai, C. J., Sinha, N., & Nussinov, R. (2000). Folding and binding cascades: Dynamic landscapes and population shifts. *Protein Science, 9,* 10–19.

Lee, X., Thompson, A., Zhang, Z., Ton–that, H., Biesterfeldt, J., Ogata, C., et al. (1998). Structure of the complex of Maclura pomifera agglutinin and the T-antigen disaccharide, Galbeta1,3GalNAc. *The Journal of Biological Chemistry, 273,* 6312–6318.

Manoj, N., Srinivas, V. R., Surolia, A., Vijayan, M., & Suguna, K. (2000). Carbohydrate specificity and salt-bridge mediated conformational change in acidic winged bean agglutinin. *Journal of Molecular Biology, 302,* 1129–1137.

Meagher, J. L., Winter, H. C., Ezell, P., Goldstein, I. J., & Stuckey, J. A. (2005). Crystal structure of banana lectin reveals a novel second sugar binding site. *Glycobiology, 15,* 1033–1042.

Merritt, E. A., Kuhn, P., Sarfaty, S., Erbe, J. L., Holmes, R. K., & Hol, W. G. (1998). The 1.25 A resolution refinement of the cholera toxin B-pentamer: Evidence of peptide backbone strain at the receptor-binding site. *Journal of Molecular Biology, 282,* 1043–1059.

Ngai, P. H., & Ng, T. B. (2007). A lectin with antifungal and mitogenic activities from red cluster pepper (Capsicum frutescens) seeds. *Applied Microbiology and Biotechnology, 74,* 366–371.

Prabu, M. M., Sankaranarayanan, R., Puri, K. D., Sharma, V., Surolia, A., Vijayan, M., et al. (1998). Carbohydrate specificity and quaternary association in basic winged bean lectin: X-ray analysis of the lectin at 2.5 A resolution. *Journal of Molecular Biology, 276,* 787–796.

Prabu, M. M., Suguna, K., & Vijayan, M. (1999). Variability in quaternary association of proteins with the same tertiary fold: A case study and rationalization involving legume lectins. *Proteins, 35,* 58–69.

Pratap, J. V., Jeyaprakash, A. A., Rani, P. G., Sekar, K., Surolia, A., & Vijayan, M. (2002). Crystal structures of artocarpin, a Moraceae lectin with mannose specificity, and its complex with methyl-alpha-D-mannose: Implications to the generation of carbohydrate specificity. *Journal of Molecular Biology, 317,* 237–247.

Rabijns, A., Barre, A., Van Damme, E. J., Peumans, W. J., De Ranter, C. J., & Rouge, P. (2005). Structural analysis of the jacalin-related lectin MornigaM from the black mulberry (Morus nigra) in complex with mannose. *The FEBS Journal, 272,* 3725–3732.

Rao, K. N., Suresh, C. G., Katre, U. V., Gaikwad, S. M., & Khan, M. I. (2004). Two orthorhombic crystal structures of a galactose-specific lectin from Artocarpus hirsuta in complex with methyl-alpha-D-galactose. *Acta Crystallographica. Section D, Biological Crystallography, 60,* 1404–1412.

Ravishankar, R., Ravindran, M., Suguna, K., Surolia, A., & Vijayan, M. (1997). Crystal structure of the peanut lectin-T-antigen complex. Carbohydrate specificity generated by water bridges. *Current Science, 72,* 855–861.

Rini, J. M., & Lobsanov, Y. D. (1999). New animal lectin structures. *Current Opinion in Structural Biology, 9,* 578–584.

Rudiger, H., & Gabius, H. J. (2001). Plant lectins: Occurrence, biochemistry, functions and applications. *Glycoconjugate Journal, 18,* 589–613.

Rutenber, E., Katzin, B. J., Ernst, S., Collins, E. J., Mlsna, D., Ready, M. P., et al. (1991). Crystallographic refinement of ricin to 2.5 A. *Proteins, 10,* 240–250.

Sankaranarayanan, R., Sekar, K., Banerjee, R., Sharma, V., Surolia, A., & Vijayan, M. (1996). A novel mode of carbohydrate recognition in jacalin, a Moraceae plant lectin with a beta-prism fold. *Nature Structural Biology, 3,* 596–603.

Schneider, T. R. (2004). Domain identification by iterative analysis of error-scaled difference distance matrices. *Acta Crystallographica. Section D, Biological Crystallography, 60,* 2269–2275.

Shaanan, B., Lis, H., & Sharon, N. (1991). Structure of a legume lectin with an ordered N-linked carbohydrate in complex with lactose. *Science, 254*, 862–866.

Sharma, A., Chandran, D., Singh, D. D., & Vijayan, M. (2007). Multiplicity of carbohydrate-binding sites in beta-prism fold lectins: Occurrence and possible evolutionary implications. *Journal of Biosciences, 32*, 1089–1110.

Sharma, A., Sekar, K., & Vijayan, M. (2009). Structure, dynamics, and interactions of jacalin. Insights from molecular dynamics simulations examined in conjunction with results of X-ray studies. *Proteins, 77*, 760–777.

Sharma, A., & Vijayan, M. (2011). Influence of glycosidic linkage on the nature of carbohydrate binding in beta-prism I fold lectins: An X-ray and molecular dynamics investigation on banana lectin-carbohydrate complexes. *Glycobiology, 21*, 23–33.

Sharon, N., & Lis, H. (2004). History of lectins: From hemagglutinins to biological recognition molecules. *Glycobiology, 14*, 53R–62R.

Singh, D. D., Saikrishnan, K., Kumar, P., Surolia, A., Sekar, K., & Vijayan, M. (2005). Unusual sugar specificity of banana lectin from Musa paradisiaca and its probable evolutionary origin. Crystallographic and modelling studies. *Glycobiology, 15*, 1025–1032.

Sinha, S., Gupta, G., Vijayan, M., & Surolia, A. (2007). Subunit assembly of plant lectins. *Current Opinion in Structural Biology, 17*, 498–505.

Skehel, J. J., & Wiley, D. C. (2000). Receptor binding and membrane fusion in virus entry: The influenza hemagglutinin. *Annual Review of Biochemistry, 69*, 531–569.

Swaminathan, S., & Eswaramoorthy, S. (2000). Structural analysis of the catalytic and binding sites of Clostridium botulinum neurotoxin B. *Nature Structural Biology, 7*, 693–699.

Sweeney, E. C., Tonevitsky, A. G., Palmer, R. A., Niwa, H., Pfueller, U., Eck, J., et al. (1998). Mistletoe lectin I forms a double trefoil structure. *FEBS Letters, 431*, 367–370.

Taylor, M. E., & Drickamer, K. (2007). Paradigms for glycan-binding receptors in cell adhesion. *Current Opinion in Cell Biology, 19*, 572–577.

Tsai, C. J., Ma, B., & Nussinov, R. (1999). Folding and binding cascades: Shifts in energy landscapes. *Proceedings of the National Academy of Sciences of the United States of America, 96*, 9970–9972.

Turner, M. W. (1996). Mannose-binding lectin: The pluripotent molecule of the innate immune system. *Immunology Today, 17*, 532–540.

Van Damme, E. J. M., Barre, A., Mazard, A., Verhaert, P., & Horman, A. (1999). Characterization and molecular cloning of the lectin from *Helianthus tuberosus*. *European Journal of Biochemistry, 259*, 135–142.

Vijayan, M., & Chandra, N. (1999). Lectins. *Current Opinion in Structural Biology, 9*, 707–714.

Wang, W., Cole, A. M., Hong, T., Waring, A. J., & Lehrer, R. I. (2003). Retrocyclin, an antiretroviral theta-defensin, is a lectin. *Journal of Immunology, 170*, 4708–4716.

Wilson, B. S., Indiveri, F., Quaranta, V., Pellegrino, M. A., & Ferrone, S. (1981). Level of Ia-like antigens on human B and T lymphocytes. *Transplantation Proceedings, 13*, 1033–1034.

Wright, C. S. (1977). The crystal structure of wheat germ agglutinin at 2-2 A resolution. *Journal of Molecular Biology, 111*, 439–457.

Zhang, R. G., Scott, D. L., Westbrook, M. L., Nance, S., Spangler, B. D., Shipley, G. G., et al. (1995). The three-dimensional crystal structure of cholera toxin. *Journal of Molecular Biology, 251*, 563–573.

Zhang, G. Q., Sun, J., Wang, H. X., & Ng, T. B. (2009). A novel lectin with antiproliferative activity from the medicinal mushroom Pholiota adiposa. *Acta Biochimica Polonica, 56*, 415–421.

Zhang, R. G., Westbrook, M. L., Westbrook, E. M., Scott, D. L., Otwinowski, Z., Maulik, P. R., et al. (1995). The 2.4 A crystal structure of cholera toxin B subunit pentamer: Choleragenoid. *Journal of Molecular Biology, 251*, 550–562.

Ziolkowska, N. E., O'Keefe, B. R., Mori, T., Zhu, C., Giomarelli, B., Vojdani, F., et al. (2006). Domain-swapped structure of the potent antiviral protein griffithsin and its mode of carbohydrate binding. *Structure, 14*, 1127–1135.

CHAPTER FIVE

Conformational Changes of Enzymes and DNA in Molecular Dynamics: Influenced by pH, Temperature, and Ligand

Wen-Ting Chu, Qing-Chuan Zheng[1]
State Key Laboratory of Theoretical and Computational Chemistry, Institute of Theoretical Chemistry, Jilin University, Changchun, PR China
[1]Corresponding author: e-mail address: zhengqc@jlu.edu.cn

Contents

Abstract

Protein conformation, which has been a research hotspot for human diseases, is an important factor of protein properties. Recently, a series of approaches have been utilized to investigate the conformational changes under different conditions. Some of them have gained promising achievements, but it is still deficient in the detail

Advances in Protein Chemistry and Structural Biology, Volume 92
ISSN 1876-1623
http://dx.doi.org/10.1016/B978-0-12-411636-8.00005-5

researches at the atomic level. In this chapter, a series of computational examples of protein conformational changes under different pH environment, temperature, and ligand binding are described. We further show some useful methods, such as constant pH molecular dynamics simulations, molecular docking, and molecular mechanics Poisson–Boltzmann surface area/generalized Born surface area calculations. In comparison with the experimental results, the methods mentioned above are reasonable to detect and predict the interaction between residue and residue, residue and DNA, and residue and ligand. Additionally, some crucial interactions that cause protein conformational changes are discovered and discussed in this chapter. In summary, our work can give penetrating information to understand the pH-, temperature-, and ligand-induced conformational change mechanisms.

1. INTRODUCTION

Proteins are the most versatile macromolecules in living systems and serve crucial functions in essentially all biological processes. They can catalyze chemical reactions, transport and store small molecules such as oxygen, provide mechanical support and immune protection, generate movement, transmit nerve impulses, and control growth and differentiation (Berg, Tymoczko, & Stryer, 2008). Generally, there is a close relationship between the protein sequence and structure, and protein structure and function, which makes the research of the protein structural change more and more concerned (Bork & Koonin, 1998; Chothia & Lesk, 1986).

In most cases, the structures will be influenced by some factors, such as pH environment, temperature, and ligand binding. These factors lead to the protein structural changes from different aspects; these structural changes will be detected effectively at the atomic level by computational method, for example, molecular dynamics (MD). Nowadays, MD simulations methods are widely used to obtain the conformational change information with kinetic and thermodynamic information on the time evolution (Adcock & McCammon, 2006; Cheatham & Kollman, 2000; Karplus & McCammon, 2002; Moraitakis, Purkiss, & Goodfellow, 2003; Norberg & Nilsson, 2002). The trajectories are determined by numerically solving the Newton's equations of motion for a system of interacting particles, where forces between the particles and potential energy are defined by molecular mechanics force fields (Alder & Wainwright, 1959). The MD simulations are performed at the atomic level, which is clear to observe the interactions established and disappeared. In addition, the interaction changes between protein and ligand through the MD trajectories can be determined by calculating free binding energy with molecular mechanics Poisson–Boltzmann surface area/generalized Born surface area (MM-PBSA/GBSA) method (Hou, Wang, Li, & Wang, 2011a, 2011b). Consequently, MD simulation is a useful method to find out what causes the conformational changes of protein under different conditions.

In this chapter, we try to answer whether protein changes its structure to lose ligand in acidic solution, what changes of protein caused by high temperature will influence the DNA binding, and what are the structural differences between the protein structures with different small molecular ligands. Our aim of this work is to provide a brief review on the structural changes of protein during MD simulations, which are caused by different conditions.

Section 2 focuses on the structural changes of insect odorant-binding proteins (insect OBPs) induced by different pH solutions. By performing and analyzing the constant pH molecular dynamics (CpHMD) simulations of two kinds of insect OBPs (CquiOBP1 and AaegOBP1), different structures in low pH are obtained (Chu, Wu, et al., 2012; Chu, Zhang, et al., 2012). Molecular docking method is also used in this section to get protein–ligand complexes. Large conformational changes are found in the active site, and pH-sensing triad is shown to be important for ligand binding.

Section 3 includes the stability of DNA by binding to the protein. In order to find out how the protein–DNA interaction changes at different temperature, the MD simulations and binding free energy analyses of two kinds of hyperthermophilic proteins (Sso7d and Cren7) complexed with DNA duplex are performed (Chen, Zheng, Yu, et al., 2012; Chen, Zheng, Zhang, et al., 2012). These two proteins can stabilize DNA duplex in a certain temperature range in the binary complex compared with the unbound DNA molecules.

Section 4 summarizes structural and dynamical basis of human cytochrome P4507B1 (CYP7B1) (Cui et al., 2013). A series of investigations including homology modeling, MD, and automatic docking, combined with the results of previous experimental site-directed mutagenesis studies and access channels analysis, have identified the structural features relevant to the substrate selectivity of CYP7B1.

2. CpHMD SIMULATIONS ON INSECT OBPs

Insect OBPs were considered to transport lipophilic odors and pheromones across the aqueous sensillar lymph to receptor proteins, before reaching the neuronal membrane. They play a crucial role in insect behaviors such as host-seeking, oviposition, and mating (Breer, 1997;

Kaissling & Thorson, 1980; Pelosi & Maida, 1995; Steinbrecht, 1996; Vogt, 1987; Vogt, Callahan, Rogers, & Dickens, 1999; Vogt & Riddiford, 1981. Some insects, such as *Culex quinquefasciatus*, *Anopheles gambiae*, and *Aedes aegypti*, are major vectors of malaria, filariasis, and cephalitis transmission all around the world (Barbosa, Furtado, Regis, & Leal, 2010; Chandre et al., 1998; Ishida et al., 2004; Zhou, He, Pickett, & Field, 2008). Therefore, the OBPs of these pests will be selected as the main target of pest control (Laurence & Pickett, 1982; Matsuo, Sugaya, Yasukawa, Aigaki, & Fuyama, 2007; Pelletier, Guidolin, Syed, Cornel, & Leal, 2010).

In this contribution, we focused on the OBPs of *C. quinquefasciatus* (CquiOBP1, PDB ID: 3OGN) (Mao et al., 2010) and *A. aegypti* (AaegOBP1, PDB ID: 3K1E) (Leite et al., 2009). Both of the two OBPs have pH-dependent feature, which makes them show high affinity with their ligands at high pH but show no or lower affinity at low pH (Leal et al., 2008). The C-terminus of these insect OBPs makes up part of the central cavity wall in neutral solution. An earlier research by Wogulis, Morgan, Ishida, Leal, and Wilson (2006) suggested that lowering pH might disrupt hydrogen bonds in insect OBPs, especially in the pH-sensing triad (His23–Tyr54–Val125 in CquiOBP1; Arg23–Tyr54–Ile125 in AaegOBP1), and the ligand release in lower pH was likely due to this. Binding assays by Leal et al. (2008) obviously showed that its ligand mosquito oviposition pheromone (MOP, (5R,6S)-6-acetoxy-5-hexadecanolide) bound to CquiOBP1 at pH 7 with high affinity, but with very low affinity at pH 5. In 2009, Leite et al. (2009) reported that the loss of affinity had something to do with the disruption of hydrogen bond network between the conserved pH-sensing triad. Later in 2010, Mao et al. (2010) hypothesized that the disruption would be caused by the displacement of C-terminus from the central cavity. However, there has not been a detailed study for the conformational change and the mechanism of ligand release in these two insect OBPs at low pH.

Our aim is to discover the structures of CquiOBP1 and AeagOBP1 at acidic (pH 5) and neutral (pH 7) environment with the CpHMD method, respectively, and find out which leads to the structural changes. Furthermore, using docking studies, we can detect which structure shows higher affinity to the ligand MOP.

2.1. CpHMD method

Traditionally, MD simulations have employed constant protonation states for titratable residues, which have many drawbacks. CpHMD is a useful theoretical method to simulate the structural change of protein according to acidic or basic solution. CpHMD method can be largely classified into two categories, discrete and continuous. Earlier continuous protonation models include a grand canonical MD algorithm developed by Mertz and Pettitt (1994) in 1994 and a method introduced by Baptista, Martel, and Petersen (1998) in 1997. Later, Börjesson and Hünenberger (2001, 2004) developed a continuous protonation variable model using the explicit solvent. The Brooks group further developed the continuous protonation state model recently (Khandogin & Brooks, 2005, 2007; Khandogin, Chen, & Brooks, 2006; Lee, Salsbury, & Brooks, 2004). In addition, discrete protonation state methods have also been developed for the protein pH–dependent study (Baptista, Teixeira, & Soares, 2002; Bürgi, Kollman, & van Gunsteren, 2002; Dlugosz & Antosiewicz, 2005; Machuqueiro & Baptista, 2006, 2009; Mongan, Case, & McCammon, 2004; Walczak & Antosiewicz, 2002). Discrete CpHMD methods combine MD and Monte Carlo (MC) sampling (Meng & Roitberg, 2010; Williams, De Oliveira, & McCammon, 2010) and use Metropolis criterion to accept or reject the protonation change. The Baptista group (Baptista et al., 2002; Machuqueiro & Baptista, 2006, 2009) used Poisson–Boltzmann (PB) equation to calculate protonation energies. Walczak and Antosiewicz (Dlugosz & Antosiewicz, 2005; Walczak & Antosiewicz, 2002) performed Langevin dynamics between MC steps. Bürgi et al. (2002) calculated the transition energy by thermodynamic integration method.

In the year 2004, discrete CpHMD method was first implemented in Amber by Mongan et al. (2004). It is a model using generalized Born (GB) implicit solvation and belonging to the discrete protonation state models. In this method, GB model (Bashford & Case, 2000; Onufriev, Bashford, & David, 2000) is used in protonation state transition energy as well as solvation free energy calculations. Therefore, solvent models in conformational and sampling of protonation state are consistent, making computational cost less than other methods. This method only changes protonation states of titratable residues, which can gain or release a "ghost" proton unrealistically.

2.2. Comparison of the structures of CquiOBP1 and AaegOBP1 in acidic solution and neutral solution

The CpHMD simulations of CquiOBP1 and AaegOBP1 have similar results. Taking the results of AaegOBP1, for example, after 20-ns MD simulations, a comparison of stable structures of wild–type AaegOBP1 at pH 7

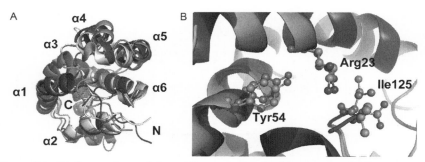

Figure 5.1 (A) Comparison of three 3D structures of wild-type AaegOBP1 at pH 7.0 (deep blue) and at pH 5.0 (magenta) after CpHMD, and the crystal structure of AaegOBP1 (PDB ID: 3K1E) at pH 8.5 (white). The N- and C-termini and six helices are labeled. (B) The structures of the pH-sensing triad (Arg23-Tyr54-Ile125) at pH 7 (deep blue) and at pH 5 (magenta). Carbon atoms are shown in deep blue (pH 7) and magenta (pH 5). Nitrogen and oxygen atoms are colored blue and red, respectively. (For interpretation of the references to color in this figure legend, the reader is referred to the online version of this chapter.)

and pH 5 is shown in Fig. 5.1. From this figure, we can easily find that there is an obvious difference in the loop regions between the two structures (at pH 7 and pH 5), especially N- and C-termini loops. A hydrophobic tunnel is formed between helices $\alpha 4$ and $\alpha 5$, and it moves a lot from the position at pH 5 with respect to that at pH 7 (Fig. 5.1A). Consequently, we can conjecture that the posture and direction of ligand binding should change with the transformation of the hydrophobic tunnel. Figure 5.1B illustrates the comparison of the significant pH–sensing triad (Arg23–Tyr54–Ile125) at pH 7 and pH 5. The figure demonstrates that the C-terminus residue, Ile125, has big conformational change at pH 5 compared to pH 7. That is, the protein structure will be affected by the acidic environment at pH 5, and the CpHMD method can capture the effect of pH value properly.

Moreover, backbone root mean square deviation (RMSD) analyses demonstrate that all the insect OBPs simulations appear to equilibrate after 3 ns. Large structural change of both CquiOBP1 and AaegOBP1 between pH 7 and pH 5 can be easily found in Fig. 5.2. Apparently, in CquiOBP1, the value of RMSD at pH 5 is stabilized near 4.0 Å, fairly higher than that at pH 7, which is straight at 2.0 Å (Fig. 5.2A). A similar situation can be found in the AaegOBP1 simulations. The RMSD value at pH 5 is stabilized near 4.0 Å, fairly higher than that at pH 7 (2.0 Å) (Fig. 5.2B). The RMSD values of the stable structures at pH 7 and pH 5 are 1.54 and 3.95 Å, with respect to the crystal structure. The RMSD results of the CquiOBP1 and AeagOBP1 show that both of their structures at pH 7 are closer to the crystal structure

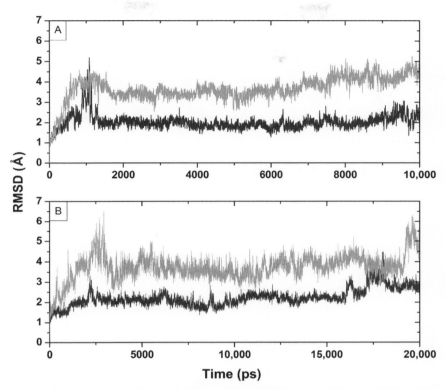

Figure 5.2 RMSD values of CquiOBP1 (A) and AaegOBP1 (B) at pH 7 (deep blue) and pH 5 (magenta) during the whole molecular dynamics simulations, respectively. The RMSD value was calculated with respect to the first frame of the CpHMD simulations. (For interpretation of the references to color in this figure legend, the reader is referred to the online version of this chapter.)

than their structures at pH 5. The large conformational shift of the two proteins at pH 5 may influence the ligand binding to protein. Additionally, there is little change of the secondary structure during all the MD simulations at pH 7 and pH 5. According to our results, lowering the pH disrupts the tertiary structure of these two insect OBPs, rather than their secondary structure. It is consistent with the conclusion of Leal et al. (Leite et al., 2009).

2.3. Importance of the H-bonds in pH-sensing triad

It is reported that the pH-sensing triad plays an important role in the binding and releasing of ligand (Leite et al., 2009; Mao et al., 2010). In CquiOBP1, this triad includes His23-Tyr54-Val125, but it turns to be Arg23-Tyr54-Ile125 in

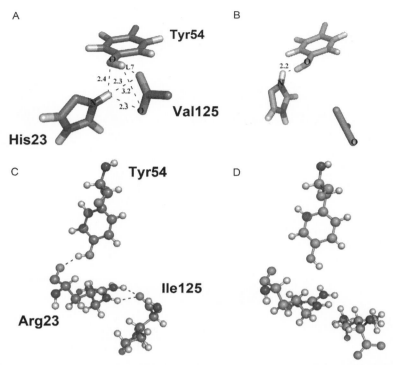

Figure 5.3 Conformation comparison of the pH-sensing triad of CquiOBP1 (His23, Tyr54, and Val125; (A) pH 7; (B) pH 5) and AaegOBP1 (Arg23, Tyr54, and Ile125; (C) pH 7; (D) pH 5). H-bonds are labeled with black-dashed line. (For color version of this figure, the reader is referred to the online version of this chapter.)

AaegOBP1. We have detected the relationship between the pH–sensing triad and the conformational change of protein in different pH environment by using a series of analysis methods, as illustrated in Fig. 5.3.

In the case of CquiOBP1, the triad seems to rotate during the simulation at pH 5 compared to the conformation of CquiOBP1 at pH 7. This will result in the increase of the distance through the carboxyl oxygen atom of Val125, the δ-nitrogen atom of His23, and the hydroxyl hydrogen atom of Tyr54. It is easy to find that the occupancy percentage of these H–bonds decreases and even vanishes at pH 5 (Table 5.1). Concretely, there are five H–bonds in the pH–sensing triad at pH 7 structure of CquiOBP1 (Fig. 5.3A and B) compared to the four H–bonds at pH 8.2. However, only one of them exists during the MD trajectory at pH 5, which is the H–bond 54@OH–23@ND1 (Fig. 5.3C and D). Then we calculated the distance between pH–sensing triad during the MD simulations (carboxyl oxygen atom of Val125, the δ-nitrogen atom of His23, and the hydroxyl oxygen

Table 5.1 The properties of H-bonds within the pH-sensing triad of CquiOBP1 at pH 7.0, including occupied percentage, distance, and angle

H-bond	Occupied (%)	Distance (Å)	Angle (°)
pH 7			
125@OXT-54@OH	44.46	2.688	19.74
125@O-54@OH	40.62	2.695	20.43
125@O-23@ND1	39.38	2.804	25.02
125@OXT-23@ND1	31.08	2.810	25.66
54@OH-23@ND1	11.46	2.883	39.10
pH 5			
54@OH-23@ND1	21.90	2.864	32.54
125@O-23@ND1	12.62	2.778	21.33
23@O-54@OH	6.90	2.847	28.10

atom of Tyr54). At pH 5, residue Val125 tends to depart from His23 and Tyr54, but the relative position between His23 and Tyr54 keeps steady. The H-bond (125@O-23@ND1) only stands for merely 2 ns. Then with the increase in the distance between them, this H-bond is disrupted. Additionally, four H-bonds between Val125 and Arg5, and Val125 and Arg6 are formed at pH 5. Asp7 and Glu9 can form some important H-bonds to the Arg5 at pH 7, making N-terminus stable. With the protonation of Asp7 and Glu9, these H-bonds are disrupted at pH 5, making the carboxyl group of Val125 turn its direction to form four H-bonds to Arg5 and Arg6. This causes the movement of C-terminus at pH 5, which makes part of binding pocket in neutral pH environment.

In the case of AaegOBP1, Table 5.2 demonstrates the H-bonds data analyzed by Amber. Both the H-bond between oxygen of Arg23 and hydroxyl hydrogen of Tyr54, and the H-bond between carboxyl oxygen of Ile125 and two amino hydrogen atoms of Arg23 exist continuously and the distance between them keeps close during the simulation at pH 7. The H-bond at pH 5 only exists in the beginning of MD simulation, with much less occupied percentage than that at pH 7. According to these H-bond analyses, we can conjecture that the hydroxyl of Tyr54 and the carboxylate of Ile125 change their directions to Arg23, which leads to the H-bond disappearance. These results are consistent with the conclusion that the H-bonds in pH sensing will be disrupted at low pH (Leite et al., 2009). Moreover, it is

Table 5.2 The properties of H-bonds within the pH-sensing triad of AaegOBP1 at pH 7.0, including occupied percentage, distance, and angle

H-bond	Occupied (%)	Distance (Å)	Angle (°)
pH 7			
23@O-54@HH	29.31	2.819	27.91
125@O-23@HH21	28.96	2.824	25.86
125@O-23@HH22	24.52	2.839	23.72
pH 5			
125@O-23@HH12	6.07	2.823	21.24

evident that all of the three H-bonds are related to Arg23, which seems to be the center of the pH-sensing triad. In order to find out the importance of this residue, the H-bonds of the three mutants (R23L, Y54F, and R23L/Y54F) of AaegOBP1 at pH 7 are analyzed. The R23L model has only one H-bond in the beginning of the simulation, and this H-bond disappears with the increasing distance between Tyr54 and Ile125. In the simulation of Y54F model, we find that the situation of Ile125 is held by two H-bonds between Arg23 and Ile125, and both the H-bonds exist continuously to the end of CpHMD simulation. But during the R23L/Y54F mutant simulation, the structure shows big changes in the beginning of MD simulation as no H-bond exists. Otherwise, we compared the H-bonds between the residue 23 and other residues of wild-type AaegOBP1 and mutation model R23L. It is easy to find out that, in wild type of AaegOBP1, there is a strong H-bond net around the Arg23. But in the simulation of R23L, the residue 23 forms only two H-bonds with Met19 and Ile27. As a result, the Arg23 plays a significant role in keeping protein tertiary structural stabilization.

2.4. Reduction of the affinity of ligand MOP in acidic solution

Autodock 4.0 was used to dock ligand MOP to the structures of CquiOBP1 and AaegOBP1 at pH 7 and pH 5, respectively. In the case of CquiOBP1, the binding free energy of MOP to CquiOBP1 is -8.13 kcal mol^{-1} at pH 7 and -6.36 kcal mol^{-1} at pH 5, respectively. It shows that MOP pulls its long lipid "tail" into the tunnel between helices $\alpha4$ and $\alpha5$ at pH 7. The ligand holds its lactone head in the central cavity, similar to the crystal complex structure at pH 8.2. However, when it comes to the structure at pH 5, the ligand bound to the protein with a completely different pose, its head stretches out of the protein at a different direction and its tail stays in the central cavity. It should be noticed that the position of tunnel

($\alpha4$ and $\alpha5$) undergoes a movement away from the original site at pH 5, which may influence the binding posture of MOP. Also, the change in loop between helices $\alpha3$ and $\alpha4$ combined with C-terminus contributes to this. We can find that the movement of the hydrophobic tunnel tends to hinder the interaction with the tail of ligand MOP. Additionally, the distance between the lactone head of MOP and C-terminus is increasing at pH 5. These phenomena result from the protonation of some important acidic titratable residues in low pH. Based on the significant difference of ligand binding at pH 7 and pH 5, we can conjecture that the ligand should be released through another tunnel between $\alpha3$ and $\alpha4$ at pH 5. The flexible loop between $\alpha3$ and $\alpha4$ is from Phe59 to Leu76, similar to the conclusion of ligand release paths in BmorPBP making by Gräter, de Groot, Jiang, and Grubmüller (2006).

In the case of AaegOBP1, the docking results demonstrate that AaegOBP1 can bind to MOP at pH 7, with binding energy of $-5.09\,\text{kcal mol}^{-1}$. On the other hand, it has no affinity to MOP at pH 5, as binding energy is $4.17\,\text{kcal mol}^{-1}$. In order to further understand the mechanism of pH-dependent binding of AaegOBP1, we select other four ligands: decanal, geranylacetone, nonanal, and octanal, and contrast these four ligands to MOP. Compared with MOP, they all have a long lipid tail except a lactone ring head. The four ligands have lower affinity to AaegOBP1 at pH 5, which indicates that the four ligands are less sensitive to pH than MOP. The differences between the binding results of MOP and other four ligands may be derived from the lactone ring head. Figure 5.4A depicts that the lactone ring head of MOP is close to the position of pH-sensing triad of wild-type AaegOBP1 at pH 7. In addition, the binding results of mutation models suggest that the pH-sensing triad is indispensable to MOP binding, but it seems to have no effect on the other four ligands binding, as shown in Fig. 5.4B. This implicates that the lactone ring head of MOP has an important interaction with pH-sensing triad of AaegOBP1. Therefore, AaegOBP1 acts as a selective filter of odor molecule, sensitive to the ligands like MOP with lactone ring head. And the interaction between ligand head and pH-sensing triad will be disrupted with the decrease in solution pH value. It is probably relevant to the moving of binding pocket made by the C-terminus change at pH 5, which blocks MOP binding to the protein.

Both CquiOBP1 and AaegOBP1 have high binding affinity with ligand MOP at pH 7, CquiOBP1 has lower binding affinity with MOP at pH 5, but AaegOBP1 loses its binding affinity in the same pH. That seems to have something with the number of H-bonds in pH-sensing triad (Section 2.3). At pH 5, all the three H-bonds disappear in AaegOBP1, which may be relative to the MOP binding affinity loss.

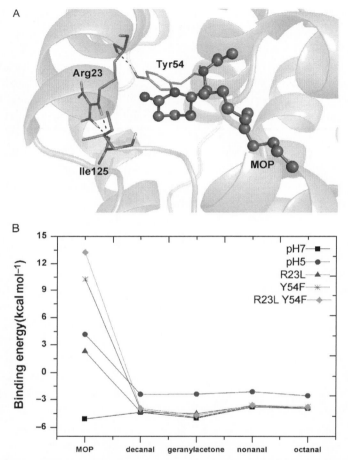

Figure 5.4 (A) Docking result of MOP to AaegOBP1 using Autodock at pH 7.0. Substrate MOP is shown in the deep blue ball and stick style, and pH-sensing triad is shown in the yellow stick style. All the oxygen atoms are colored in red and nitrogen atoms are colored in blue. (B) Binding energy plot of the five kinds of ligands binding to five other different proteins. (For interpretation of the references to color in this figure legend, the reader is referred to the online version of this chapter.)

3. THERMAL STABILIZATION OF HYPERTHERMOPHILIC PROTEIN Sso7d AND Cren7 COMPLEX WITH DNA

A large amount of similar DNA–binding proteins, which are believed to play a significant role in protecting its DNA against thermal denaturation, have been expressed and focused on the recent reports (Agha–Amiri & Klein, 1993; Baumann, Knapp, Lundbäck, Ladenstein, & Härd, 1994;

Luscombe, Austin, Berman, & Thornton, 2000). One of them is Sso7d, a 7-kDa protein of 63 amino acids, belonging to small DNA-binding proteins family isolated from *S. solfataricus* (Choli, Henning, Wittmann-Liebold, & Reinhardt, 1988). Sso7d is endowed with extreme thermal, acid, and chemical stability as well as DNA-binding properties, which make the protein an attractive system for structural, thermodynamic, and DNA-binding studies (Andrea et al., 2011; Gao et al., 1998; Gera, Hussain, Wright, & Rao, 2011). The 3D structures of Sso7d–DNA complexes have been determined by X-ray crystallography (PDB ID: 1BNZ and 1BF4) (Gao et al., 1998), revealing that Sso7d binds DNA nonspecifically by partial intercalation of the hydrophobic side chains of Val26 and Met29 (McAfee, Edmondson, Datta, Shriver, & Gupta, 1995). Meanwhile, the DNA molecules in those complexes undergo conformational changes (Dostál, Chen, Wang, & Welfle, 2004). Impressively, although Sso7d can significantly distort duplex DNA structures, the protein structure in the complexes is similar to that of the unbound Sso7d (Napoli et al., 2002). Previous studies also showed that Sso7d promotes the renaturation of complement DNA strands at high temperatures and then facilitates the stabilization of DNA duplexes (Guagliardi, Napoli, Rossi, & Ciaramella, 1997).

Another hyperthermophilic protein, Cren7, is also able to compact DNA and plays a vital role in DNA stabilization at high growth temperatures (Choli, Wittmann-Liebold, & Reinhardt, 1988; Grote, Dijk, & Reinhardt, 1986; Kimura, Kimura, Davie, Reinhardt, & Dijk, 1984; Sandman & Reeve, 2005; Todorova & Atanasov, 2004). Cren7 is discovered in *Sulfolobus shibatae* (Guo et al., 2008) and bears significant resemblance in both structure and biochemical properties to Sul7 (Sac7d and Sso7d), although the two proteins are unrelated at the amino acid sequence level (Bell & White, 2010; Guo et al., 2008). Cren7 binds to DNA without obvious sequence preference and significantly increases the melting temperature of dsDNA against thermal denaturation (Guo et al., 2008). The 3D structures of Cren7–DNA complex (PDB ID: 3LWH and 3LWI) (Zhang, Gong, Guo, Jiang, & Huang, 2010) show that the Cren7 protein binds DNA nonspecifically in the minor groove of DNA and binds the DNA double helix through the intercalation of the hydrophobic side chains of Leu28 (Zhang et al., 2010). Same as Sso7d, the conformation of interacting partners undergoes a significant conformational change upon binding. In addition, structural studies and mutational data show that several residues play a critical role in stabilizing the protein–DNA complex by hydrogen bonds and hydrophobic interactions (Feng, Yao, & Wang, 2010; Guo et al., 2008; Zhang et al., 2010).

In our work, we attempt to characterize and explore the role of these two hyperthermophilic proteins in enhancing the stabilization of DNA duplexes via protein–DNA interactions by using MD simulations at different temperatures and MM-PBSA approach (Bahadur, Kannan, & Zacharias, 2009; Gohlke, Kiel, & Case, 2003; Hou et al., 2011a, 2011b; Lee, Kim, & Seok, 2010; Liu & Yao, 2009).

3.1. MM-PBSA method

The MM-PBSA method (Hou et al., 2011a, 2011b; Srinivasan, Cheatham, Cieplak, Kollman, & David, 1998; Swanson, Henchman, & McCammon, 2004) using a triplet-trajectory analysis was carried out to calculate the binding free energies for both the Sso7d– and Cren7–DNA complexes. The MM-PBSA method and the nmode module implemented in Amber11 (Case et al., 2010) were employed to estimate the binding free energies. In this method, the binding free energies (ΔG_{bind}) are computed by the following equations:

$$\Delta G_{bind} = G_{complex} - G_{protein} - G_{DNA} \tag{5.1}$$

Here, $G_{complex}$, $G_{protein}$, and G_{DNA} are the free energies of complex, protein, and DNA, respectively. The free energy ($G_{x=complex,protein,DNA}$) of each species can be estimated by using the MM-PBSA methods:

$$G_{x=complex,protein,DNA} = E_{MM} + G_{solv} - T\Delta S \tag{5.2}$$

$$E_{MM} = E_{ele} + E_{vdw} + E_{int} \tag{5.3}$$

$$G_{solv} = G_{pb} + G_{nonp} \tag{5.4}$$

Here, E_{MM} is the gas phase molecular mechanical energy, G_{solv} is the solvation free energy, and E_{ele}, E_{vdw}, and E_{int} are the electrostatic energy, the van der Waals interaction energy, and the internal energy, respectively. The solvation free energy, G_{solv}, can be computed as the sum of electrostatic solvation energy (G_{pb}) and nonelectrostatic solvation energy (G_{nonp}). The polar component (the electrostatic solvation energy) is computed by using the PB (Bashford & Case, 2000; Luo, David, & Gilson, 2002) in Amber 11. The dielectric constants were set to 1 and 80 (Milev et al., 2003; Yang, Zhu, Wang, & Chen, 2010) for the solute and the surrounding solvent, respectively, in our calculations, and the ionic strength was set to 0.1 M. The nonpolar contribution (the nonelectrostatic solvation energy) is estimated by the equation:

$$G_{nonp} = \gamma SASA + \beta \tag{5.5}$$

Here, the γ and β, two empirical constants, were set as $0.00542\,\text{kcal mol}^{-1}\,\text{Å}^{-2}$ and $0.92\,\text{kcal mol}^{-1}$ (Cheatham, Srinivasan, Case, & Kollman, 1998; Sitkoff, Sharp, & Honig, 1994), respectively, and SASA is the solvent accessible surface area determined by a probe radius of 1.4 Å. The solute entropy S (Gouda, Kuntz, Case, & Kollman, 2002) is estimated by normal mode analysis, using the normal module analysis (Case, 1994) implemented in the Amber11.

3.2. Structural changes of protein and DNA at different temperatures

To explore the specific structural changes of the binary complexes, the back-bone RMSD values referenced to the corresponding starting structures are calculated (Tables 5.3 and 5.4). First in Sso7d (Table 5.3), protein is still quite stable at 360 K. This result is consistent with the experimental observation and computational studies for the Sso7d protein (Merlino, Graziano, & Mazzarella, 2004; Priyakumar, Ramakrishna, Nagarjuna, & Reddy, 2010; Xu, Su, Chen, & Wang, 2011). At other two high temperatures (420 and 480 K), the maximum average RMSD value was only 3.8 Å, indicating the protein does not undergo major structural transitions in the timescale of the simulations. However, with temperature increasing, the mean RMSD values of the binary complexes with respect to the X-ray crystal structures rise from 1.5 to 6.5 Å. This leads to the conclusion that the large deviations of the binary complexes may result from the instabilities of DNA double helix. Therefore, we focused on the behaviors of DNA in the complexes in comparison to the unbound DNA during the simulations. From analysis of the change in RMSD values (Table 5.3), we get the conclusion that the changes in the structure of DNA molecules give rise to the RMSD fluctuations of the binary complexes. Table 5.3 also shows the RMSD of the bound DNA and the unbound DNA with respect to the canonical A- and B-DNA (Arnott & Hukins, 1972) to illustrate the influence of the protein Sso7d on the bending of the DNA. Because of the kink of DNA by intercalation of the side chains of Val26 and Met29 of Sso7d, the RMSD values of both the bound and the unbound DNA are reasonably larger compared with the values from the X-ray crystal structures. Judging from the change in RMSD values, we can make out that the bound DNA in the complexes is close to the canonical A–DNA than to the B–DNA, while the unbound DNA is vice versa.

Then in Cren7, RMSD values have similar results as that in Sso7d (Table 5.4). The relatively larger average values were obtained at the three

Table 5.3 Average RMSD values (Å) of the Sso7d, the Sso7d–DNA binary complexes, and the DNA structures with respect to the X-ray crystal structures and canonical A- and B-forms

	300 K	360 K	360 K^{-1}	420 K	480 K
1BNZ					
Protein (alone)	1.79±0.20	1.81±0.22		1.93±0.27	3.32±0.77
Protein (complexed)	1.30±0.17	1.29±0.22	1.53±0.35	1.83±0.48	3.25±0.75
Protein–DNA	1.57±0.23	3.43±0.25	3.07±0.67	4.48±0.37	6.52±1.19
DNA (complexed)–xtal	1.66±0.39	3.96±0.31	3.52±0.86	6.42±0.58	9.27±1.97
DNA (complexed)–B-form	5.89±0.32	6.20±0.24	6.17±0.32	7.91±0.43	11.36±1.91
DNA (complexed)–A-form	3.41±0.21	5.63±0.28	5.14±0.90	7.67±0.51	9.37±1.56
DNA (alone)–xtal	4.55±0.54	5.08±0.86		7.11±1.35	21.88±11.75
DNA (alone)–B-form	3.86±0.43	4.01±0.65		7.19±1.27	22.80±12.37
DNA (alone)–A-form	5.80±0.5	5.98±0.61		7.65±1.33	21.51±11.80
1BF4					
Protein (complexed)	1.26±0.14	1.19±0.19	1.21±0.19	1.30±0.18	3.80±1.19
Protein–DNA	1.44±0.19	2.42±0.26	2.56±0.59	3.43±0.45	6.47±0.88
DNA (complexed)–xtal	1.40±0.31	3.20±0.44	2.90±0.75	4.77±0.73	6.83±0.66
DNA (complexed)–B-form	4.93±0.25	5.67±0.28	5.55±0.41	6.20±0.51	7.68±0.66
DNA (complexed)–A-form	3.92±0.26	4.64±0.33	4.50±0.40	6.13±0.54	7.35±0.57
DNA (alone)–xtal	3.90±0.56	3.67±0.58		5.45±0.65	10.46±0.88
DNA (alone)–B-form	3.56±0.41	3.47±0.40		4.35±0.51	11.10±0.97
DNA (alone)–A-form	5.41±0.51	5.32±0.58		6.75±0.67	10.12±0.42

Table 5.4 Average RMSD values (Å) of the Cren7, the Cren7–DNA binary complexes, and the DNA structures with respect to the X-ray crystal structures and canonical A- and B-DNA

	300 K	350 K	400 K	450 K
3LWH				
Protein–DNA	1.44 ± 0.15	2.33 ± 0.27	2.78 ± 0.41	3.91 ± 0.64
Protein (complex)	0.89 ± 0.14	1.06 ± 0.19	1.69 ± 0.47	2.10 ± 0.38
DNA (complexed)–xtal	1.85 ± 0.27	3.21 ± 0.62	3.52 ± 0.63	5.26 ± 1.13
DNA (complex)–B-form	4.35 ± 0.19	6.01 ± 0.36	6.07 ± 0.47	6.80 ± 0.99
DNA (complex)–A-form	3.10 ± 0.25	4.34 ± 0.38	4.36 ± 0.59	5.58 ± 0.99
DNA (alone)–xtal	4.06 ± 0.40	4.69 ± 0.61	9.06 ± 1.76	20.47 ± 7.36
DNA (alone)–B-form	3.39 ± 0.26	4.72 ± 0.51	8.62 ± 1.70	21.29 ± 8.04
DNA (alone)–A-form	5.23 ± 0.42	6.18 ± 0.56	9.36 ± 1.45	19.91 ± 7.22
Protein (alone)	1.41 ± 0.37	1.91 ± 0.28	2.55 ± 0.44	3.25 ± 0.64
3LWI				
Protein–DNA	1.50 ± 0.18	1.81 ± 0.27	2.15 ± 0.44	3.63 ± 0.32
Protein (complex)	1.13 ± 0.19	1.27 ± 0.24	1.75 ± 0.42	1.97 ± 0.29
DNA (complexed)–xtal	1.70 ± 0.28	2.24 ± 0.45	3.72 ± 0.88	9.44 ± 0.92
DNA (complex)–B-form	5.19 ± 0.31	5.70 ± 0.32	5.68 ± 0.49	9.51 ± 0.50
DNA (complex)–A-form	2.85 ± 0.33	3.28 ± 0.37	3.95 ± 0.74	8.22 ± 0.84
DNA (alone)–xtal	3.98 ± 0.49	4.81 ± 0.72	6.09 ± 0.68	9.64 ± 1.68
DNA (alone)–B-form	2.28 ± 0.49	3.28 ± 0.51	4.24 ± 0.99	9.42 ± 1.47
DNA (alone)–A-form	4.39 ± 0.55	4.97 ± 0.59	6.12 ± 0.57	10.37 ± 2.01
Protein (alone)	1.56 ± 0.36	2.27 ± 0.33	2.39 ± 0.50	2.80 ± 0.50

high temperatures (350, 400, and 450 K), implying that the structures of the two Cren7–DNA complexes undergo constant evolution at high temperatures, which is similar with the results of Sso7d. Obviously, there are no remarkable changes for Cren7 protein in the timescale of the simulations. However, similar calculations for the DNA within the two complexes show that the DNA molecules exhibit larger deviations in the two complexes. In a word, the large deviations of the two complexes originate from the instabilities of the DNA double helix. Furthermore, we calculated the RMSD values of the DNA in MD simulations of DNA alone and protein–DNA complexes versus the canonical A- and B-DNA of the same sequence to explore the influence of the Cren7 protein on the bending of the DNA. These values indicate that the two DNA molecules in the complexes are closer to the A-DNA than to the B-DNA, while the free DNA molecules are vice versa. These deviations are likely to result from the kink of DNA by the intercalation of the side chains of L28 in Cren7, which is in agreement with the study of Zhang et al. (2010).

3.3. Structural stability and conformational transitions of DNA molecules

The criteria for Watson–Crick-type base pairing are about 3.2 Å between N1 and N3 atoms, and the distance longer than 5 Å shows the disruption of base pair (Bueren-Calabuig, Giraudon, Galmarini, Egly, & Gago, 2011; Priyakumar, Harika, & Suresh, 2010). In Sso7d, all base pairs remain unchanged in the Sso7d–DNA binary complexes at 300 K. In contrast, the three specific base pairs (T2A7 and A7T2 in the 1BNZ, and G7C2 in the 1BF4) in the two unbound DNA strands undergo partial base opening at 360 K, and the probabilities of the distance exhibit nonzero between N1 and N3 atoms beyond 5 Å. At 420 K, a higher temperature, almost all base pairs of the DNA undergo base opening. All base pairs of DNA molecules in the binary complexes are lost at a relative extreme temperature of 480 K. Similar things are observed in Cren7. The A7T2 base pair of the unbound DNA strands from 3LWH starts undergoing base opening at 350 K displaying nonzero probabilities beyond 5 Å. At 400 K, all base pairs of the unbound DNA strands from 3LWH and two specific base pairs (C2G7 and G3C6) of the unbound DNA strands from 3LWI undergo base opening, while only one specific base pair (A7T2) exhibits nonzero probabilities beyond 5 Å in 3LWH. At 450 K, all base pairs of the DNA undergo base opening even in the two Cren7–DNA complexes. These analyses above show that both Sso7d and Cren7 facilitate stabilization of the DNA in the complexes over a certain temperature range. The results are similar to that of the studies on Sac7d–DNA complexes (Priyakumar et al., 2010).

Then, the pseudorotation angle (P) parameter (Cremer & Pople, 1975; Harvey & Prabhakaran, 1986; Saenger, 1984) was used to classify the sugar pucker conformation as follows: A-like sugars include the P values from $-30°$ to $90°$ and the B-like sugars include values between $90°$ and $210°$ (Tolstorukov, Jernigan, & Zhurkin, 2004). The probability distributions

Table 5.5 Percentage (%) of A-like (A) and B-like (B) conformations for the 10 individual nucleotides and the total 10 nucleotides in both complementary strands of the Sso7d-bound DNA at 300 and 360 K

	1BNZ						1BF4					
	300 K		360 K		360 K^{-1}		300 K		360 K		360 K^{-1}	
	A	B	A	B	A	B	A	B	A	B	A	B
D2a	29.7	70.0	26.7	72.1	25.3	73.4	59.6	38.4	57.8	41.8	57.7	41.3
D2b	0.1	99.7	4.1	94.3	3.7	93.4	0.6	95.1	0.8	94.3	0.8	93.8
D3a	39.6	58.0	34.0	65.8	35.2	64.1	1.6	93.6	3.0	89.2	3.2	88.4
D3b	0.7	98.3	0.5	90.9	0.5	90.3	95.1	4.9	96.1	3.8	96.8	3.1
D4a	0.5	93.2	26.1	72.8	27.8	70.1	45.2	54.9	43.1	57.0	42.5	56.7
D4b	93.8	4.8	93.5	5.6	92.9	6.3	60.1	39.6	70.8	28.4	68.7	30.6
D5a	31.1	68.7	81.7	18.2	83.3	16.4	52.8	47.0	62.2	37.2	60.3	38.6
D5b	89.8	9.7	85.8	14.1	83.5	15.3	7.4	86.8	7.2	90.0	7.2	90.7
D6a	65.6	34.2	99.0	0.9	89.7	1.2	0.7	99.2	2.4	95.8	2.2	96.1
D6b	10.2	89.2	5.0	93.1	6.7	92.5	2.5	96.7	0.1	99.6	0.2	99.3
D7a	1.5	98.0	1.1	86.2	1.3	88.5	1.7	98.2	0.2	99.3	0.7	98.7
D7b	7.6	92.2	1.2	97.3	1.7	96.8	8.9	90.7	2.1	96.9	3.2	95.1
Total	34.2	64.5	40.2	57.6	42.4	56.1	33.3	65.8	34.4	63.2	35.6	61.7

of sugar pseudorotation angles for both DNA strands are shown in Tables 5.5 and 5.6.

In Sso7d, the nucleotides surrounding the interaction sites exhibit similar conformational behavior in the two complexes. The three nucleotides (D6b, D7a, and D7b) in the 1BNZ and the five nucleotides (D5b, D6a, D6b, D7a, and D7b) in the 1BF4, which adopt the B-like sugar conformations in the binary complexes, locate far away from the kinking sites. The three nucleotides (D4b, D5b, and D6a) in the 1BNZ and the two nucleotides (D3b and D4b) in the 1BF4, which sample the A–like sugar conformations, are close to the kinking sites. The impact of temperature on the structural transition of DNA is mainly contributed by D4a, D5a, and D6a in the 1BNZ as well as by D4b and D5a in the 1BF4.

Similarly, in Cren7, three nucleotides (D3a, D3b, and D4b) in 3LWH and the two nucleotides (D3a and D5a) in 3LWI, which are close to the kink sites, take the A–like sugar conformations, while the four nucleotides (D2a,

Table 5.6 Percentage (%) of A-like (A) and B-like (B) conformations for the 12 individual nucleotides and the total 12 nucleotides in both complementary strands of the Cren7-bound DNA at 300 and 350 K

	3LWH				3LWI			
	300		350		300		350	
	A	B	A	B	A	B	A	B
D2a	7.4	92.6	9.3	90.6	8.1	91.8	8.7	91.2
D2b	56.3	43.7	40.7	59.3	21.3	78.7	60.1	38.9
D3a	61.8	38.0	52.3	47.6	68.5	30.7	68.2	31.7
D3b	60.3	39.3	68.9	30.1	34.6	65.4	90.4	8.7
D4a	29.2	70.8	26.9	73.0	34.2	65.8	39.6	60.4
D4b	94.9	2.1	93.8	5.0	66.9	33.0	9.3	90.7
D5a	28.2	71.8	85.2	14.7	75.2	24.8	61.5	38.5
D5b	76.1	23.7	3.1	96.8	58.2	41.7	34.8	65.1
D6a	3.8	96.2	84.6	15.1	4.6	95.3	13.3	86.7
D6b	29.2	70.7	4.1	95.8	28.6	71.4	30.0	69.9
D7a	36.4	63.5	65.6	33.4	44.4	55.6	40.7	59.2
D7b	20.4	79.6	19.8	80.1	10.5	89.5	12.3	87.7
Total	42.0	57.7	44.1	55.6	37.9	62.0	40.4	59.5

D4a, D6b, and D7b) in 3LWH and the six nucleotides (D2a, D4a, D6a, D6b, D7a, and D7b) in 3LWI, which locate far away from the kink sites, adopt the B-like sugar conformations. The effect of temperature on the conformational transition of DNA is mainly contributed by the five nucleotides (D2b, D5a, D5b, D6a, and D7a) in 3LWH and the four nucleotides (D2b, D3b, D4b, and D5b) in 3LWI. These structural transitions of nucleotides in both Sso7d and Cren7 may be relevant to the hydrophobic residues.

3.4. Free energy analysis of protein–DNA complexes

The stability of binary complexes at 300 and 360 K in Sso7d and 350 K in Cren7 allows us to analyze binding free energies and structural properties in detail. As shown in Tables 5.7 and 5.8, the binding free energies (ΔG_{bind}) at 360 K in Sso7d and 350 K in Cren7 are lower than those at 300 K, indicating the binding is stronger at 360 K in Sso7d and 350 K in Cren7, respectively.

Table 5.7 Binding free energies (kcal mol^{-1}) contributed by enthalpy and entropy at 300 and 360 K from triplet-trajectory analysis for the two Sso7d–DNA complexes

	1BNZ			1BF4		
	300 K	360 K	360 K^{-1}	300 K	360 K	360 K^{-1}
E_{ele}	−2393.2	−2481.8	−2480.9	−2295.5	−2321.6	−2322.9
E_{vdw}	−88.9	−86.5	−87.8	−87.1	−85.4	−85.3
G_{nonp}	−8.8	−8.7	−8.7	−7.9	−7.4	−7.3
G_{pb}	2426.8	2509.8	2509.7	2331.6	2351.7	2352.5
ΔG_{np}[a]	−97.7	−95.1	−96.5	−95.0	−92.8	−92.6
ΔG_{pb}[b]	33.6	28.0	28.8	36.1	30.1	29.6
ΔE_{MM}	−2488.8	−2581.5	−2582.3	−2389.6	−2420.6	−2420.9
ΔG_{solv}	2418.0	2501.1	2501.0	2323.7	2344.3	2345.2
ΔH	−70.8	−80.4	−81.3	−65.9	−76.3	−75.7
$-T\Delta S$	39.4	45.8	45.8	42.3	46.7	47.5
ΔG_{bind}[c]	−31.4	−34.6	−35.5	−23.7	−29.6	−28.2

[a]$\Delta G_{np} = E_{vdw} + G_{nonp}$.
[b]$\Delta G_{pb} = E_{ele} + G_{pb}$.
[c]$\Delta G_{bind} = \Delta G_{np} + \Delta G_{pb} + E_{int} - T\Delta S$.

In order to characterize and identify the key residues of the protein–DNA interaction interface in Sso7d and Cren7, per-residue free energy decomposition was performed. In the case of Sso7d, the favorable interactions are mainly made by eight amino acids (Lys7, Tyr8, Lys9, Lys22, Trp24, Val26, Met29, and Arg43) in the binary complexes. Polar interactions are found to facilitate the binding for three amino acids (Lys7, Lys9, and Lys22), which are involved in ionic interactions with the phosphates. Five amino acids (Tyr8, Trp24, Val26, Met29, and Arg43) are involved in the main binding attractions by van der Waals interactions. It is obvious that the energy decomposition results are consistent with the results in the two-crystal structure analyses (1BNZ and 1BF4) (Gao et al., 1998). In the case of Cren7, seven amino acid residues (Lys24, Trp26, Leu28, Pro30, Lys31, Arg51, and Lys53) make major contributions to the binding free energy with more than 3.5 kcal mol^{-1} free energy for 3LWH and 3LWI. These residues are very important for the Cren7–DNA binding, which are in good agreement with the previous experimental identification (Feng et al., 2010; Zhang et al., 2010). Compared with the decomposition

Table 5.8 Binding free energies (kcal mol^{-1}) contributed by enthalpy and entropy at 300 and 350 K from triplet-trajectory analysis for the two Cren–DNA complexes

	3LWH		3LWI	
	300	350	300	350
E_{ele}	-2661.3 ± 4.75	-2753.8 ± 3.69	-2625.0 ± 4.01	-2632.2 ± 4.62
E_{vdw}	-76.5 ± 0.87	-85.9 ± 0.93	-90.7 ± 0.96	-80.9 ± 0.92
G_{nonp}	-8.2 ± 0.06	-8.8 ± 0.19	-8.5 ± 0.22	-8.1 ± 0.23
G_{pb}	2667.1 ± 3.08	2757.6 ± 4.33	2648.4 ± 3.81	2642.4 ± 4.21
E_{int}	4.6 ± 1.21	3.3 ± 1.43	5.9 ± 1.38	3.2 ± 1.27
ΔG_{np}[a]	-84.7 ± 0.81	-94.7 ± 1.31	-99.2 ± 1.47	-89.0 ± 1.11
ΔG_{pb}[b]	5.8 ± 0.27	3.8 ± 1.12	23.4 ± 1.94	10.2 ± 1.67
ΔE_{MM}	-2733.2 ± 3.82	-2836.4 ± 4.25	-2709.8 ± 3.35	-2716.3 ± 4.92
ΔG_{solv}	2659.0 ± 5.19	2748.9 ± 4.61	2639.9 ± 3.78	2634.3 ± 4.59
ΔH	-74.3 ± 1.82	-87.6 ± 1.34	-70.0 ± 1.96	-82.0 ± 2.44
$-T\Delta S$	32.1 ± 0.68	42.8 ± 0.77	35.3 ± 0.82	41.1 ± 0.71
ΔG_{bind}[c]	-42.2 ± 1.24	-48.6 ± 1.85	-34.7 ± 1.29	-40.9 ± 1.63

[a]$\Delta G_{np} = E_{vdw} + G_{nonp}$.
[b]$\Delta G_{pb} = E_{ele} + G_{pb}$.
[c]$\Delta G_{bind} = \Delta G_{np} + \Delta G_{pb} + E_{int} - T\Delta S$.

energies of residues at 300 K, the decomposition energies of the three amino acid residues (Trp26, Leu28, and Pro30) are increased at 350 K, indicating a larger contribution of conserved residues to the stability of the two complexes. Although the decomposition energy of Lys24, Lys31, Arg51, and Lys53 is not increased at 350 K, it has a significantly favorable contribution to the binding free energy.

4. SUBSTRATE SELECTIVITY AND MAJOR ACTIVE-SITE ACCESS CHANNELS OF HUMAN CYTOCHROME P4507B1

The cytochrome P450s (CYPs) are ubiquitous heme-containing mixed function oxygenases that catalyze the hydroxylation of nonactivated hydrocarbon compounds, dealkylation, epoxidation, and dehydrogenation reactions involved in oxidative metabolism (Kühnel et al., 2008; Meng, Zheng, & Zhang, 2009; Yamashita, Feng, Yoshida, Itoh, & Hashida, 2011).

P4507B1 (CYP7B1), one of the CYP7 family members, is widely expressed in the brain, particularly in the hippocampus, and also in the liver and kidney, albeit at much lower levels (Martin, Bean, Rose, Habib, & Seckl, 2001; Myant & Mitropoulos, 1977; Noshiro, Nishimoto, Morohashi, & Okuda, 1989; Noshiro & Okuda, 1990). It appears to contribute to bile acid synthesis through 7a-hydroxylation of hydroxycholesterol. Recent studies show that mutations in the CYP7B1 gene are directly responsible for spastic paraplegia type 5, which is known as a progressive neuropathy (Erichsen, Koht, Stray-Pedersen, Abdelnoor, & Tallaksen, 2009; Setchell et al., 1998; Tsaousidou et al., 2008). The enzyme is a steroid cytochrome P450 7a-hydroxylase that provides the primary metabolic routes for neurosteroids dehydroepiandrosterone (DHEA) and pregnenolone, cholesterol derivatives 25-hydroxycholesterol (25-HOChol) and 27-hydroxycholesterol, and other steroids including estrogen receptor ligands such as 5α-androstane-3b,17β-diol (anediol) and 5α-androstene-3b,17β-diol (enediol) (Chalbot & Morfin, 2006; Goizet et al., 2009; Li & Bigelow, 2010; Martin et al., 2001; Pettersson, Lundqvist, & Norlin, 2010; Stiles, McDonald, Bauman, & Russell, 2009; Tang, Eggertsen, Chiang, & Norlin, 2006; Tsaousidou et al., 2008; Yau et al., 2003). DHEA is one of the most abundant neurosteroids and performs many important functions and is widely under investigation as a representative neurosteroid in the brain (Martin et al., 2001; Rose et al., 1997; Tang et al., 2006). Similarly, 25-HOChol, which is hydroxylated by CYP7B1 as a part of the cholesterol degradative pathway, can be viewed as a representative cholesterol derivative in the liver (Goizet et al., 2009; Martin et al., 2001). Therefore, the most representative substrates including 25-HOChol, DHEA, anediol, and enediol were chosen as models to investigate the structural features relevant to the substrate selectivity of CYP7B1 and also to save computational time.

Previous experimental work has examined the kinetic characterization of CYP7B1 activity, but intermolecular interactions involved in the catalytic mechanism are still unknown at the atomic level (Martin et al., 2001; Rose et al., 1997). In the absence of an X-ray crystal structure, computational approaches provide a series of methods such as homology modeling, MD, and automatic docking that in combination with previous experimental site-directed mutagenesis studies and access channels analysis allow the structural features relevant to the substrate selectivity of CYP7B1 to be investigated.

4.1. Establish the initial model for CYP7B1 and four ligand complexes

The primary sequence of CYP7B1 was obtained from the Swiss–Prot database (accession number O75881.2), and BLAST searching revealed a high sequence identity between the CYP7B1 and CYP7A1 (41%) that allowed a straightforward sequence alignment. The crystal structure of CYP7A1 is available from the Protein Data Bank (PDB ID: 3SN5) (Stiles et al., 2009). The initial model of CYP7B1 was built by using Discovery Studio 2.5 with residues from 38 to 506. The heme iron is covalently bound to the side chain thiolate group of the conserved cysteine residue (Cys449) (Stiles et al., 2009). This model was refined by MD simulation. It was clear that the RMSD of the system remained in equilibrium during the last 4 ns, which indicated that the model was stable and could be used in further docking studies. The self-compatibility score for this protein was 199.61, which is higher than the low score (96.3801) and close to the top score (214.179). The statistical score of the Ramachandran plot shows that 94.3% are in the most favored regions, 4.0% in the additional allowed regions, and 1.7% in generously allowed regions. In summary, the above results indicate that the homology model is reliable.

Experimental data have shown that CYP7B1 can catalyze substrates with medium chain lengths, such as 25-HOChol (Martin et al., 2001). Four substrates (Fig. 5.5) were docked into the active site of the enzyme using the CDOCKER protocol of Discovery Studio 2.5. The set of 40 conformations in each docking system were divided into two groups. In the "A" conformation, the two hydrophobic methyl groups (C18 and C19) are directed toward the heme plane, whereas in the "B" conformation the two hydrophobic methyl groups extend back toward the heme plane (Fig. 5.6). Among the 10 conformations of CYP7B1–25-HOChol complex structures, both binding modes were found; the highest score conformations of the two types were selected because they cover all the representative conformations of A and B. According to the bound conformation in the crystal structure of CYP7A1 (PDB ID: 3SN5), two hydrophobic methyl groups of the substrate cholest-4-en-3-one extend back toward the heme plane. Because the structure of cholest-4-en-3-one is similar to that of 25-HOChol, the B conformation is considered to be most closely matched with the experimental results. All the conformations of CYP7B1–enediol, CYP7B1–DHEA, and CYP7B1–anediol complex structures show only B conformation binding mode.

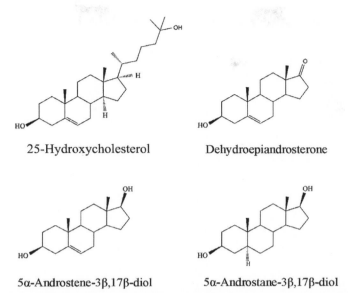

Figure 5.5 Molecular structures of the four substrates of CYP7B1: 25-hydroxycholesterol (25-HOChol), dehydroepiandrosterone (DHEA), 5α-androstane-3b,17β-diol (anediol), and 5α-androstene-3b,17β-diol (enediol).

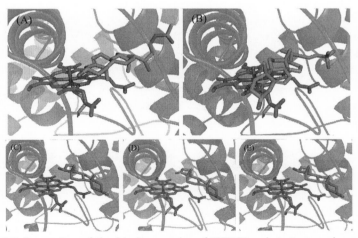

Figure 5.6 Stereoview of the binding modes of CYP7B1–substrate complexes for (A) 25-HOChol-A conformation, (B) 25-HOChol-B conformation, (C) enediol-B conformation, (D) DHEA-B conformation, and (E) anediol-B conformation, respectively. The complex structures are shown in transparent ribbon representation (helixes in red, beta strands in yellow, and loops and turns in green), and substrates and CpdI are shown as stick representations. (For interpretation of the references to color in this figure legend, the reader is referred to the online version of this chapter.)

4.2. MD simulations and total interaction energy calculations

After 5-ns MD simulations of each CYP7B1–ligand complex, the binding free energies calculated by MM–GBSA and the statistic interaction energies of four substrates with a series of residues at the active site were performed, as shown in Table 5.9. 25-HOChol was found to be the substrate with the highest binding affinity. The binding free energies for the A and B binding modes of 25-HOChol in the complexes are -30.38 and -33.84 kcal mol^{-1}, and the interaction energies are -68.82 and -81.99 kcal mol^{-1}, respectively. On the basis of these results, we can conjecture that CYP7B1 may allow multiple conformations of 25-HOChol to be accommodated and that the B conformation may be the major binding mode (Itoh, Takemura, Shimoi, & Yamamoto, 2010). The results of the analysis show excellent agreement with experimental data in all cases mentioned earlier (Martin et al., 2001).

To gain a deeper insight into the interactions between the substrates and the enzyme, residues with interaction energies lower than -1 kcal mol^{-1} are listed in Table 5.10. The analysis of interaction energies highlights 15 residues that are consistently involved in both binding modes of the 25-HOChol substrate within the active site: Phe114, Ser115, Leu118, Leu119, Arg267, Gly288, Phe289, Trp291, Ala292, Ser366, Thr367, Ile369, Leu488, Phe489, and Gly490. Arg267, Thr367, and Leu488 are found to form significant hydrogen bonds with the substrate in more than 80% frames in the simulation trajectories, which could fix the substrate in a favorable position for further metabolic reaction. Ala292 is positioned in proximity to the C3–OH of the substrate, and the corresponding residue Ala283 in CYP8A1 is important in an early hydroxylation step (Chiang, Yeh, Wang, & Chan, 2006). The hydrophobic residues, particularly Leu118, Gly288, Trp291, Leu488, and Phe489, collectively form a hydrophobic cavity and provide hydrophobic interactions to stabilize the substrate. Phe489 is observed to adopt a favorable position to merge the active site with adjacent channel by a rotation of its side chain and subsequently has indirect impact on substrate binding. In general, the main binding attractions come from the following seven residues: Arg267, Trp291, Ala292, Ser366, Thr367, Leu488, and Phe489, with strong interaction and significant roles in positioning the substrates by restructuring the active-site volume.

For the other three selected substrates (enediol, DHEA, and anediol), the results listed in Table 5.10 indicate that the following are the key residues that contribute to the binding of the three substrates: Phe114, Ser115, Leu118, Trp291, Ala292, Ala295, Asn296, Thr367, Phe489, and Gly490. Residues Ala292, Asn296, and Thr369 are found to form hydrogen bonds

Table 5.9 Binding free energies (kcal mol^{-1}) for four substrates in the complexes

System	ΔE_{ele}	ΔE_{vdw}	ΔE_{MM}[a]	ΔG_{GB}	$\Delta G_{nonpolar}$	$\Delta G_{MM\text{-}GB/SA}$[b]	$-T\Delta S$	$\Delta\Delta G_{TOT}$[c]
25-HOChol-B	−10.01	−61.98	−71.99	25.86	−6.25	−52.38	18.54	−33.84
25-HOChol-A	−11.64	−60.17	−71.81	30.58	−6.55	−47.78	17.40	−30.38
Enediol	−16.47	−41.80	−58.27	27.60	−4.82	−35.49	15.03	−20.46
DHEA	−11.11	−42.78	−53.89	24.26	−4.81	−34.44	15.39	−19.05
Anediol	−9.85	−31.74	−41.59	24.70	−4.78	−21.67	15.65	−6.02

[a] $\Delta E_{MM} = \Delta E_{ele} + \Delta E_{vdw}$.
[b] $\Delta G_{MM\text{-}GB/SA} = \Delta E_{MM} + \Delta G_{GB} + \Delta G_{nonpolar}$.
[c] $\Delta\Delta G_{TOT} = \Delta E_{MM} + \Delta G_{GB} + \Delta G_{nonpolar} - T\Delta S$.

Table 5.10 The total interaction energies (kcal mol^{-1}) of the substrate with the each key residue in the active site for four substrates, respectively

Interaction energies	25-HOChol-B	25-HOChol-A	Enediol	DHEA	Anediol
Phe114	−1.36	−3.41	−3.85	−3.64	−1.24
Ser115	−1.20	−2.19	−2.15	−1.79	−2.23
Leu118	−2.45	−2.55	−2.00	−2.18	−2.53
Leu119	−2.29	−1.40	−0.16	−0.22	−0.45
Arg267	−9.10	−1.09	0.42	0.23	0.40
Gly288	−2.30	−2.66	−0.29	0.56	−0.11
Phe289	−2.48	−1.13	−0.28	0.36	−0.17
Trp291	−3.01	−3.22	−3.24	−2.79	−1.12
Ala292	−1.74	−1.17	−1.99	−1.32	−1.29
Ala295	−0.20	−0.07	−1.30	−2.20	−1.77
Asn296	−0.97	−0.96	−3.61	−3.69	−3.44
Ser366	−3.22	−1.38	−0.16	−0.31	−0.63
Thr367	−7.99	−2.26	−1.59	−0.94	−0.42
Ile369	−0.97	−1.39	−1.69	−1.39	−1.41
Leu488	−5.21	−4.88	−0.76	−0.61	−0.36
Phe489	−5.40	−5.55	−4.73	−4.21	−2.76
Gly490	−1.33	−1.33	−1.21	−0.99	−0.62

with all three substrates. Trp291 and Ala295 in CYP7B1 are expected to play similar important roles to that of the corresponding residue Trp283 in CYP7A1, which is considered, on the basis of site-directed mutagenesis studies, to form a "lid" to keep cholesterol near the heme (Mast et al., 2005). In summary, for all four substrates, the crucial residues with hydrophobic side chains in the active site, particularly Trp291 and Phe489, are important for CYP7B1 selectivity and may have a great impact on the active-site architecture and substrate binding.

4.3. Analysis of access channels in CYP7B1–small molecule complex structures

Conformational changes caused by multiple ligand binding are considered to be essential for ligand channeling in CYPs. In addition, the access channels

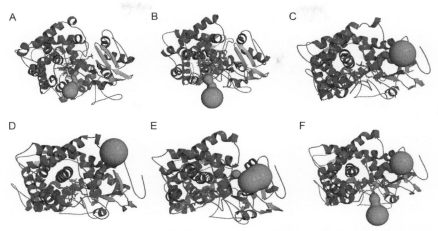

Figure 5.7 Access channels identified from the average structures of CYP7B1–substrate complexes for (A) A conformation of CYP7B1–25-HOChol complex, (B) B conformation of CYP7B1–25-HOChol complex, (C) CYP7B1–enediol complex, (D) CYP7B1–DHEA complex, and (E) CYP7B1–anediol complex, respectively. (F) Path 1 and path 2 channels are shown in the complex as the different access/egress routes for substrates. Channels are shown as cyan spheres, the complex structures are shown in ribbon representation (helixes in red, beta strands in yellow, and loops and turns in green), and substrates and CpdI are shown in stick representations. (For interpretation of the references to color in this figure legend, the reader is referred to the online version of this chapter.)

could guide the substrates directly to the position of the active site (Krishnamoorthy, Gajendrarao, Thangapandian, Lee, & Lee, 2010). All the potential substrate access channels can be clustered into two groups, which are designated in Fig. 5.7A–E. Path 1 represents the large channel located between the B-helix, the B–B′ loop, and the B–C′ loop; path 2 represents the minor channel that lies between the A′, B, and E helices.

For 25-HOChol, path 1 is obtained in both A and B conformations of the complex structures. Other substrates including DHEA, enediol, and anediol show path 2 channels. This observation suggests that similar substrates with large substituents at the 17 position could enter the active site of CYP7B1 from path 1 and then reorient for more favorable binding within the active site just like 25-HOChol. Substrates that are hydroxy substituted or keto substituted at the 17 position may follow path 2 and the binding conformational states could associate with enediol. Several hydrophobic residues, including aromatic residues, are observed to lie close to the major access channel. Phe111, Phe114, Leu118, Leu119, Phe123, Ile125, Phe219, and Trp291, which lie close to path 1, are suggested to make hydrophobic interactions with the hydrocarbon chain of 25-HOChol. Shorter ligands such as enediol that are hydroxy substituted or keto substituted at the 17 position are more polar and may use path 2, which is surrounded by more hydrophilic residues. Thus, it can be inferred that CYP7B1 could provide different access/egress routes for substrates that are either more or less hydrophobic, and the two pathways may play different roles in each case, which is consistent with the results obtained with CYP2E1 enzyme reported by Porubsky, Battaile, and Scott (2010).

For further analysis, another two 20-ns MD simulations were performed for two representative CYP7B1–substrate complexes: CYP7B1–25-HOChol complex and CYP7B1–enediol complex. The Phe111-His139 pair in the side chains of the CYP7B1–25-HOChol complex is observed to project out of the active site during the simulation, elucidating an extension motion in path 1. However, the Phe223-Phe489 pair is practically constant in the CYP7B1–25-HOChol complex (Fig. 5.8). In the 25-HOChol-bound structure, this aromatic ring of Phe489 forms a barrier between the active-site void and path 2, whereas in the enediol-bound structure, the plane of the aromatic ring rotates about 90° to connect the two voids. This may suggest that Phe489 can adopt either position, depending on the occupancy of the ligand. Minor

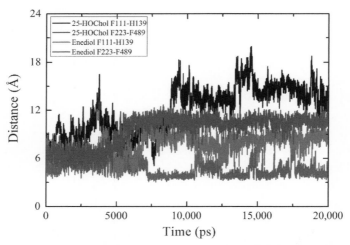

Figure 5.8 Distance between F111(CZ)-H139(NE2) and F223(CE2)-F480(CZ) residues in the two complexes indicated extension motions of the path 1 channel of the CYP7B1–25-HOChol complex and the path 2 channel of the CYP7B1–enediol complex, respectively. (For color version of this figure, the reader is referred to the online version of this chapter.)

rotations are observed in other residues, such as Phe111 and Phe289, to accommodate the ligands.

All the above structural analyses and comparisons suggest that the CYP7B1 enzyme appears to operate through two paths depending on the substrate; long-chain steroids enter the enzyme through the large channel (path 1), and path 2 may be the potential channel used for other truncated steroids such as enediol. In summary, mapping the dissimilar chain lengths of substrates reveals different access/egress routes, which can subsequently lead to the substrate binding indirectly as a result of impairment in catalytic function.

5. SUMMARY AND CONCLUSION

Proteins change their structures under the influence of environment and then that changes will have an effect on the properties of proteins. The conformational changes of proteins mainly embody on the changes of the interactions between some crucial residues. In order to find out what causes the protein conformational changes, a representative collection of proteins, insect OBPs, hyperthermophilic proteins, and human cytochrome P450 were selected for investigation, by using MD simulations, docking, and free energy calculations methods.

CpHMD simulations and docking methods were carried out to detect the conformational change of CquiOBP1 and AaegOBP1 in different pH solution. The results of CpHMD show that, from pH 7 to pH 5, the conformational changes of the two OBPs are mainly due to the decreasing of H-bonds number in pH-sensing triad, which leads to the movement of C-terminus in low pH solution. The Arg23 plays vital role in the maintenance of the structural stability of AaegOBP1. In addition, docking results indicate that MOP bound to the protein with its tail stretch into the tunnel between α4 and α5 at pH 7, similar to the condition in the crystal structure. But at pH 5, the ligand turns its direction with its head out of central cavity, and the binding free energy increases with it. Thus, another tunnel between α3 and α4, which is formed at pH 5, will be important for the ligand MOP release. All the findings above have paved the way for the future investigation of pH-induced ligand-releasing mechanism.

Then we investigated the thermal stability and structural transitions of DNA depending on temperature by performing MD simulations on Sso7d– and Cren7–DNA complexes at different temperatures. The results of both complexes indicate that DNAs alone in systems are more prone to denature than those in complexes with temperature increasing. In addition to the thermal stability, the DNA molecules in the two complexes also undergo B-like to A-like form transitions with increased temperature; however, the transitions only occur in a part of the nucleotides. The MM–PBSA results indicate that the binding affinity for both complexes shows an increasing trend as temperatures rise. Furthermore, the extensive interactions between the two proteins and DNA phosphate backbones are the reasons nonspecific DNA-binding protein is proposed. Some residues (Lys7, Tyr8, Lys9, Lys22, Trp24, Val26, Met29, and Arg43 in Sso7d; Lys24, Trp26, Leu28, Pro30, Lys31, Arg51, and Lys53 in Cren7) are critical to the binding to DNA in the binary complexes.

To study protein–ligand and receptor–ligand interactions, a major hurdle in computational investigation of CYP7B1 complex is the absence of a crystal structure of the protein. The complex model of CYP7B1 was constructed by using human CYP7A1 as a template and utilized docking four ligands to the protein. The calculated dynamic properties indicate that those residues with hydrophobic side chains in the active site, particularly Trp291 and Phe489, are important for CYP7B1 selectivity and have a significant impact on the active-site architecture and substrate binding. The hydrophobic residues, including the Phe cluster, that lie above the active site collectively form a hydrophobic cavity and provide hydrophobic interactions to stabilize the substrate. Two major access channels are presented for the entry of substrates with dissimilar chain lengths, as proposed on the basis of fatty acid binding to CYP2E1. Moreover, the Phe cluster, particularly Phe489, is proposed mainly to merge the active site with the adjacent channel to the surface and accommodate substrate binding in a reasonable orientation.

ACKNOWLEDGMENTS

This work is supported by Natural Science Foundation of China and Specialized Research Fund for the Doctoral Program of Higher Education (Grant Nos. 21273095, 20903045, 21203072, and 20070183046).

REFERENCES

Adcock, S. A., & McCammon, J. A. (2006). Molecular dynamics: Survey of methods for simulating the activity of proteins. *Chemical Reviews, 106*(5), 1589–1615.
Agha-Amiri, K., & Klein, A. (1993). Nucleotide sequence of a gene encoding a histone-like protein in the archaeon Methanococcus voltae. *Nucleic Acids Research, 21*(6), 1491.
Alder, B., & Wainwright, T. (1959). Studies in molecular dynamics. I. General method. *The Journal of Chemical Physics, 31*(2), 459–466.
Andrea, B., Ottavia, S., Roberto, C., Ivana, A., Paola, F., Simone, C., et al. (2011). Hydration studies on the archaeal protein Sso7d using NMR measurements and MD simulations. *BMC Structural Biology, 11*, 44–53.

Arnott, S., & Hukins, D. (1972). Optimised parameters for A-DNA and B-DNA. *Biochemical and Biophysical Research Communications*, 47(6), 1504–1509.

Bahadur, R. P., Kannan, S., & Zacharias, M. (2009). Binding of the bacteriophage P22 N-peptide to the boxB RNA motif studied by molecular dynamics simulations. *Biophysical Journal*, 97(12), 3139–3149.

Baptista, A. M., Martel, P. J., & Petersen, S. B. (1998). Simulation of protein conformational freedom as a function of pH: Constant-pH molecular dynamics using implicit titration. *Proteins: Structure, Function, and Bioinformatics*, 27(4), 523–544.

Baptista, A. M., Teixeira, V. H., & Soares, C. M. (2002). Constant-pH molecular dynamics using stochastic titration. *The Journal of Chemical Physics*, 117, 4184–4200.

Barbosa, R. M. R., Furtado, A., Regis, L., & Leal, W. S. (2010). Evaluation of an oviposition-stimulating kairomone for the yellow fever mosquito, *Aedes aegypti*, in Recife, Brazil. *Journal of Vector Ecology*, 35(1), 204–207.

Bashford, D., & Case, D. A. (2000). Generalized Born models of macromolecular solvation effects. *Annual Review of Physical Chemistry*, 51(1), 129–152.

Baumann, H., Knapp, S., Lundbäck, T., Ladenstein, R., & Härd, T. (1994). Solution structure and DNA-binding properties of a thermostable protein from the archaeon *Sulfolobus solfataricus*. *Nature Structural & Molecular Biology*, 1(11), 808–819.

Bell, S. D., & White, M. F. (2010). Archaeal chromatin organization. *Bacterial Chromatin*, 205–217.

Berg, J. M., Tymoczko, J. L., & Stryer, L. (2008). *Biochemistry* (5th ed.). New York: WH Freeman.

Börjesson, U., & Hünenberger, P. H. (2001). Explicit-solvent molecular dynamics simulation at constant pH: Methodology and application to small amines. *The Journal of Chemical Physics*, 114, 9706–9719.

Börjesson, U., & Hünenberger, P. H. (2004). pH-dependent stability of a decalysine α-helix studied by explicit-solvent molecular dynamics simulations at constant pH. *The Journal of Physical Chemistry. B*, 108(35), 13551–13559.

Bork, P., & Koonin, E. V. (1998). Predicting functions from protein sequences—Where are the bottlenecks? *Nature Genetics*, 18(4), 313–318.

Breer, H. (1997). *Molecular mechanisms of pheromone reception in insect antennae*. New York: Chapman & Hall.

Bueren-Calabuig, J. A., Giraudon, C., Galmarini, C. M., Egly, J. M., & Gago, F. (2011). Temperature-induced melting of double-stranded DNA in the absence and presence of covalently bonded antitumour drugs: Insight from molecular dynamics simulations. *Nucleic Acids Research*, 39(18), 8248–8257.

Bürgi, R., Kollman, P. A., & van Gunsteren, W. F. (2002). Simulating proteins at constant pH: An approach combining molecular dynamics and Monte Carlo simulation. *Proteins: Structure, Function, and Bioinformatics*, 47(4), 469–480.

Case, D. A. (1994). Normal mode analysis of protein dynamics. *Current Opinion in Structural Biology*, 4(2), 285–290.

Case, D., Darden, T., Cheatham, T., III., Simmerling, C., Wang, J., Duke, R., et al. (2010). *AMBER 11*. San Francisco: University of California.

Chalbot, S., & Morfin, R. (2006). Dehydroepiandrosterone metabolites and their interactions in humans. *Drug Metabolism and Drug Interactions*, 22(1), 1–23.

Chandre, F., Darriet, F., Darder, M., Cuany, A., Doannio, J., Pasteur, N., et al. (1998). Pyrethroid resistance in *Culex quinquefasciatus* from West Africa. *Medical and Veterinary Entomology*, 12(4), 359–366.

Cheatham, T. E., III., & Kollman, P. A. (2000). Molecular dynamics simulation of nucleic acids. *Annual Review of Physical Chemistry*, 51(1), 435–471.

Cheatham, T. E., III., Srinivasan, J., Case, D. A., & Kollman, P. A. (1998). Molecular dynamics and continuum solvent studies of the stability of polyG-polyC and polyA–polyT DNA duplexes in solution. *Journal of Biomolecular Structure & Dynamics*, 16(2), 265–280.

Chen, L., Zheng, Q. C., Yu, L. Y., Chu, W. T., Zhang, J. L., Xue, Q., et al. (2012). Insights into the thermal stabilization and conformational transitions of DNA by hyperthermophile protein Sso7d: Molecular dynamics simulations and MM-PBSA analysis. *Journal of Biomolecular Structure & Dynamics*, 30(6), 716–727.

Chen, L., Zheng, Q. C., Zhang, J. L., Yu, L. Y., Chu, W. T., Xue, Q., et al. (2012). Influence of hyperthermophilic protein Cren7 on the stability and conformation of DNA: Insights from molecular dynamics simulation and free energy analysis. *The Journal of Physical Chemistry. B*, 116(41), 12415–12425.

Chiang, C. W., Yeh, H. C., Wang, L. H., & Chan, N. L. (2006). Crystal structure of the human prostacyclin synthase. *Journal of Molecular Biology*, 364(3), 266–274.

Choli, T., Henning, P., Wittmann–Liebold, B., & Reinhardt, R. (1988). Isolation, characterization and microsequence analysis of a small basic methylated DNA-binding protein from the Archaebacterium, *Sulfolobus solfataricus*. *Biochimica et Biophysica Acta*, 950(2), 193–203.

Choli, T., Wittmann–Liebold, B., & Reinhardt, R. (1988). Microsequence analysis of DNA-binding proteins 7a, 7b, and 7e from the archaebacterium *Sulfolobus acidocaldarius*. *The Journal of Biological Chemistry*, 263(15), 7087–7093.

Chothia, C., & Lesk, A. M. (1986). The relation between the divergence of sequence and structure in proteins. *The EMBO Journal*, 5(4), 823.

Chu, W. T., Wu, Y. J., Zhang, J. L., Zheng, Q. C., Chen, L., Xue, Q., et al. (2012). Constant pH molecular dynamics (CpHMD) and mutation studies: Insights into AaegOBP1 pH-induced ligand releasing mechanism. *Biochimica et Biophysica Acta*, 1824(7), 913–918.

Chu, W. T., Zhang, J. L., Zheng, Q. C., Chen, L., Wu, Y. J., Xue, Q., et al. (2012). Constant pH molecular dynamics (CpHMD) and molecular docking studies of CquiOBP1 pH-induced ligand releasing mechanism. *Journal of Molecular Modeling*, 19(3), 1301–1309.

Cremer, D., & Pople, J. (1975). A general definition of ring puckering coordinates. *Journal of the American Chemical Society*, 97(6), 1354–1358.

Cui, Y. L., Zhang, J. L., Zheng, Q. C., Niu, R. J., Xu, Y., Zhang, H. X., et al. (2013). Structural and dynamic basis of human cytochrome P450 7B1: A survey of substrate selectivity and major active site access channels. *Chemistry: A European Journal*, 19(2), 549–557.

Dlugosz, M., & Antosiewicz, J. (2005). Effects of solute-solvent proton exchange on polypeptide chain dynamics: A constant-pH molecular dynamics study. *The Journal of Physical Chemistry. B*, 109(28), 13777–13784.

Dostál, L., Chen, C. Y., Wang, A. H. J., & Welfle, H. (2004). Partial B-to-A DNA transition upon minor groove binding of protein Sac7d monitored by Raman spectroscopy. *Biochemistry*, 43(30), 9600–9609.

Erichsen, A. K., Koht, J., Stray-Pedersen, A., Abdelnoor, M., & Tallaksen, C. M. E. (2009). Prevalence of hereditary ataxia and spastic paraplegia in southeast Norway: A population-based study. *Brain*, 132(6), 1577–1588.

Feng, Y., Yao, H., & Wang, J. (2010). Crystal structure of the crenarchaeal conserved chromatin protein Cren7 and double-stranded DNA complex. *Protein Science*, 19(6), 1253–1257.

Gao, Y. G., Su, S. Y., Robinson, H., Padmanabhan, S., Lim, L., McCrary, B. S., et al. (1998). The crystal structure of the hyperthermophile chromosomal protein Sso7d bound to DNA. *Nature Structural & Molecular Biology*, 5(9), 782–786.

Gera, N., Hussain, M., Wright, R. C., & Rao, B. M. (2011). Highly stable binding proteins derived from the hyperthermophilic Sso7d scaffold. *Journal of Molecular Biology*, 409(4), 601–616.

Gohlke, H., Kiel, C., & Case, D. A. (2003). Insights into protein–protein binding by binding free energy calculation and free energy decomposition for the Ras-Raf and Ras-RalGDS complexes. *Journal of Molecular Biology*, 330(4), 891–914.

Goizet, C., Boukhris, A., Durr, A., Beetz, C., Truchetto, J., Tesson, C., et al. (2009). CYP7B1 mutations in pure and complex forms of hereditary spastic paraplegia type 5. *Brain*, *132*(6), 1589–1600.

Gouda, H., Kuntz, I. D., Case, D. A., & Kollman, P. A. (2002). Free energy calculations for theophylline binding to an RNA aptamer: Comparison of MM-PBSA and thermodynamic integration methods. *Biopolymers*, *68*(1), 16–34.

Gräter, F., de Groot, B. L., Jiang, H., & Grubmüller, H. (2006). Ligand-release pathways in the pheromone-binding protein of *Bombyx mori*. *Structure*, *14*(10), 1567–1576.

Grote, M., Dijk, J., & Reinhardt, R. (1986). Ribosomal and DNA binding proteins of the thermoacidophilic archaebacterium *Sulfolobus acidocaldarius*. *Biochimica et Biophysica Acta*, *873*(3), 405–413.

Guagliardi, A., Napoli, A., Rossi, M., & Ciaramella, M. (1997). Annealing of complementary DNA strands above the melting point of the duplex promoted by an archaeal protein. *Journal of Molecular Biology*, *267*(4), 841–848.

Guo, L., Feng, Y., Zhang, Z., Yao, H., Luo, Y., Wang, J., et al. (2008). Biochemical and structural characterization of Cren7, a novel chromatin protein conserved among Crenarchaea. *Nucleic Acids Research*, *36*(4), 1129–1137.

Harvey, S. C., & Prabhakaran, M. (1986). Ribose puckering: Structure, dynamics, energetics, and the pseudorotation cycle. *Journal of the American Chemical Society*, *108*(20), 6128–6136.

Hou, T., Wang, J., Li, Y., & Wang, W. (2011a). Assessing the performance of the MM/PBSA and MM/GBSA methods: I. The accuracy of binding free energy calculations based on molecular dynamics simulations. *Journal of Chemical Information and Modeling*, *51*(1), 69–82.

Hou, T., Wang, J., Li, Y., & Wang, W. (2011b). Assessing the performance of the molecular mechanics/Poisson Boltzmann surface area and molecular mechanics/generalized Born surface area methods. II. The accuracy of ranking poses generated from docking. *Journal of Computational Chemistry*, *32*(5), 866–877.

Ishida, Y., Chen, A. M., Tsuruda, J. M., Cornel, A. J., Debboun, M., & Leal, W. S. (2004). Intriguing olfactory proteins from the yellow fever mosquito, *Aedes aegypti*. *Naturwissenschaften*, *91*(9), 426–431.

Itoh, T., Takemura, H., Shimoi, K., & Yamamoto, K. (2010). A 3D model of CYP1B1 explains the dominant 4-hydroxylation of estradiol. *Journal of Chemical Information and Modeling*, *50*(6), 1173–1178.

Kaissling, K. E., & Thorson, J. (1980). *Insect olfactory sensilla: Structural, chemical and electrical aspects of the functional organization* (Vol. 261). Amsterdam: Elsevier.

Karplus, M., & McCammon, J. A. (2002). Molecular dynamics simulations of biomolecules. *Nature Structural & Molecular Biology*, *9*(9), 646–652.

Khandogin, J., & Brooks, C. L., III. (2005). Constant pH molecular dynamics with proton tautomerism. *Biophysical Journal*, *89*(1), 141.

Khandogin, J., & Brooks, C. L. (2007). Linking folding with aggregation in Alzheimer's β-amyloid peptides. *Proceedings of the National Academy of Sciences of the United States of America*, *104*(43), 16880–16885.

Khandogin, J., Chen, J., & Brooks, C. L., III. (2006). Exploring atomistic details of pH-dependent peptide folding. *Proceedings of the National Academy of Sciences of the United States of America*, *103*(49), 18546–18550.

Kimura, M., Kimura, J., Davie, P., Reinhardt, R., & Dijk, J. (1984). The amino acid sequence of a small DNA binding protein from the archaebacterium *Sulfolobus solfataricus*. *FEBS Letters*, *176*(1), 176–178.

Krishnamoorthy, N., Gajendrarao, P., Thangapandian, S., Lee, Y., & Lee, K. W. (2010). Probing possible egress channels for multiple ligands in human CYP3A4: A molecular modeling study. *Journal of Molecular Modeling*, *16*(4), 607–614.

Kühnel, K., Ke, N., Cryle, M. J., Sligar, S. G., Schuler, M. A., & Schlichting, I. (2008). Crystal structures of substrate-free and retinoic acid-bound cyanobacterial cytochrome P450 CYP120A1. *Biochemistry, 47*(25), 6552–6559.

Laurence, B. R., & Pickett, J. A. (1982). Erythro-6-acetoxy-5-hexadecanolide, the major component of a mosquito oviposition attractant pheromone. *Journal of the Chemical Society, Chemical Communications, 1*, 59–60.

Leal, W. S., Barbosa, R. M. R., Xu, W., Ishida, Y., Syed, Z., Latte, N., et al. (2008). Reverse and conventional chemical ecology approaches for the development of oviposition attractants for *Culex* mosquitoes. *PLoS One, 3*(8), e3045.

Lee, J., Kim, J. S., & Seok, C. (2010). Cooperativity and specificity of Cys2His2 zinc finger protein-DNA interactions: A molecular dynamics simulation study. *The Journal of Physical Chemistry. B, 114*(22), 7662–7671.

Lee, M. S., Salsbury, F. R., & Brooks, C. L. (2004). Constant-pH molecular dynamics using continuous titration coordinates. *Proteins: Structure, Function, and Bioinformatics, 56*(4), 738–752.

Leite, N. R., Krogh, R., Xu, W., Ishida, Y., Iulek, J., Leal, W. S., et al. (2009). Structure of an odorant-binding protein from the mosquito *Aedes aegypti* suggests a binding pocket covered by a pH-sensitive "Lid" *PLoS One, 4*(11), e8006.

Li, A., & Bigelow, J. C. (2010). The 7-hydroxylation of dehydroepiandrosterone in rat brain. *Steroids, 75*(6), 404–410.

Liu, H., & Yao, X. (2009). Molecular basis of the interaction for an essential subunit PA-PB1 in influenza virus RNA polymerase: Insights from molecular dynamics simulation and free energy calculation. *Molecular Pharmaceutics, 7*(1), 75–85.

Luo, R., David, L., & Gilson, M. K. (2002). Accelerated Poisson–Boltzmann calculations for static and dynamic systems. *Journal of Computational Chemistry, 23*(13), 1244–1253.

Luscombe, N. M., Austin, S. E., Berman, H. M., & Thornton, J. M. (2000). An overview of the structures of protein-DNA complexes. *Genome Biology, 1*(1), reviews001.

Machuqueiro, M., & Baptista, A. M. (2006). Constant-pH molecular dynamics with ionic strength effects: Protonation-conformation coupling in decalysine. *The Journal of Physical Chemistry. B, 110*(6), 2927–2933.

Machuqueiro, M., & Baptista, A. M. (2009). Molecular dynamics at constant pH and reduction potential: Application to cytochrome c 3. *Journal of the American Chemical Society, 131*(35), 12586–12594.

Mao, Y., Xu, X., Xu, W., Ishida, Y., Leal, W. S., Ames, J. B., et al. (2010). Crystal and solution structures of an odorant-binding protein from the southern house mosquito complexed with an oviposition pheromone. *Proceedings of the National Academy of Sciences of the United States of America, 107*(44), 19102–19107.

Martin, C., Bean, R., Rose, K., Habib, F., & Seckl, J. (2001). cyp7b1 catalyses the 7alpha-hydroxylation of dehydroepiandrosterone and 25-hydroxycholesterol in rat prostate. *The Biochemical Journal, 355*(Pt 2), 509.

Mast, N., Graham, S. E., Andersson, U., Bjorkhem, I., Hill, C., Peterson, J., et al. (2005). Cholesterol binding to cytochrome P450 7A1, a key enzyme in bile acid biosynthesis. *Biochemistry, 44*(9), 3259–3271.

Matsuo, T., Sugaya, S., Yasukawa, J., Aigaki, T., & Fuyama, Y. (2007). Odorant-binding proteins OBP57d and OBP57e affect taste perception and host-plant preference in *Drosophila sechellia*. *PLoS Biology, 5*(5), e118.

McAfee, J. G., Edmondson, S. P., Datta, P. K., Shriver, J. W., & Gupta, R. (1995). Gene cloning, expression, and characterization of the Sac7 proteins from the hyperthermophile *Sulfolobus acidocaldarius*. *Biochemistry, 34*(31), 10063–10077.

Meng, Y., & Roitberg, A. E. (2010). Constant pH replica exchange molecular dynamics in biomolecules using a discrete protonation model. *Journal of Chemical Theory and Computation, 6*(4), 1401–1412.

Meng, X. Y., Zheng, Q. C., & Zhang, H. X. (2009). A comparative analysis of binding sites between mouse CYP2C38 and CYP2C39 based on homology modeling, molecular dynamics simulation and docking studies. *Biochimica et Biophysica Acta, 1794*(7), 1066–1072.

Merlino, A., Graziano, G., & Mazzarella, L. (2004). Structural and dynamic effects of α-Helix deletion in Sso7d: Implications for protein thermal stability. *Proteins: Structure, Function, and Bioinformatics, 57*(4), 692–701.

Mertz, J. E., & Pettitt, B. M. (1994). Molecular dynamics at a constant pH. *International Journal of High Performance Computing Applications, 8*(1), 47–53.

Milev, S., Gorfe, A. A., Karshikoff, A., Clubb, R. T., Bosshard, H. R., & Jelesarov, I. (2003). Energetics of sequence-specific protein-DNA association: Binding of integrase Tn916 to its target DNA. *Biochemistry, 42*(12), 3481–3491.

Mongan, J., Case, D. A., & McCammon, J. A. (2004). Constant pH molecular dynamics in generalized Born implicit solvent. *Journal of Computational Chemistry, 25*(16), 2038–2048.

Moraitakis, G., Purkiss, A. G., & Goodfellow, J. M. (2003). Simulated dynamics and biological macromolecules. *Reports on Progress in Physics, 66*(3), 383.

Myant, N., & Mitropoulos, K. (1977). Cholesterol 7 alpha-hydroxylase. *Journal of Lipid Research, 18*(2), 135–153.

Napoli, A., Zivanovic, Y., Bocs, C., Buhler, C., Forterre, P., & Ciaramella, M. (2002). DNA bending, compaction and negative supercoiling by the architectural protein Sso7d of Sulfolobus solfataricus. *Nucleic Acids Research, 30*(12), 2656–2662.

Norberg, J., & Nilsson, L. (2002). Molecular dynamics applied to nucleic acids. *Accounts of Chemical Research, 35*(6), 465–472.

Noshiro, M., Nishimoto, M., Morohashi, K., & Okuda, K. (1989). Molecular cloning of cDNA for cholesterol 7α-hydroxylase from rat liver microsomes: Nucleotide sequence and expression. *FEBS Letters, 257*(1), 97–100.

Noshiro, M., & Okuda, K. (1990). Molecular cloning and sequence analysis of cDNA encoding human cholesterol 7α-hydroxylase. *FEBS Letters, 268*(1), 137–140.

Onufriev, A., Bashford, D., & David, A. (2000). Modification of the generalized Born model suitable for macromolecules. *The Journal of Physical Chemistry. B, 104*(15), 3712–3720.

Pelletier, J., Guidolin, A., Syed, Z., Cornel, A. J., & Leal, W. S. (2010). Knockdown of a mosquito odorant-binding protein involved in the sensitive detection of oviposition attractants. *Journal of Chemical Ecology, 36*(3), 245–248.

Pelosi, P., & Maida, R. (1995). Odorant-binding proteins in insects. *Comparative Biochemistry and Physiology. Part B, Biochemistry & Molecular Biology, 111*(3), 503–514.

Pettersson, H., Lundqvist, J., & Norlin, M. (2010). Effects of CYP7B1-mediated catalysis on estrogen receptor activation. *Biochimica et Biophysica Acta, 1801*(9), 1090–1097.

Porubsky, P. R., Battaile, K. P., & Scott, E. E. (2010). Human cytochrome P450 2E1 structures with fatty acid analogs reveal a previously unobserved binding mode. *The Journal of Biological Chemistry, 285*(29), 22282–22290.

Priyakumar, U. D., Harika, G., & Suresh, G. (2010). Molecular simulations on the thermal stabilization of DNA by hyperthermophilic chromatin protein Sac7d, and associated conformational transitions. *The Journal of Physical Chemistry. B, 114*(49), 16548–16557.

Priyakumar, U. D., Ramakrishna, S., Nagarjuna, K., & Reddy, S. K. (2010). Structural and energetic determinants of thermal stability and hierarchical unfolding pathways of hyperthermophilic proteins, Sac7d and Sso7d. *The Journal of Physical Chemistry. B, 114*(4), 1707–1718.

Rose, K. A., Stapleton, G., Dott, K., Kieny, M. P., Best, R., Schwarz, M., et al. (1997). Cyp7b, a novel brain cytochrome P450, catalyzes the synthesis of neurosteroids 7α-hydroxy dehydroepiandrosterone and 7α-hydroxy pregnenolone. *Proceedings of the National Academy of Sciences of the United States of America, 94*(10), 4925–4930.

Saenger, W. (1984). *Principles of nucleic acid structure* (Vol. 7). New York: Springer-Verlag.

Sandman, K., & Reeve, J. N. (2005). Archaeal chromatin proteins: Different structures but common function? *Current Opinion in Microbiology, 8*(6), 656–661.

Setchell, K., Schwarz, M., O'Connell, N. C., Lund, E. G., Davis, D. L., Lathe, R., et al. (1998). Identification of a new inborn error in bile acid synthesis: Mutation of the oxysterol 7alpha-hydroxylase gene causes severe neonatal liver disease. *The Journal of Clinical Investigation, 102*(9), 1690–1703.

Sitkoff, D., Sharp, K. A., & Honig, B. (1994). Accurate calculation of hydration free energies using macroscopic solvent models. *The Journal of Physical Chemistry, 98*(7), 1978–1988.

Srinivasan, J., Cheatham, T. E., III., Cieplak, P., Kollman, P. A., & David, A. (1998). Continuum solvent studies of the stability of DNA, RNA, and phosphoramidate-DNA helices. *Journal of the American Chemical Society, 120*(37), 9401–9409.

Steinbrecht, R. A. (1996). Are odorant-binding proteins involved in odorant discrimination? *Chemical Senses, 21*(6), 719–727.

Stiles, A. R., McDonald, J. G., Bauman, D. R., & Russell, D. W. (2009). CYP7B1: One cytochrome P450, two human genetic diseases, and multiple physiological functions. *The Journal of Biological Chemistry, 284*(42), 28485–28489.

Swanson, J. M. J., Henchman, R. H., & McCammon, J. A. (2004). Revisiting free energy calculations: A theoretical connection to MM/PBSA and direct calculation of the association free energy. *Biophysical Journal, 86*(1), 67–74.

Tang, W., Eggertsen, G., Chiang, J. Y. L., & Norlin, M. (2006). Estrogen-mediated regulation of CYP7B1: A possible role for controlling DHEA levels in human tissues. *The Journal of Steroid Biochemistry and Molecular Biology, 100*(1), 42–51.

Todorova, R., & Atanasov, B. (2004). The role of the salt concentration, proton, and phosphate binding on the thermal stability of wild and cloned DNA-binding protein Sso7d from *Sulfolobus solfataricus. International Journal of Biological Macromolecules, 34*(1), 135–147.

Tolstorukov, M. Y., Jernigan, R. L., & Zhurkin, V. B. (2004). Protein-DNA hydrophobic recognition in the minor groove is facilitated by sugar switching. *Journal of Molecular Biology, 337*(1), 65–76.

Tsaousidou, M. K., Ouahchi, K., Warner, T. T., Yang, Y., Simpson, M. A., Laing, N. G., et al. (2008). Sequence alterations within CYP7B1 implicate defective cholesterol homeostasis in motor-neuron degeneration. *American Journal of Human Genetics, 82*(2), 510–515.

Vogt, R. (1987). *The molecular basis of pheromone reception: Its influence on behavior. Pheromone biochemistry.* Orlando, FL: Academic Press, pp. 385–431.

Vogt, R., Callahan, F., Rogers, M., & Dickens, J. (1999). Odorant binding protein diversity and distribution among the insect orders, as indicated by LAP, an OBP-related protein of the true bug Lygus lineolaris (Hemiptera, Heteroptera). *Chemical Senses, 24*(5), 481–495.

Vogt, R. G., & Riddiford, L. M. (1981). Pheromone binding and inactivation by moth antennae. *Nature, 293*, 161–163.

Walczak, A. M., & Antosiewicz, J. M. (2002). Langevin dynamics of proteins at constant pH. *Physical Review E, 66*(5), 051911.

Williams, S. L., De Oliveira, C. A. F., & McCammon, J. A. (2010). Coupling constant pH molecular dynamics with accelerated molecular dynamics. *Journal of Chemical Theory and Computation, 6*(2), 560–568.

Wogulis, M., Morgan, T., Ishida, Y., Leal, W. S., & Wilson, D. K. (2006). The crystal structure of an odorant binding protein from *Anopheles gambiae*: Evidence for a common ligand release mechanism. *Biochemical and Biophysical Research Communications, 339*(1), 157–164.

Xu, X., Su, J., Chen, W., & Wang, C. (2011). Thermal stability and unfolding pathways of Sso7d and its mutant F31A: Insight from molecular dynamics simulation. *Journal of Biomolecular Structure & Dynamics, 28*(5), 717–727.

Yamashita, F., Feng, C., Yoshida, S., Itoh, T., & Hashida, M. (2011). Automated information extraction and structure–activity relationship analysis of cytochrome P450 substrates.
Journal of Chemical Information and Modeling, 51(2), 378–385.

Yang, B., Zhu, Y., Wang, Y., & Chen, G. (2010). Interaction identification of Zif268 and
TATAZF proteins with GC-/AT-rich DNA sequence: A theoretical study. *Journal of
Computational Chemistry, 32*(3), 416–428.

Yau, J., Rasmuson, S., Andrew, R., Graham, M., Noble, J., Olsson, T., et al. (2003).
Dehydroepiandrosterone 7-hydroxylase cyp7b: Predominant expression in primate
hippocampus and reduced expression in Alzheimer's disease. *Neuroscience, 121*(2),
307–314.

Zhang, Z., Gong, Y., Guo, L., Jiang, T., & Huang, L. (2010). Structural insights into the
interaction of the crenarchaeal chromatin protein Cren7 with DNA. *Molecular Microbiology, 76*(3), 749–759.

Zhou, J. J., He, X. L., Pickett, J., & Field, L. (2008). Identification of odorant-binding proteins of the yellow fever mosquito *Aedes aegypti*: Genome annotation and comparative
analyses. *Insect Molecular Biology, 17*(2), 147–163.

CHAPTER SIX

Protein Functional Dynamics in Multiple Timescales as Studied by NMR Spectroscopy

Gabriel Ortega*, Miquel Pons†, Oscar Millet*,1
*Structural Biology Unit, CIC bioGUNE, Derio, Spain
†Biomolecular NMR Laboratory, Organic Chemistry Department, University of Barcelona, Barcelona, Spain
¹Corresponding author: e-mail address: omillet@cicbiogune.es

Contents

Abstract

Protein functional dynamics are defined as the atomic thermal fluctuations or the segmental motions that are essential for the function of the biomolecule. NMR is a very versatile technique that allows obtaining quantitative information from these processes at atomic resolution. This review is focused on the use of ^{15}N spin relaxation methods to study functional dynamics although the connections with other NMR methods and biophysical techniques will be briefly mentioned. In the first part of the chapter, methodological aspects will be considered, while a set of selected cases will be described in more detail in the second part.

1. FUNCTIONAL DYNAMICS AND NMR

1.1. Biomolecular motions: From thermal noise to functional dynamics

It is commonly accepted that a static view of a protein, represented by the time–averaged structure, is not a proper descriptor since, when not at absolute zero, proteins do not just adopt one structure but are continuously sampling a plethora of conformations. Understanding the bases of such dynamic processes is important not only for a better description of the protein entity in itself but also for the profound implications that protein plasticity has for the function of the molecule.

Defining *protein dynamics* as the changes in the protein's conformation over time stresses the need to differentiate motional events that may occur at multiple timescales. Often, a correlation exists between the timescale of the motion and the number of atoms involved (Fig. 6.1). Fastest librations occur at the bond geometry and they can be detected by laser spectroscopy at the femtosecond timescale (Silva, Murkin, & Schramm, 2011), time-resolved X-ray crystallography (Moffat, 2001), and NMR spectroscopy at the picosecond timescale (Abu-Abed, Millet, MacLennan, & Ikura, 2004;

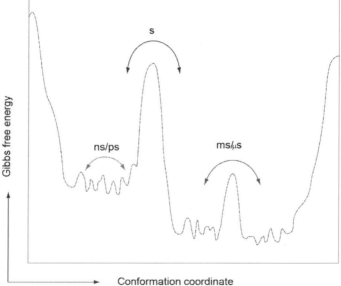

Figure 6.1 A hierarchy of conformations. Due to their intrinsic flexibility, proteins sample a range of thermodynamically accessible conformations within a hierarchy of timescales. The population of each state is determined by its Gibbs free energy, while the interconversion between states is owed to the energy barriers that separate them. Three examples of the continuum of timescales are depicted in the figure for clarity: fast bond librations occurring in the picosecond (ps) to nanosecond (ns) timescale, segmental motions involving a larger number of atoms occurring in the microsecond (μs)–millisecond (ms) timescale, and large domain rearrangements occurring in slow timescales (s).

Palmer, 2001) among other techniques. Considering the range of molecular weights normally found in proteins (between 5 kDa and 1 MDa), protein tumbling in solution is associated to correlation times that fall in the nanosecond timescale, where many side chain motions and segmental reorientations of the protein also occur. Such motions involve very small activation energies and many states will be accessible at room temperature. Slower motions are associated with larger activation energies between the exchanging conformations (Fig. 6.2), typically as a result of steric hindrance or by the existence of some stabilizing interactions. Examples of these restricted motions in proteins are the tryptophan ring flips (Gall, Cross, DiVerdi, & Opella, 1982), disulfide bond rearrangements (Rosengren, Daly, Plan, Waine, & Craik, 2003), and proline *cis–trans* isomerization (Wedemeyer, Welker, & Scheraga, 2002). Finally, on much slower timescales (second or slower) sometimes reside the process of protein folding/unfolding, allosteric motions, and enzyme catalysis that can also be monitored by mass spectrometry (Eyles & Kaltashov, 2004) and molecular dynamics (Karplus, 2003), among other techniques.

When considering the amplitude of motion, changes in the conformational space are also intrinsically related to the number of atoms involved (Frauenfelder, Sligar, & Wolynes, 1991). Computational methods based on principal component analysis and coarse-grained simulations can predict and characterize collective motions for large proteins at relatively large timescales (Bahar, Lezon, Yang, & Eyal, 2010; Bahar & Rader, 2005). Structural biology techniques can provide information about the instances of the motional trajectory, complementing the information obtained from molecular dynamic simulations (Adcock & McCammon, 2006). For folded proteins, segmental motions can be divided into hinge motions and shear motions depending on the orientation of the main axis of motion (Gerstein, Lesk, & Chothia, 1994) although this is a rather limited view, since motions found in nature are complex and often need multiple axes for a proper description. For instance, disorder-to-order and order-to-disorder transitions observed after ligand binding showed no clear evidence of correlated motion between the implicated atoms (Homans, 2005). This is always the case in intrinsically disordered proteins, where flexibility is consubstantial and the amplitudes of such dynamics are often governed by the physical laws of polymers (Tompa et al., 2009).

Protein dynamics affect and/or regulate many processes such as signaling/regulation (Smock & Gierasch, 2009), sensing intracellular ion concentration (Millet, Bernado, Garcia, Rizo, & Pons, 2002), protein folding

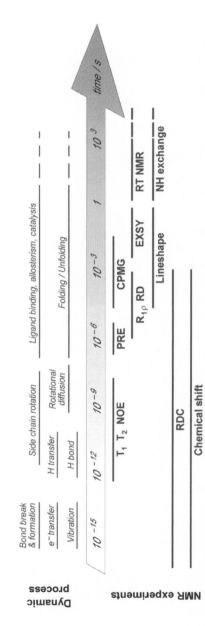

Figure 6.2 Timescales involved in protein dynamics and NMR observable. Spin relaxation experiment covers most of the timescales where the effective protein internal (and overall) motions take place. Faster librations can be detected by nuclear spin relaxation measurements (R_1, R_2, NOE) or by analyzing the dynamic contribution of the residual dipolar couplings (RDC). The line-shape analysis and the relaxation dispersion experiments account for motions involving a larger number of atoms and are most relevant in biology. The slower motions can be characterized in NMR using EXSY and ZZ exchange experiments. (For color version of this figure, the reader is referred to the online version of this chapter.)

(Neudecker, Lundstrom, & Kay, 2009), solvent–induced conformational rearrangements (Ortega et al., 2011), and protein thermostability (Fayos, Pons, & Millet, 2005; Tadeo, Lopez-Mendez, Castano, Trigueros, & Millet, 2009). Enzymes also constitute striking examples of functionally coupled dynamics since they are dynamic machines: conformational changes are often required to recognize the substrate and to generate the Michaelis complex, as well as to accelerate the chemical reaction and to decrease the affinity for the product (Bosco et al., 2010; Popovych, Sun, Ebright, & Kalodimos, 2006). Enzymes and proteins can undergo excursions to lowly populated excited states (Korzhnev & Kay, 2008), favoring conformational selection versus induced–fit mechanisms (Velyvis, Yang, Schachman, & Kay, 2007).

Albeit its importance, protein motions are difficult to be characterized experimentally. As a consequence of the high versatility of the technique, NMR is a particularly well–suited technique to study protein dynamics (Mittermaier & Kay, 2006), and a plethora of NMR observables can provide information about motions in a broad range of timescales with atomic resolution (Kleckner & Foster, 2011). Thus, not surprisingly, a continuous effort over years has been made to develop new experiments and labeling strategies (Gardner & Kay, 1998; Sprangers, Velyvis, & Kay, 2007) for the study of protein dynamics in solution (Sheppard, Sprangers, & Tugarinov, 2010).

1.2. NMR experiments for the characterization of protein motions

The aim of this section is to briefly describe the standard methods used for the NMR characterization of protein motions in different timescales. Only the most accepted methods (restricted to amide backbone relaxation) are discussed here. Experimental details of the experiments are preferentially discussed and the section is most intended to be a user guide.

1.2.1 Relaxation events in the ps–ns timescale

For a nucleus of spin $1/2$, there are two independent relaxation observables: the longitudinal (R_1) and the transversal (R_2) relaxation rates, describing the return of the magnetization to the equilibrium after the pulse perturbation. Specifically, R_1 is defined as the return rate of magnetization parallel to the magnetic field, being responsible for the restitution of the Boltzmann populations. It is determined using an inversion–recovery experiment, where R_1 is obtained from the exponential fitting of the peak intensity versus the recovery time. In proteins, backbone relaxation is typically collected from the amide ^{15}N relaxation (Kay, Torchia, & Bax, 1989). Due to spectral complexity, a bidimensional experiment based on a ^{1}H-^{15}N-HSQC scheme is used and the intensity is obtained from the peak volume integration (Fig. 6.3). A train of pulses is applied during the recovery delay to avoid interference from the ^{1}H-^{15}N dipole–dipole cross correlation. In theory, the intensity at long recovery times should result in a nonzero final value (proportional to the equilibrium magnetization), but this contribution is cancelled out by the phase cycle and the intensity exponentially decays to zero with the increasing recovery time.

The transversal relaxation rate monitors the loss of the phase coherence between spins. For amide groups in proteins, the R_2 rate is obtained using a very similar pulse sequence as for R_1, but with the inversion-recovery period substituted by a Carr–Purcell–Meiboom–Gill (CPMG) train of pulses (Fig. 6.3). To minimize overheating effects on the sample, pulses during the CPMG are usually softer than the pulses used in the preparation period. This will result in lower effective B_1 fields, and peaks at large offsets from the carrier may undergo incomplete inversion during the CPMG. An alternative experiment to be used is the $R_{1\rho}$, where a spin-lock substitutes the CPMG train. The pure transversal relaxation rate is related to the experimental $R_{1\rho}$ by the following expression:

$$R_{1\rho} = R_1 \cos^2\theta + R_2 \sin^2\theta$$

where $\theta = \mathrm{arccot}(\Omega/v)$, v is the ^{15}N spin-lock field strength and Ω is the resonance offset from the spin-lock carrier. When using the $R_{1\rho}$ experiment, the phase of the continuous wave (CW) field is randomly altered to avoid the introduction of artifacts due to cross correlation (Korzhnev, Skrynnikov, Millet, Torchia, & Kay, 2002).

When also considering the amide proton directly bound to the nitrogen, the R_1 and R_2 rates are normally complemented with the heteronuclear ^{1}H-^{15}N-NOE, which contains trivial structural information but effectively reports on the dynamics at the same timescale. The heteronuclear NOE is measured according to the pulse sequence also shown in Fig. 6.3. Typically, a train of 120° pulses is used for the proton saturation and the experiment is repeated with and without NOE enhancement. For this experiment, it is very important that the equilibrium magnetization is completely restored before pulsing again (i.e., $\geq 5\cdot(R_1)^{-1}$), so delays between scans must be very long (≥ 10s).

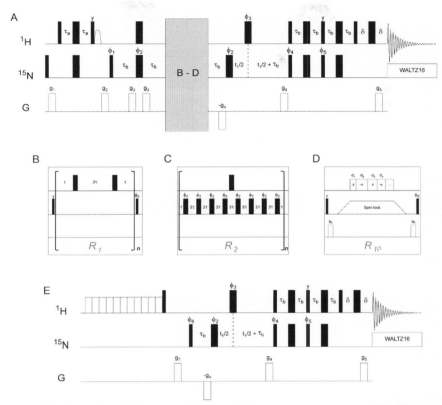

Figure 6.3 Pulse sequences for the measurement of the R_1, R_2, $R_{1\rho}$, and ^{1}H-^{15}N-NOE relaxation rates. The basic pulse sequence shown in (A) is used for the determination of the relaxation rates for R_1 (module B), R_2 (module C), and $R_{1\rho}$ (module D). The heteronuclear NOE experiment is measured using the pulse sequence shown in (E). The incremental time units are $4 \cdot \tau \cdot n$ for R_1, $16 \cdot \tau \cdot n$ for R_2, and τ for the $R_{1\rho}$ experiment. Narrow and wide solid bars represent the $\pi/2$ and π pulses, respectively, applied with an x phase unless otherwise indicated. For the $R_{1\rho}$ pulse sequence, adiabatic pulses (ramps in the trapezoidal figure) are used to transfer magnetization from the z-axis to the spin-lock field. In the $R_{1\rho}$ experiment, cross-correlation suppression of the cross correlation is achieved by a ^{1}H CW field, with random phase alternation (between x and $-x$, with d_i of about 10 ms on average). A 1.5-kHz WALTZ16 decoupling train of pulses is used during acquisition. Values for the delays are $\tau_a < 1/4J_{NH}$ (typically 2.2 ms), $\tau_b = 1/4J_{NH}$ (2.7 ms), and $d = 1.3$ ms. The spin lock is applied along the x-axis during a total time $\tau + 2\eta$. The selective pulse in the proton channel has a sine shape (1 ms), centered on the water position. Proton saturation is achieved with a train of 120° pulses (5 kHz, total duration of 5 s) prior to nitrogen excitation (E). The phase cycle employed is $\phi_1 = 4(x)$, $4(-x)$ for the R_1 experiment and $\phi_1 = x$, $-x$ for the R_2 experiment; $\phi_2 = x$ for the R_1 and NOE experiments and $\phi_2 = 4(x)$, $4(-x)$ for the R_2 experiment; $\phi_3 = 2(x)$, $2(-x)$; $\phi_4 = 2(x)$, $2(-x)$; $\phi_5 = 2(y)$, $2(-y)$; $\phi_6 = y$, $-y$; and $\phi_{rec} = x$, $-x$; $-x$, x; $-x$, x; x, $-x$ for the R_1 and NOE, while $\phi_{rec} = x$, $-x$; $-x$, x for the R_2. Quadrature detection is achieved using the sensitivity enhancement approach using g4 and g5 for coherence selection.

The phenomenological relaxation rates are connected to the intramolecular dynamics via the relaxation mechanisms with the lattice. For the nitrogen backbone nucleus, proton–nitrogen dipole–dipole and ^{15}N chemical shift anisotropy are the main relaxation mechanisms acting, so the relaxation rates can be related to the spectral density function according to the following expressions:

$$R_1 = d[J(\omega_H - \omega_N) + 3J(\omega_N) + 6J(\omega_H + \omega_N)] + cJ(\omega_N)$$

$$R_2 = \frac{d}{2}[4J(0) + J(\omega_H - \omega_N) + 3J(\omega_N) + 6J(\omega_H) + 6J(\omega_H + \omega_N)]$$

$$+ \frac{c}{6}[4J(0) + 3J(\omega_N)]$$

$$\text{NOE} = 1 + \frac{\gamma_H}{\gamma_N}\frac{d}{R_1}[6J(\omega_H + \omega_N) - J(\omega_H - \omega_N)]$$

where d corresponds to the prefactor for the dipolar mechanism:

$$d = \frac{1}{4}\left(\frac{\mu_0}{4\pi}\right)\frac{\hbar^2\gamma_N^2\gamma_H^2}{r_{HN}^6}$$

and c is equivalent to the prefactor for the chemical shift anisotropy mechanism:

$$c = \frac{(\omega_N \Delta\omega)^2}{3}$$

and γ_N and γ_H are the gyromagnetic ratios, r_{NH} is the vibrationally averaged H—N bond length, $\Delta\omega$ is the nitrogen CSA, and ω_N is the Larmor frequency of ^{15}N (in rad^{-1}). These equations have been obtained from the Bloch–Wangsness–Redfield theory (Redfield, 1978) and link the phenomenological observable (R_1, R_2, and NOE) with the molecular motions at discrete frequencies: 0, ω_N, ω_H, $\omega_H + \omega_N$, ω_H, and $\omega_H - \omega_N$. Notice that the gyromagnetic ratio of nitrogen is negative and therefore $\omega_H + \omega_N < \omega_H < \omega_H - \omega_N$. More information on the spectral density function can be obtained by increasing the number of sampled frequencies. Experimentally, this can be done by combining relaxation data at different magnetic fields or by including new independent observable. For a proper comparison of data from different magnets, special care must be taken to ensure a proper temperature calibration in all magnets (Orekhov, Pervushin, Korzhnev, & Arseniev, 1995).

The relationship between bond vector motions and the spin relaxation data is far from trivial. Among the different analyses proposed in the literature, the spectral density mapping (SDM) is the simplest one because it does not introduce any model for the interpretation of the data and, therefore, it is unbiased (Peng & Wagner, 1995). In particular, it is applicable to characterize the complex dynamics of intrinsically disordered proteins. Analysis of SDM of globular proteins is straightforward since the shape of the function allows discriminating between motions that are faster, slower, and of similar timescale than the overall tumbling (Krizova, Zidek, Stone, Novotny, & Sklenar, 2004; Lefevre, Dayie, Peng, & Wagner, 1996). In case a large number of relaxation rates are available, the analytical solution for the equation system becomes accessible and the value of the spectral density function can be obtained at all the frequencies under consideration. When a single magnetic field is used, a data set of at least six independent relaxation observables is required. However, when the number of experimental data is limited, an approximated analysis (reduced SDM) is also possible (Farrow, Zhang, Szabo, Torchia, & Kay, 1995). Reduced SDM makes use of the quasi-constant value that the spectral density values has at high frequencies and assumes that $J(\omega_H + \omega_N) \approx J(\omega_H) \approx J(\omega_H - \omega_N) \approx J(0.87\omega_H)$. Any case, if the SDM approach is to be considered, it is always very convenient to collect data at multiple fields since the increase in the number of variables is usually lower than the increase in the number of experimental rates.

As an alternative to the SDM, data can be fitted to a model where a specific shape for the motional correlation function is given. In this context, the model-free formalism is the preferred approach for spin relaxation data analysis of globular proteins. In the model-free formalism (Lipari & Szabo, 1982), a reduced expression describing the autocorrelation of a vector non-rigidly anchored to the surface of a rigid object is assumed. The only assumption is the independence between the overall tumbling and local motions affecting the vector. This minimalistic description depends on a small number of parameters. The values of these parameters are obtained from the relaxation data. In the original implementation, apart from the global correlation time (τ_c), two more motional parameters were considered: the generalized order parameter (S^2) and the correlation time τ_e, which account for the amplitude and the timescale of the internal motions, respectively (Table 6.1). Extensions of the model have also included new parameters to account for local motions in other (slower) timescales (Clore et al., 1990; d'Auvergne & Gooley, 2008a, 2008b). The spectral density function is obtained as the Fourier transform of the correlation function, and the

Table 6.1 The model-free formalism

Variant[a]	Spectral density function	Fitted parameters[b]
LS-2	$J(\omega) = \frac{S^2 \cdot \tau_c}{1+\omega^2 \cdot \tau_c^2} + (1-S^2)\frac{\tau'}{1+\omega^2 \cdot \tau'^2}$ $\tau'^{-1} = \tau_c^{-1} + \tau_e^{-1}$	S^2, τ_e
LS-3	$J(\omega) = \frac{S^2 \cdot \tau_i}{1+\omega^2 \cdot \tau_i^2} + (1-S^2)\frac{\tau'}{1+\omega^2 \cdot \tau'^2}$ $\tau'^{-1} = \tau_i^{-1} + \tau_e^{-1}$	S^2, τ_e, τ_i
LS-4	$J(\omega) = \frac{S_f^2 \cdot S_s^2 \cdot \tau_c}{1+\omega^2 \cdot \tau_c^2} + \left(1-S_f^2\right)\frac{\tau'_f}{1+\omega^2 \cdot \tau'_f} + S_f^2\left(1-S_s^2\right)\frac{\tau'_s}{1+\omega^2 \cdot \tau'^2_s}$ $\tau'^{-1}_f = \tau_c^{-1} + \tau_f^{-1}$ $\tau'^{-1}_s = \tau_c^{-1} + \tau_s^{-1}$	$S_f^2, S_s^2, \tau_f, \tau_s$

Motional parameters and expressions for the spectral density function.
[a]Number of fitted parameters in the model: LS-2, LS-3, and LS-4 fit 2, 3, and 4 parameters, respectively.
[b]S^2 is the generalized order parameter; S_f^2 and S_s^2 are the specific order parameters accounting for the fast and slow motions, respectively; τ_c is the overall correlation time; τ_e is the timescale of the internal motions; and τ_f and τ_s account for the timescales of the fast and slow internal motions, respectively.

relaxation rates are calculated as a function of the values of the parameters. A minimization algorithm provides the optimal values of the parameters that fit the experimental results, assuming a given form of the autocorrelation function. A statistical test is used to select the simplest model that explains the observed relaxation rates. The best model for the local motion component may vary between residues, but the description of the global motion should be common, if the assumption of a rigid object is correct. The hydrodynamic motion of rigid objects can be accurately simulated and, for rigid proteins of known structure, a comparison between experimental relaxation rates and the hydrodynamic predictions is a good check of the initial assumptions (Bernado, Garcia de la Torre, & Pons, 2002; Garcia de la Torre, Huertas, & Carrasco, 2000).

1.2.2 Relaxation events in the μs–ms timescale

NMR is a technique especially well suited for the study of molecular motions in the microsecond–millisecond timescale (Mittermaier & Kay, 2006). Random variations in the spin environment result in loss of coherence analogous to that resulting from spin–spin relaxation. Thus, exchange processes in these timescales result in an additional contribution to transverse relaxation and line broadening.

$$R_2 = R_2^0 + R_{ex}$$

where R_2^0 is the line width in the absence of exchange and R_{ex} accounts for the exchange contribution of the process. Thus, disentangling the two components of the line shape can provide a lot of kinetic, thermodynamic, and structural information of the motional event. Depending on the timescale of the motions under study, two experiments are typically used to that end: the Carr–Purcell–Meiboom–Gill relaxation dispersion (CPMG-RD) experiment (for slow-intermediate processes with k_{ex} between 40 and 4000 s^{-1}) and the rotating frame relaxation dispersion (RF-RD) experiment (for fast processes with k_{ex} between 10,000 and 50,000 s^{-1}) (Loria, Berlow, & Watt, 2008).

The CPMG-RD experiment is based on the refocusing of the exchange-induced line broadening after the application of spin–echo elements to transverse magnetization during a fixed delay period (Fig. 6.4). In the absence of exchange, the magnetization is fully recovered after a CPMG element of duration τ–$180°$–τ, except for the loss due to transversal relaxation (which is of the order of $\approx \exp(-2\tau \cdot R_2)$). More important, in the absence of motional processes, the signal recovered is totally independent of the number of $180°$ pulses inserted in the period 2τ. However, when conformational chemical exchange is active, there is a certain probability for the magnetization to exchange between two different chemical shifts during the dephasing and refocusing delays of the CPMG. Because of the stochastic nature of the exchange process, the exchanged signal will not be refocused back, resulting in a neat loss of intensity at the end of the CPMG. The key element is that the intensity drop is proportional to the total effective time that the magnetization has been under free precession, ultimately determined by the number of π pulses included in the relaxation delay (Fig. 6.4). A relaxation dispersion profile is obtained by measuring the signal intensity as a function of the number of π pulses included in a fixed time CPMG module (expressed as the effective field in Hz, ν_{CPMG}).

For amide groups in proteins, the CPMG-RD experiment is measured using a similar strategy as for the transversal relaxation rate, with the pulse sequence shown in Fig. 6.4 (Loria, Rance, & Palmer, 1999). After the preparation period, a mixture of in-phase ($N_{x,y}$) and antiphase ($N_{x,y}H_z$) transversal magnetization is generated. The two coherences decay at different rates. Moreover, heteronuclear coupling during the free precession delays interconverts the two and, therefore, the effective transverse relaxation would still depend on the separation between the π pulses, even in the absence of exchange. To overcome this problem, the CPMG period is split into two complementary periods separated by a U-sequence element that interconverts in–phase and antiphase terms. Thus, at the end of the second CPMG period, the in–phase and antiphase magnetizations have been active

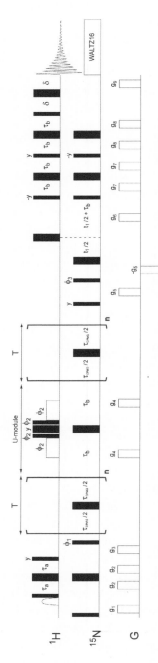

Figure 6.4 Pulse sequence for the relaxation dispersion experiment of backbone amide groups. The dispersion profile can be obtained by altering the number of π pulses in the CPMG module (τ_{CPMG} is the space between two π pulses). The total CPMG time is divided into two periods (each one of duration = T) separated by the U-module that ensures the conversion between the in-phase and antiphase magnetizations. The ^{1}H π pulse in the U-module is of composite nature and it is flanked by two water-selective pulses (square shape, 1.5 ms). A 1.5-kHz WALTZ16 decoupling train of pulses is used during acquisition. The selective pulse in the proton channel has an e-BURP shape (6 ms), centered on the water position. Values for the delays are $\tau_a < 1/4J_{NH}$ (typically 2.2 ms), $\tau_b = 1/4J_{NH}$ (2.7 ms), and $d = 1.3$ ms. The phase cycle employed is $\phi_1 = x, -x; \phi_2 = 2(x), 2(-x); \phi_3 = 4(y), 4(-y);$ and $\phi_{rec} = x, -x; x, -x; -x, x; -x, x.$ Quadrature detection is achieved using the sensitivity enhancement approach using the g6 and g9 for coherence selection.

for the same amount of time regardless of the number of π pulses, thus ensuring a proper baseline in the relaxation dispersion profile.

Assuming that the exchange process occurs between two states (A and B),

$$A \frac{k_A}{k_B} B$$

Carver and Richards' equation provides an analytical solution to the relaxation dispersion profiles. The equation provides the exchange contribution to the line shape (R_{ex}, Fig. 6.5), as a function of the exchange rate k_{ex} ($k_{ex} = k_A + k_B$), the populations of the two states (p_A and p_B), and their chemical shift difference: $\Delta \omega = \omega_A - \omega_B$. Actually, only the magnitude can be obtained from the data fitting, while some elegant additional experiments

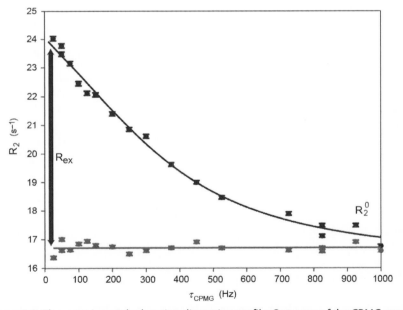

Figure 6.5 The experimental relaxation dispersion profile. Outcome of the CPMG experiment (Fig. 6.4) for two amide groups showing effective (black) and flat (red) relaxation dispersion profiles. The solid lines correspond to the fitting to the Carver–Richards equation (black line) and a linear regression (red line). The limit value at infinite τ_{CPMG} corresponds to the line broadening in the absence of exchange (R_2^0). The difference between the R_2 values when $\tau_{CPMG} \rightarrow \infty$ and when $\tau_{CPMG} \rightarrow 0$ is the exchange contribution to the line broadening (R_{ex}). (For interpretation of the references to color in this figure legend, the reader is referred to the online version of this chapter.)

allow obtaining the sign of the chemical shift change (Korzhnev, Neudecker, Mittermaier, Orekhov, & Kay, 2005; Skrynnikov, Dahlquist, & Kay, 2002). For fast exchange, as compared to the chemical shift differences, the following expression applies:

$$R_{ex} \approx \frac{p_A(1-p_A)\cdot k_{ex}}{1+(k_{ex}/\Delta\omega)^2}$$

This expression has the advantage of simplicity but it is only approximate. On the other hand, combining data from multiple magnetic fields allows obtaining the timescale of the chemical exchange as well as offers a much more robust fitting of the parameters (in particular, the separation between the populations and the chemical shift change) (Millet, Loria, Kroenke, Pons, & Palmer, 2000).

Exchange events in the 20–100 μs time window can be studied with the rotating frame relaxation dispersion. In a similar way to CPMG-RD, RF-RD alters the effective transverse magnetization in the lab frame using spin–lock fields in the range of 1000 to more than 50,000 Hz (Watt, Shimada, Kovrigin, & Loria, 2007). This experimental observable can be measured with the pulse sequence shown in Fig. 6.3. In the presence of chemical exchange ($R_{ex} > 0$), the relaxation in the rotating frame ($R_{1\rho}$) is increased because the nucleus spends time in different chemical environments during the spin–lock period. As the field strength is increased, the effective chemical shift difference between the states is decreased and the exchange contribution is attenuated, in a process analogous to the suppression of exchange contributions during a CPMG period when the separation of the π pulses is decreased. In fact, the effect of the CPMG sequence is often described in terms of the applied effective field. The motional parameters are related to the experimental setup according to the expression:

$$R_{ex} = \frac{p_A p_B \Delta\omega^2 k_{ex}}{k_{ex}^2 + \omega_{eff}^2}$$

where ω_{eff} is the effective field strength of the applied spin lock. Although the experiment depends on the same parameters as for the CPMG-RD, here the analysis is limited because exchange phenomena are always in the fast regime, where the exchange populations and the chemical shift differences become convoluted and cannot be estimated independently in a proper way. Another drawback of the method is that the high spin–lock powers used may produce undesired heating effects of the sample.

2. EXAMPLES OF FUNCTIONAL DYNAMICS

In this section, a few examples are discussed on how NMR relaxation data can be successfully applied for the characterization of proteins dynamics.

2.1. Interdomain orientations in periplasmic binding proteins

Periplasmic binding proteins (PBPs) constitute a large family of receptors that recognize a plethora of small molecules and ions in gram-negative bacteria (Quiocho & Ledvina, 1996). Nutrients like maltose can cross the outer membrane through the homotrimeric maltoporin to reach the periplasm, where it binds to the respective PBP. The binding event produces a conformational change in the protein that activates the interaction with the periplasmic loop regions of the transport complex, allowing the substrate translocation to the cytoplasm (Bordignon, Grote, & Schneider, 2010). Alternatively, the binary complex can bind to the Tar receptor, triggering the bacterial chemotactic cascade (Manson, Armitage, Hoch, & Macnab, 1998).

PBPs represent a paradigm for functional dynamics since binding is associated with a large closure motion that traps the ligand in the cleft between the two protein domains (the so-called *Venus flytrap mechanism*) (Vyas, Vyas, & Quiocho, 1983, 1988). In many cases, structural data and computational studies have provided clear descriptions for the conformational trajectory (Millet, Hudson, & Kay, 2003; Shilton, Flocco, Nilsson, & Mowbray, 1996) and have identified the hinge composition as a crucial element in the closure mechanism (Bermejo, Strub, Ho, & Tjandra, 2010; Borrok, Kiessling, & Forest, 2007).

The *Venus flytrap mechanism* is also currently being tested for the design of better biosensors for monitoring blood glucose levels in diabetic patients, over 100 million people worldwide. The two conformations of the PBP serve as an on/off switch that can be used to detect bound ligand in a reagent-free manner. For that purpose, glucose/galactose-binding protein and ribose-binding protein (RBP) (that also binds glucose) have been engineered using site-directed mutagenesis to incorporate fluorescent or bioluminescent probes that can effectively report for the binding event.

In our laboratory, we have investigated the dynamic properties of these two PBPs using NMR spectroscopy. *Escherichia coli* glucose/galactose-binding protein (GGBP) and *E. coli* RBP have a low sequence identity (only a 24%) but they share a very high homology in the structure (Vyas, Vyas, & Quiocho, 1991), constituted by two globular domains of similar size, connected by a three-strand hinge (Vyas et al., 1983; Vyas, Vyas, & Quiocho, 1987). For each protein, relaxation data (T_1, T_2, and ^1H-^{15}N-NOE) were measured for the backbone amides at several magnetic fields. These data were used to estimate the correlation time, taking into consideration the rotational diffusion anisotropy (Bruschweiler, Liao, & Wright, 1995; Lee, Rance, Chazin, & Palmer, 1997). To that end, several structures of GGBP and RBP, including their solution structures (also obtained in our laboratory), were employed. For RBP in the absence of ligand, the NMR structure provided a better fitting to the relaxation data than the *E. coli* X-ray structures with open (1URP) (Bjorkman & Mowbray, 1998) or closed (2DRI) (Bjorkman et al., 1994) interdomain orientations. When both domains were analyzed independently (to pull out the contribution from the interdomain orientation), the resulting tensors were very comparable to the one obtained using the full molecule, indicating that a single conformation is sufficient to explain all the experimental data. The excellent agreement found for RBP contrasts with GGBP, where no structural model showed good agreement with the experimental relaxation data set, an indirect evidence of interdomain motions in *apo*GGBP. Early studies with ^{19}F-NMR (Luck & Falke, 1991) and disulfide trapping (Careaga, Sutherland, Sabeti, & Falke, 1995) have already suggested the existence of large amplitude motions in GGBP, in accordance with the multiple hinge angles found in the deposited X-ray structures.

To further study their intramolecular dynamics, the same relaxation rates were measured for the two proteins (GGBP and RBP) associated with glucose. Apo- and ligand-bound relaxation data were analyzed using the Lipari–Szabo formalism, which is free of assumptions regarding the motional mechanism (model free) and only requires for the fast local librations to be independent from the overall tumbling of the molecule (Lipari & Szabo, 1982). The model-free analysis (LS-2) yields order parameters that measure the extent of the angular motion in the amide vector. For RBP, the changes in the order parameter upon ligand binding ($\Delta S^2 = S^2_{apo} - S^2_{holo}$) oscillate from positive to negative values with no clear trend while, remarkably, the vast majority of residues in GGPB present positive ΔS^2 values (Fig. 6.6). The latter effect is pervasive and the resulting neat contribution in GGBP evidences an increase in flexibility for GGBP bound to the ligand as compared to free GGBP.

Several studies have identified changes in pico- to nanosecond backbone dynamics upon ligand binding, dispersed throughout the protein (Arumugam et al., 2003; Yun, Jang, Kim, Choi, & Lee, 2001), including

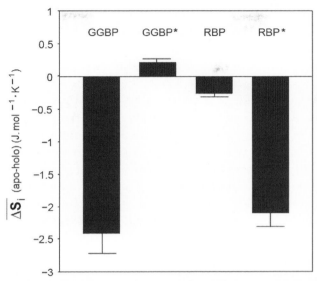

Figure 6.6 Functional dynamics in the ps–ns timescale. Overall changes in protein dynamics are expressed as the average conformational entropy upon ligand binding $\left(\overline{\Delta S_i} = \overline{S_i(\text{apo}) - S_i(\text{holo})} \right)$ for the four proteins under consideration. These values represent the average overall sampled residues. Error bars are obtained by propagation of the experimental uncertainties for the relaxation data.

several cases on protein–carbohydrate interaction (Akke, 2012) and at least one study involving a PBP (MacRaild, Daranas, Bronowska, & Homans, 2007). This effect can be produced by a raise in the fast librations for the amide bonds in *glu*GGBP that would compensate for the restriction in the freedom of motion of the domains upon ligand binding or by segmental motions in the nanosecond timescale, affecting the order parameter and breaking the required independence between local and overall tumbling. Further analysis of the data using different models (Skrynnikov, Millet, & Kay, 2002) evidenced that nanosecond dynamics are also contributing to the order parameter. Such motions would be consistent with the multiple interdomain closure angles observed in the pool of ligand-bound structures available for GGPB, with differences in the closure angle of up to 9° (Borrok et al., 2007).

It is important to ascertain whether the motions identified in *glu*RBP are a reflection of thermal motion (see Section 1.1) or if they are of a much more complex nature, underscoring a concerted motion of the protein. To test this hypothesis, the two proteins were engineered at the hinge level. Based on the changes in the C^α torsion angle between *apo*GGBP and *glu*GGBP, the amino acids G109, T110, and, to a minor extent, Y111 and V293 were identified as the main contributions to the interdomain dynamics. For RBP, changes in the torsion angle were concentrated in residues V263 and K266. According to sequence alignment, V293 in GGBP and V263 in RBP correspond to the same residue. K266 (RBP) is very close to the C–terminus of the molecule and is no further considered. Taking everything into consideration, two new proteins were generated by site–directed mutagenesis: G109A/T110S-GGBP (GGBP*) that replaces two residues from RBP in the first strand of the hinge and the mirror substitution in RBP (A102G/S103T-RBP, RBP*). Relaxation data were collected for the two mutant proteins, with and without monosaccharide. GGBP* abrogates all evidences for the pervasive fast dynamics. Remarkably, the opposite is also true and RBP* rescues the net positive contribution in the ΔS^2 profile, much comparable to wild–type GGBP.

From a functional point of view, the open–to–close transition is thermodynamically coupled with the intrinsic ligand affinity (Table 6.2), constituting a paradigm of functional dynamics. The natural substrate affinities were measured by isothermal titration calorimetry for wild–type and mutant proteins. Very high affinities were observed for D–ribose (RBP) and D–glucose (GGBP), in line with previously reported affinity constants measured by other techniques (Aqvist & Mowbray, 1995; Binnie, Zhang, Mowbray, & Hermodson, 1992). The two hinge residue mutations drastically altered the apparent sugar binding constant, and GGBP* binds D–glucose with much *weaker* affinity. RBP* conversely binds D–ribose with 10–fold *stronger* affinity than wild–type RBP. The thermodynamic parameters extracted from the ITC data reveal that the differences in affinity measured are caused exclusively by changes in the entropy of the system, in full agreement with the NMR relaxation data analysis.

2.2. Allosteric motions in *E. coli* dihydrofolate reductase

The enzyme dihydrofolate reductase (DHFR) reduces dihydrofolic acid to tetrahydrofolic acid. This is a critical cellular role since the enzyme processes the only source of tetrahydrofolate in the cell (Bhabha, Tuttle, Martinez-Yamout, & Wright, 2011). More specifically, the enzyme catalyzes the reduction of 7,8–dihydrofolate to 5,6,7,8–tetrahydrofolate using NADPH as a cofactor and electron donor (Matthews et al., 1977). The kinetic mechanism involves rebinding NADPH to assist the release of

Table 6.2 Affinity between the periplasmic binding protein and the natural ligand, as determined by isothermal titration calorimetry

Protein	Wild type[c]			Mutant[c]		
	K_D (nM)	$\Delta H°$ (kcal mol^{-1})	$T\Delta S°$ (kcal mol^{-1})	K_D (nM)	$\Delta H°$ (kcal mol^{-1})	$T\Delta S°$ (kcal mol^{-1})
RBP[a]	170±9	−11.0±0.1	−1.4±0.1	28.0±4	−11.0±0.2	−0.3±0.05
GGBP[b]	290±20	−21.0±0.3	−11.7±0.4	237×10³±1530	−21±0.5	−15.8±0.6

[a] Affinity for D-ribose.
[b] Affinity for D-glucose.
[c] Experiment carried out at 310 K.

the tetrahydrofolate product. The *E. coli* DHFR represents an important example of functional dynamics since NMR investigations have undoubtedly established the linkage between protein dynamics and catalytic function. This has been possible because of the nature and the timescales of the observed motions and because of the availability of a large amount of structural and mechanistic data that have provided a detailed framework to interpret the motions.

The enzyme has been thoroughly studied using NMR experiments in combination with the available structural information. In solution, the enzyme adopts five major intermediates: the enzyme bound to the cofactor (E:NADPH), the Michaelis complex (E:NADPH:DHF), the ternary complex with the product (E:NADP+:THF), the binary complex with the product (E:THF), and the ternary complex with the reduced cofactor (E:NADPH: THF) (Fierke, Johnson, & Benkovic, 1987). The structures of all the kinetic intermediates have been obtained by X-ray crystallography (Sawaya & Kraut, 1997; Venkitakrishnan et al., 2004) and chemical shift differences have also been assigned for the different conformations (Osborne, Venkitakrishnan, Dyson, & Wright, 2003). Interestingly, in two of the kinetic instances (the Michaelis complex and E:NADPH), the Met20 loop adopts a closed conformation, packing tightly against the nicotinamide ring of the bound cofactor. In the other three substates, Met20 adopts an occluded conformation, leaving the nicotinamide ring outside from the binding pocket and preventing it from binding to the active site.

The five conformational substates of *E. coli* DHFR have been characterized using CPMG-RD experiments (Boehr, McElheny, Dyson, & Wright, 2006). Dispersion data for the amide nitrogen, measured at two frequencies, were fitted to a two-site exchange model. The analysis showed that many of the residues exhibiting exchange contributions are directly or indirectly associated with the Met20 loop, where excursions toward an excited state of higher energy (and therefore low population) were found. Remarkably, the comparison of the chemical shift values obtained from dynamics experiments ($\Delta\omega$ values obtained from the fitting) against the differences of chemical shift obtained by the chemical shift perturbation analysis (Osborne et al., 2003) always gave excellent linear correlations. This extraordinary result means that each intermediate in the catalytic cycle samples a lowly populated excited state, whose conformation resembles the ground state of the following (or the preceding) intermediate.

The data obtained from relaxation dispersion directly report on the catalytic cycle of the enzyme, and the higher energy conformations observed in the relaxation dispersion experiment appear to play a direct role in catalysis. Because the excited states resemble the ground states for the following inter-mediate in the catalytic cycle, the binding of ligands to the enzyme must occur through conformational selection instead of the more canonical induced-fit mechanism. Thus, ligand association modulates the energy land-scape, funneling the enzyme through its reaction cycle along a preferred kinetic path: a minor conformational substate resembling the ligand-bound or induced conformation is already in solution and ligand binds to the minor substate, causing a shift in the equilibrium to make the ligand-bound con-formation the new major substate.

The dynamics in the picosecond to nanosecond timescale has also been investigated for DHFR using NMR spectroscopy (Epstein, Benkovic, & Wright, 1995; McElheny, Schnell, Lansing, Dyson, & Wright, 2005). In contrast to the dynamics on the μs–ms timescale, dynamics in the ps–ns timescale appears to be dominated by the overall protein conformation and the nature of bound ligands has little effect on the protein motion in this timescale. Moreover, the energetics of the different states in the energy land-scape has also been investigated using the same experimental technique (Boehr, McElheny, Dyson, & Wright, 2010). Results show that the dynam-ics in the substate and product binary complexes are very different and governed by quite different kinetic and thermodynamic parameters. These differences are difficult to rationalize from the sole information from the high-resolution structure and suggest that dynamics in DHFR are exqui-sitely tuned for every intermediate during the catalytic cycle.

2.3. Self-association and supramolecular proenzymes

Weak protein homo- or heteroassociation is increasingly recognized as an important functional element in many regulatory processes. Weak binding is associated with energies not much higher than kT, allowing its efficient modulation by a number of factors. From an experimental point of view, weak interactions are difficult to detect due to the low population of the bound species in the usual concentrations employed and the presence of fast exchange between free and bound forms. Weak homo–oligomerization processes are a special case as, in the general case, the participating species (monomer, dimer, etc.) may not be studied and structurally characterized independently. Fast equilibration on the NMR timescale leads to average parameters, and, in the case of weak oligomerization, the average value may not be very different from that of the major (monomeric species). Thus, chemical shift perturbations as a function of protein concentration may fail to detect weak oligomerization processes using moderate concentration variations. In contrast, methods based on NMR relaxation are very sensitive to the presence of even small amounts of species with a very different rotational correlation time.

Concentration-dependent ^{15}N relaxation has been used by a number of groups (Bernado, Akerud, Garcia de la Torre, Akke, & Pons, 2003; Fushman, Cahill, & Cowburn, 1997; Jensen et al., 2008). The sensitivity of the methods stems from the nonlinear increase of the contribution of the spectral density function at the origin ($J(0)$) to relaxation as the molecular weight of the species increases. Furthermore, in contrast to methods that detect a limited number of probes located at the oligomer interface, the effect on global tumbling is sensed by all the nuclei in the molecule including those that are far from the interface. The presence of multiple probes contributes to increase the precision by which the relaxation rates can be measured and to minimize the effect of local perturbations. In particular, changes in local dynamics or chemical exchange that may blur the global self-association picture take place preferentially close to the oligomerization interface.

A third advantage of using relaxation measurements sensitive to the global tumbling is due to the presence of motional anisotropy. The global shape of monomer and dimers can potentially be substantially different: a dimer made of two quasi-spherical monomers will have a cylindrical symmetry. The interaction of two elongated monomers may give an even larger anisotropy or, in contrast, result in a more globular dimer. Anisotropic tumbling causes ^{15}N relaxation rates to depend on the orientation of the [15]N—H bond vector with respect to the principal axes of the rotational diffusion tensor. Thus, residue-specific relaxation rates not only provide information on the self-association thermodynamics but also on the structure of the oligomers. This is in contrast, for example, to other hydrodynamic parameters that can also be measured by NMR, such as translational diffusion in which a single diffusion coefficient is extracted, even if it can be measured on different resolved NMR signals.

Incidentally, the large effects of self-association on NMR relaxation may significantly affect the analysis of the internal dynamics of apparently monomeric proteins if weak self-association phenomena are overlooked (Schurr, Babcock, & Fujimoto, 1994). When the 3D structure of the protein is known, a comparison between experimental ^{15}N relaxation rates and the predictions from HYDRONMR provides a good indication of possible self-association effects. The atomic element radius is an adjustable parameter in HYDRONMR. The best fit between predictions and experiment is found for a value of 3.3 Å, much larger values of the atomic element radius are often associated to undetected self-association (Bernado et al., 2002).

Other relaxation-based NMR observables are sensitive reporters of self-association. Paramagnetic relaxation enhancement exploits the very efficient relaxation caused by unpaired electrons. Pseudo-self-association is studied by mixing paramagnetic and diamagnetic forms eventually combined with selective ^{15}N-labeling of the diamagnetic form allowing the detection of ultra-weak self-association with $K_D > 15$ mM (Tang, Ghirlando, & Clore, 2008).

RDC provide an alternative method to determine shape anisotropy by NMR: knowing the structure of the monomer and assuming that the oligomer has axial symmetry, which would often be the case (Jain, Wyckoff, Raetz, & Prestegard, 2004). However, this approach requires that the oligomer is the dominant species, a rigid system, and is not perturbed by the interaction with the alignment media. In the case of weak complexes, extrapolation of the residual couplings in the monomeric and dimeric species can be obtained by comparing aligned samples at various protein concentrations (Lee et al., 2010; Ortega-Roldan et al., 2009). Low-resolution shape information of oligomers in solution can be also extracted from small-angle X-ray scattering (SAXS). The range of protein concentrations that can be studied by SAXS and NMR is very similar and the two techniques are complementary. Shape information arises from the pairwise distance distribution function of scattering elements (Koch, Vachette, & Svergun, 2003). The scattering from a mixture is a linear combination from the contributions of its components; thus, if the scattering curves are known, dissociation constants can be obtained by fitting scattering data obtained at several concentrations. If enough data are available and the shapes of the components are very different, it is possible to extract both the dissociation constants and the scattering curves of the components by decomposing the concentration-dependent scattering curves (Bernado, Mylonas, Petoukhov, Blackledge, & Svergun, 2007; Vestergaard et al., 2007). Single value decomposition (Williamson, Craig, Kondrashkina, Bailey-Kellogg, & Friedman, 2008) and multivariate curve resolution combined with alternating least squares (MCV-ALS) (Navea, Tauler, & de Juan, 2006) can, in principle, be used to study complex oligomerization processes, including the analysis of the number of species contributing to the observed data. Low-resolution structures of the monomer and dimer species of a low-molecular-weight protein tyrosine phosphatase (lmwPTP) in good agreement with the corresponding X-ray structures were obtained *ab initio* by MCV–ALS (Blobel, Bernado, Svergun, Tauler, & Pons, 2009).

The lmPTP are ubiquous in mammalian cells (Chernoff & Li, 1985; Waheed, Laidler, Wo, & Van Etten, 1988) but are also present in prokaryotes (Lescop et al., 2006; Xu, Xia, & Jin, 2006). They constitute a paradigmatic case of self-association that has been studied using various biophysical methods.

Tabernero, Evans, Tishmack, Van Etten, and Stauffacher (1999) first observed lmwPTP dimers in the crystals of a mutant of bovine lmwPTP. Dimer formation was confirmed in solution also in the wild-type protein using ultracentrifugation and NMR (Akerud, Thulin, Van Etten, & Akke, 2002). Further insight into the oligomerization of lmwPTP could be obtained by combining the known 3D X-ray structures of monomer and dimer with ^{15}N relaxation obtained at various concentrations. The linkage was provided by hydrodynamic calculations using HYDRONMR (Garcia de la Torre et al., 2000). The additional insight clearly demonstrated that lmwPTP oligomerization goes beyond dimerization, involving at least one additional oligomeric species that was compatible with an isotropic tetramer. Supporting this analysis, the few residues with relaxation rates that could not be accounted for by the best model, based on exclusive relaxation by dipolar and chemical shift anisotropy mechanisms, were clustered in the regions that would form the interface of a tetrahedral tetramer formed by the perpendicular interaction between two dimers. Further support to the formation of a higher molecular weight species could be obtained using [129]Xe-NMR (Blobel et al., 2007).

Due to its high polarizability, xenon gas is soluble in water and [129]Xe chemical shifts are sensitive reporters of its chemical environment. In particular, Xe has the capacity to interact with hydrophobic clefts present in protein oligomers and [129]Xe chemical shifts are a very sensitive protein oligomerization reporter. By comparing ^{15}N relaxation data and [129]Xe chemical shifts, it could be shown that Xe interacts weakly with lmwPTP monomer and dimer but shows a very high affinity for the higher oligomer. [129]Xe NMR was also used to show that lmwPTP oligomerization is sensitive to crowding conditions.

Complex oligomerization paralleling that observed in bovine lmwPTP was also observed in YwlE, an lmwPTP from *Bacillus subtilis* (Blobel et al., 2009). Thus, lmwPTP oligomerization is an evolutionary conserved feature suggesting it is functionally relevant, although the *in vitro* dissociation constants are in the millimolar range. Macromolecular crowding in the cyto-plasm, however, may result in a much higher tendency to self-associate (Elcock, 2010). In this context, the formation of higher oligomers (tetra-mers) may represent a strategy to modulate the effective dissociation con-stant in the crowded cytoplasm. Larger species are more sensitive to excluded-volume effects and higher stoichiometries enhance the excluded-volume corrections to relating concentrations to thermodynamic activities (Minton & Wilf, 1981).

In the structure of the lmwPTP dimer of bovine PTP, Tyr131 and Tyr132 from each monomer are inserted into the active site of the other. Therefore, the active site is blocked and the dimer form would be enzymat-ically inactive (Fig. 6.7). These tyrosine residues, when phosphorylated, increase the activity of the enzyme and the phosphorylated forms are them-selves dephosphorylated by lmwPTP. Tabernero suggested that the dimer found in the crystal could represent a transient state between the dephos-phorylation of the tyrosines and the release of the unphosphorylated mono-mer and also hypothesized that dimerization could be a regulatory mechanism for lmwPTP (Tabernero et al., 1999). Dimerization leading to inactivation of other phosphatases had been previously observed (Weiss & Schlessinger, 1998). On the other hand, in YwlE, phenylalanine is present in the dimerization site, suggesting that dimerization is indepen-dent from tyrosine phosphorylation (Blobel et al., 2009).

Oligomerization of lmwPTP has been suggested to constitute the basis of a supramolecular proenzyme (Bernado et al., 2003). Proenzymes are inactive

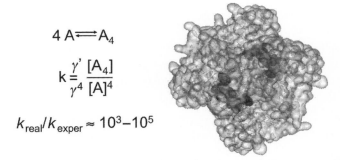

$$4\,A \rightleftharpoons A_4$$

$$k = \frac{\gamma'}{\gamma^4}\frac{[A_4]}{[A]^4}$$

$$k_{real}/k_{exper} \approx 10^3 - 10^5$$

Figure 6.7 Equilibrium oligomerization of low-molecular-weight protein tyrosine phos-phatase (lmPTP). The lmPTP in solution is best represented as the equilibrium of differ-ent species: monomer, dimer, and tetramer. NMR relaxation experiments allowed extracting the populations of each of the species (left). A structural model for the asso-ciation, consistent with the experimental data, is shown in the right. (For color version of this figure, the reader is referred to the online version of this chapter.)

enzymatic species that can be accumulated waiting for a triggering event that converts them into active forms. The accumulation of the inactive form ensures a rapid response independent from protein synthesis. In classical pro-enzymes, the triggering event is the irreversible cleavage of peptide bonds and, therefore, once activated, the enzyme cannot return to its resting state. On the contrary, a supramolecular proenzyme is maintained in its inactive form by noncovalent interactions. In the case of the lmwPTP, competition for the binding site by a substrate containing a phosphorylated tyrosine is the triggering event, releasing active monomeric phosphatase from the dimer. Thus, in the initial signaling events, kinases will phosphorylate its substrates without the competition of lmwPTP, which is initially in its inactive olig-omeric form. Once the concentration of phosphorylated substrate increases, lmwPTP oligomers disassemble, releasing the active phosphatase in a neg-ative feedback loop. The result is a signaling pulse. Once the substrate has been dephosphorylated, lmwPTP oligomers are reformed and the phospha-tase returns to its resting state, waiting for the next signaling pulse.

3. CONCLUDING REMARKS

In this review, we have explored the use of nitrogen amide backbone relaxation in the description of protein dynamics at multiple timescales, a matter particularly elusive. Special care has been paid off to the description of the technical aspects during the experimental implementation and the posterior analysis of the experiments. It has also been emphasized that the quantitative information that can be extracted from relaxation data heavily depends on the timescale of the measurement. Even though these techniques are rather well known and nowadays routinely used, we have included some recent applications, discussed in detail as case studies, to demonstrate that this approach is still very innovative and to highlight its future potential.

For folded proteins, backbone amide groups are normally well dispersed in the ^{1}H-^{15}N-HSQC-based experiments and they can be readily assigned for small- and intermediate-sized systems. For very large proteins (MW ≥ 40 kDa), the signal overlap in the spectra and the intrinsic line broadening of the peaks make this kind of approach unfeasible and the use of other probes (like the methyl groups) is highly recommended instead. For unfolded proteins, backbone relaxation analysis constitutes a very pow-erful tool and it can be used in combination with other biophysical data (SAXS, fluorescence anisotropy, etc.) for the characterization of the ensem-ble of conformations that best describes the protein behavior in solution.

ACKNOWLEDGMENTS

Support was provided from The Department of Industry, Tourism and Trade of the Government of the Autonomous Community of the Basque Country, from the Innovation Technology Department of the Bizkaia County, from the Ministerio de Economía y Competititvidad (CTQ2009-10353/BQU, CTQ2012-32183, and CSD2008-00005), MICINN-FEDER (BIO2010-15683 and ICTS RMN de Barcelona), the "Generalitat de Catalunya" (2009SGR1352), and EU 7th FP BioNMR (Contract 261863). G. O. acknowledges a fellowship from the Ministerio de Ciencia y Tecnología.

REFERENCES

Abu-Abed, M., Millet, O., MacLennan, D. H., & Ikura, M. (2004). Probing nucleotide-binding effects on backbone dynamics and folding of the nucleotide-binding domain of the sarcoplasmic/endoplasmic-reticulum Ca^{2+}-ATPase. *The Biochemical Journal*, *379*, 235–242.

Adcock, S. A., & McCammon, J. A. (2006). Molecular dynamics: Survey of methods for simulating the activity of proteins. *Chemical Reviews*, *106*, 1589–1615.

Akerud, T., Thulin, E., Van Etten, R. L., & Akke, M. (2002). Intramolecular dynamics of low molecular weight protein tyrosine phosphatase in monomer–dimer equilibrium studied by NMR: A model for changes in dynamics upon target binding. *Journal of Molecular Biology*, *322*, 137–152.

Akke, M. (2012). Conformational dynamics and thermodynamics of protein–ligand binding studied by NMR relaxation. *Biochemical Society Transactions*, *40*, 419–423.

Aqvist, J., & Mowbray, S. L. (1995). Sugar recognition by a glucose/galactose receptor. Evaluation of binding energetics from molecular dynamics simulations. *The Journal of Biological Chemistry*, *270*, 9978–9981.

Arumugam, S., Gao, G., Patton, B. L., Semenchenko, V., Brew, K., & Van Doren, S. R. (2003). Increased backbone mobility in beta-barrel enhances entropy gain driving binding of N-TIMP-1 to MMP-3. *Journal of Molecular Biology*, *327*, 719–734.

Bahar, I., Lezon, T. R., Yang, L. W., & Eyal, E. (2010). Global dynamics of proteins: Bridging between structure and function. *Annual Review of Biophysics*, *39*, 23–42.

Bahar, I., & Rader, A. J. (2005). Coarse-grained normal mode analysis in structural biology. *Current Opinion in Structural Biology*, *15*, 586–592.

Bermejo, G. A., Strub, M. P., Ho, C., & Tjandra, N. (2010). Ligand-free open–closed transitions of periplasmic binding proteins: The case of glutamine-binding protein. *Biochemistry*, *49*, 1893–1902.

Bernado, P., Akerud, T., Garcia de la Torre, J., Akke, M., & Pons, M. (2003). Combined use of NMR relaxation measurements and hydrodynamic calculations to study protein association. Evidence for tetramers of low molecular weight protein tyrosine phosphatase in solution. *Journal of the American Chemical Society*, *125*, 916–923.

Bernado, P., Garcia de la Torre, J., & Pons, M. (2002). Interpretation of 15N NMR relaxation data of globular proteins using hydrodynamic calculations with HYDRONMR. *Journal of Biomolecular NMR*, *23*, 139–150.

Bernado, P., Mylonas, E., Petoukhov, M. V., Blackledge, M., & Svergun, D. I. (2007). Structural characterization of flexible proteins using small-angle X-ray scattering. *Journal of the American Chemical Society*, *129*, 5656–5664.

Bhabha, G., Tuttle, L., Martinez-Yamout, M. A., & Wright, P. E. (2011). Identification of endogenous ligands bound to bacterially expressed human and E. coli dihydrofolate reductase by 2D NMR. *FEBS Letters*, *585*, 3528–3532.

Binnie, R. A., Zhang, H., Mowbray, S., & Hermodson, M. A. (1992). Functional mapping of the surface of Escherichia coli ribose-binding protein: Mutations that affect chemotaxis and transport. *Protein Science*, *1*, 1642–1651.

Bjorkman, A. J., Binnie, R. A., Zhang, H., Cole, L. B., Hermodson, M. A., & Mowbray, S. L. (1994). Probing protein–protein interactions. The ribose-binding protein in bacterial transport and chemotaxis. *The Journal of Biological Chemistry*, *269*, 30206–30211.

Bjorkman, A. J., & Mowbray, S. L. (1998). Multiple open forms of ribose-binding protein trace the path of its conformational change. *Journal of Molecular Biology*, *279*, 651–664.

Blobel, J., Bernado, P., Svergun, D. I., Tauler, R., & Pons, M. (2009). Low-resolution structures of transient protein-protein complexes using small-angle X-ray scattering. *Journal of the American Chemical Society*, *131*, 4378–4386.

Blobel, J., Schmidl, S., Vidal, D., Nisius, L., Bernado, P., Millet, O., et al. (2007). Protein tyrosine phosphatase oligomerization studied by a combination of 15N NMR relaxation and 129Xe NMR. Effect of buffer containing arginine and glutamic acid. *Journal of the American Chemical Society*, *129*, 5946–5953.

Boehr, D. D., McElheny, D., Dyson, H. J., & Wright, P. E. (2006). The dynamic energy landscape of dihydrofolate reductase catalysis. *Science (New York, NY)*, *313*, 1638–1642.

Boehr, D. D., McElheny, D., Dyson, H. J., & Wright, P. E. (2010). Millisecond timescale fluctuations in dihydrofolate reductase are exquisitely sensitive to the bound ligands. *Proceedings of the National Academy of Sciences of the United States of America*, *107*, 1373–1378.

Bordignon, E., Grote, M., & Schneider, E. (2010). The maltose ATP-binding cassette transporter in the 21st century—Towards a structural dynamic perspective on its mode of action. *Molecular Microbiology*, *77*, 1354–1366.

Borrok, M. J., Kiessling, L. L., & Forest, K. T. (2007). Conformational changes of glucose/galactose-binding protein illuminated by open, unliganded, and ultra-high-resolution ligand-bound structures. *Protein Science*, *16*, 1032–1041.

Bosco, D. A., Eisenmesser, E. Z., Clarkson, M. W., Wolf-Watz, M., Labeikovsky, W., Millet, O., et al. (2010). Dissecting the microscopic steps of the cyclophilin A enzymatic cycle on the biological HIV-1 capsid substrate by NMR. *Journal of Molecular Biology*, *403*, 723–738.

Bruschweiler, R., Liao, X., & Wright, P. E. (1995). Long-range motional restrictions in a multidomain zinc-finger protein from anisotropic tumbling. *Science (New York, NY)*, *268*, 886–889.

Careaga, C. L., Sutherland, J., Sabeti, J., & Falke, J. J. (1995). Large amplitude twisting motions of an interdomain hinge: A disulfide trapping study of the galactose–glucose binding protein. *Biochemistry*, *34*, 3048–3055.

Chernoff, J., & Li, H. C. (1985). A major phosphotyrosyl–protein phosphatase from bovine heart is associated with a low-molecular-weight acid phosphatase. *Archives of Biochemistry and Biophysics*, *240*, 135–145.

Clore, G. M., Szabo, A., Bax, A., Kay, L. E., Driscoll, P. C., & Gronenborn, A. M. (1990). Deviations from the simple 2-parameter model-free approach to the interpretation of n-15 nuclear magnetic-relaxation of proteins. *Journal of the American Chemical Society*, *112*, 4989–4991.

d'Auvergne, E. J., & Gooley, P. R. (2008a). Optimisation of NMR dynamic models I. Minimisation algorithms and their performance within the model-free and Brownian rotational diffusion spaces. *Journal of Biomolecular NMR*, *40*, 107–119.

d'Auvergne, E. J., & Gooley, P. R. (2008b). Optimisation of NMR dynamic models II. A new methodology for the dual optimisation of the model-free parameters and the Brownian rotational diffusion tensor. *Journal of Biomolecular NMR*, *40*, 121–133.

Elcock, A. H. (2010). Models of macromolecular crowding effects and the need for quantitative comparisons with experiment. *Current Opinion in Structural Biology*, *20*, 196–206.

Epstein, D. M., Benkovic, S. J., & Wright, P. E. (1995). Dynamics of the dihydrofolate reductase-folate complex: Catalytic sites and regions known to undergo conformational change exhibit diverse dynamical features. *Biochemistry*, *34*, 11037–11048.

Eyles, S. J., & Kaltashov, I. A. (2004). Methods to study protein dynamics and folding by mass spectrometry. *Methods, 34,* 88–99.

Farrow, N. A., Zhang, O., Szabo, A., Torchia, D. A., & Kay, L. E. (1995). Spectral density function mapping using 15N relaxation data exclusively. *Journal of Biomolecular NMR, 6,* 153–162.

Fayos, R., Pons, M., & Millet, O. (2005). On the origin of the thermostabilization of proteins induced by sodium phosphate. *Journal of the American Chemical Society, 127,* 9690–9691.

Fierke, C. A., Johnson, K. A., & Benkovic, S. J. (1987). Construction and evaluation of the kinetic scheme associated with dihydrofolate reductase from Escherichia coli. *Biochemistry, 26,* 4085–4092.

Frauenfelder, H., Sligar, S. G., & Wolynes, P. G. (1991). The energy landscapes and motions of proteins. *Science (New York, NY), 254,* 1598–1603.

Fushman, D., Cahill, S., & Cowburn, D. (1997). The main-chain dynamics of the dynamin pleckstrin homology (PH) domain in solution: Analysis of 15N relaxation with monomer/dimer equilibration. *Journal of Molecular Biology, 266,* 173–194.

Gall, C. M., Cross, T. A., DiVerdi, J. A., & Opella, S. J. (1982). Protein dynamics by solid-state NMR: Aromatic rings of the coat protein in fd bacteriophage. *Proceedings of the National Academy of Sciences of the United States of America, 79,* 101–105.

Garcia de la Torre, J., Huertas, M. L., & Carrasco, B. (2000). HYDRONMR: Prediction of NMR relaxation of globular proteins from atomic-level structures and hydrodynamic calculations. *Journal of Magnetic Resonance, 147,* 138–146.

Gardner, K. H., & Kay, L. E. (1998). The use of 2H, 13C, 15N multidimensional NMR to study the structure and dynamics of proteins. *Annual Review of Biophysics and Biomolecular Structure, 27,* 357–406.

Gerstein, M., Lesk, A. M., & Chothia, C. (1994). Structural mechanisms for domain movements in proteins. *Biochemistry, 33,* 6739–6749.

Homans, S. W. (2005). Probing the binding entropy of ligand–protein interactions by NMR. *ChemBioChem, 6,* 1585–1591.

Jain, N. U., Wyckoff, T. J., Raetz, C. R., & Prestegard, J. H. (2004). Rapid analysis of large protein-protein complexes using NMR-derived orientational constraints: The 95 kDa complex of LpxA with acyl carrier protein. *Journal of Molecular Biology, 343,* 1379–1389.

Jensen, M. R., Kristensen, S. M., Keeler, C., Christensen, H. E., Hodsdon, M. E., & Led, J. J. (2008). Weak self-association of human growth hormone investigated by nitrogen-15 NMR relaxation. *Proteins, 73,* 161–172.

Karplus, M. (2003). Molecular dynamics of biological macromolecules: A brief history and perspective. *Biopolymers, 68,* 350–358.

Kay, L. E., Torchia, D. A., & Bax, A. (1989). Backbone dynamics of proteins as studied by 15N inverse detected heteronuclear NMR spectroscopy: Application to staphylococcal nuclease. *Biochemistry, 28,* 8972–8979.

Kleckner, I. R., & Foster, M. P. (2011). An introduction to NMR-based approaches for measuring protein dynamics. *Biochimica et Biophysica Acta, 1814,* 942–968.

Koch, M. H., Vachette, P., & Svergun, D. I. (2003). Small-angle scattering: A view on the properties, structures and structural changes of biological macromolecules in solution. *Quarterly Reviews of Biophysics, 36,* 147–227.

Korzhnev, D. M., & Kay, L. E. (2008). Probing invisible, low-populated States of protein molecules by relaxation dispersion NMR spectroscopy: An application to protein folding. *Accounts of Chemical Research, 41,* 442–451.

Korzhnev, D. M., Neudecker, P., Mittermaier, A., Orekhov, V. Y., & Kay, L. E. (2005). Multiple-site exchange in proteins studied with a suite of six NMR relaxation dispersion experiments: An application to the folding of a Fyn SH3 domain mutant. *Journal of the American Chemical Society, 127,* 15602–15611.

Korzhnev, D. M., Skrynnikov, N. R., Millet, O., Torchia, D. A., & Kay, L. E. (2002). An NMR experiment for the accurate measurement of heteronuclear spin-lock relaxation rates. *Journal of the American Chemical Society, 124,* 10743–10753.

Krizova, H., Zidek, L., Stone, M. J., Novotny, M. V., & Sklenar, V. (2004). Temperature-dependent spectral density analysis applied to monitoring backbone dynamics of major urinary protein-I complexed with the pheromone 2-sec-butyl-4,5-dihydrothiazole. *Journal of Biomolecular NMR, 28,* 369–384.

Lee, L. K., Rance, M., Chazin, W. J., & Palmer, A. G., 3rd. (1997). Rotational diffusion anisotropy of proteins from simultaneous analysis of 15N and 13C alpha nuclear spin relaxation. *Journal of Biomolecular NMR, 9,* 287–298.

Lee, H. W., Wylie, G., Bansal, S., Wang, X., Barb, A. W., Macnaughtan, M. A., et al. (2010). Three-dimensional structure of the weakly associated protein homodimer SeR13 using RDCs and paramagnetic surface mapping. *Protein Science, 19,* 1673–1685.

Lefevre, J. F., Dayie, K. T., Peng, J. W., & Wagner, G. (1996). Internal mobility in the partially folded DNA binding and dimerization domains of GAL4: NMR analysis of the N-H spectral density functions. *Biochemistry, 35,* 2674–2686.

Lescop, E., Hu, Y., Xu, H., Hu, W., Chen, J., Xia, B., et al. (2006). The solution structure of Escherichia coli Wzb reveals a novel substrate recognition mechanism of prokaryotic low molecular weight protein-tyrosine phosphatases. *The Journal of Biological Chemistry, 281,* 19570–19577.

Lipari, G., & Szabo, A. (1982). Model free approach to the interpretation of nuclear magnetic relaxation in macromolecules: 1. Theory and range of validity. *Journal of the American Chemical Society, 104,* 4546–4559.

Loria, J. P., Berlow, R. B., & Watt, E. D. (2008). Characterization of enzyme motions by solution NMR relaxation dispersion. *Accounts of Chemical Research, 41,* 214–221.

Loria, J. P., Rance, M., & Palmer, A. G. (1999). A relaxation-compensated Carr-Purcell-Meiboom-Gill sequence for characterizing chemical exchange by NMR spectroscopy. *Journal of the American Chemical Society, 122,* 2331–2332.

Luck, L. A., & Falke, J. J. (1991). Open conformation of a substrate-binding cleft: 19 F NMR studies of cleft angle in the D-galactose chemosensory receptor. *Biochemistry, 30,* 6484–6490.

MacRaild, C. A., Daranas, A. H., Bronowska, A., & Homans, S. W. (2007). Global changes in local protein dynamics reduce the entropic cost of carbohydrate binding in the arabinose-binding protein. *Journal of Molecular Biology, 368,* 822–832.

Manson, M. D., Armitage, J. P., Hoch, J. A., & Macnab, R. M. (1998). Bacterial locomotion and signal transduction. *Journal of Bacteriology, 180,* 1009–1022.

Matthews, D. A., Alden, R. A., Bolin, J. T., Freer, S. T., Hamlin, R., Xuong, N., et al. (1977). Dihydrofolate reductase: X-ray structure of the binary complex with methotrexate. *Science (New York, NY), 197,* 452–455.

McElheny, D., Schnell, J. R., Lansing, J. C., Dyson, H. J., & Wright, P. E. (2005). Defining the role of active-site loop fluctuations in dihydrofolate reductase catalysis. *Proceedings of the National Academy of Sciences of the United States of America, 102,* 5032–5037.

Millet, O., Bernado, P., Garcia, J., Rizo, J., & Pons, M. (2002). NMR measurement of the off rate from the first calcium-binding site of the synaptotagmin I C2A domain. *FEBS Letters, 516,* 93–96.

Millet, O., Hudson, R. P., & Kay, L. E. (2003). The energetic cost of domain reorientation in maltose-binding protein as studied by NMR and fluorescence spectroscopy. *Proceedings of the National Academy of Sciences of the United States of America, 100,* 12700–12705.

Millet, O., Loria, J. P., Kroenke, C. D., Pons, M., & Palmer, A. G. (2000). The static magnetic field dependence of chemical exchange line broadening defines the NMR chemical shift time scale. *Journal of the American Chemical Society, 122,* 2867–2877.

Minton, A. P., & Wilf, J. (1981). Effect of macromolecular crowding upon the structure and function of an enzyme: Glyceraldehyde-3-phosphate dehydrogenase. *Biochemistry, 20,* 4821–4826.

Mittermaier, A., & Kay, L. E. (2006). New tools provide new insights in NMR studies of protein dynamics. *Science (New York, NY), 312,* 224–228.

Moffat, K. (2001). Time-resolved biochemical crystallography: A mechanistic perspective. *Chemical Reviews, 101,* 1569–1581.

Navea, S., Tauler, R., & de Juan, A. (2006). Monitoring and modeling of protein processes using mass spectrometry, circular dichroism, and multivariate curve resolution methods. *Analytical Chemistry, 78,* 4768–4778.

Neudecker, P., Lundstrom, P., & Kay, L. E. (2009). Relaxation dispersion NMR spectroscopy as a tool for detailed studies of protein folding. *Biophysical Journal, 96,* 2045–2054.

Orekhov, V. Y., Pervushin, K. V., Korzhnev, D. M., & Arseniev, A. S. (1995). Backbone dynamics of (1-71)- and (1-36)bacterioopsin studied by two-dimensional (1)H- (15)N NMR spectroscopy. *Journal of Biomolecular NMR, 6,* 113–122.

Ortega, G., Lain, A., Tadeo, X., Lopez-Mendez, B., Castano, D., & Millet, O. (2011). Halophilic enzyme activation induced by salts. *Scientific Reports, 1,* 6. http://dx.doi.org/10.1038/srep00006.

Ortega-Roldan, J. L., Jensen, M. R., Brutscher, B., Azuaga, A. I., Blackledge, M., & van Nuland, N. A. (2009). Accurate characterization of weak macromolecular interactions by titration of NMR residual dipolar couplings: Application to the CD2AP SH3-C: ubiquitin complex. *Nucleic Acids Research, 37,* e70.

Osborne, M. J., Venkitakrishnan, R. P., Dyson, H. J., & Wright, P. E. (2003). Diagnostic chemical shift markers for loop conformation and substrate and cofactor binding in dihydrofolate reductase complexes. *Protein Science, 12,* 2230–2238.

Palmer, A. G., 3rd. (2001). Nmr probes of molecular dynamics: Overview and comparison with other techniques. *Annual Review of Biophysics and Biomolecular Structure, 30,* 129–155.

Peng, J. W., & Wagner, G. (1995). Frequency spectrum of NH bonds in eglin c from spectral density mapping at multiple fields. *Biochemistry, 34,* 16733–16752.

Popovych, N., Sun, S., Ebright, R. H., & Kalodimos, C. G. (2006). Dynamically driven protein allostery. *Nature Structural and Molecular Biology, 13,* 831–838.

Quiocho, F. A., & Ledvina, P. S. (1996). Atomic structure and specificity of bacterial periplasmic receptors for active transport and chemotaxis: Variation of common themes. *Molecular Microbiology, 20,* 17–25.

Redfield, A. G. (1978). Proton nuclear magnetic resonance in aqueous solutions. *Methods in Enzymology, 49,* 253–270.

Rosengren, K. J., Daly, N. L., Plan, M. R., Waine, C., & Craik, D. J. (2003). Twists, knots, and rings in proteins. Structural definition of the cyclotide framework. *The Journal of Biological Chemistry, 278,* 8606–8616.

Sawaya, M. R., & Kraut, J. (1997). Loop and subdomain movements in the mechanism of Escherichia coli dihydrofolate reductase: Crystallographic evidence. *Biochemistry, 36,* 586–603.

Schurr, J. M., Babcock, H. P., & Fujimoto, B. S. (1994). A test of the model-free formulas. Effects of anisotropic rotational diffusion and dimerization. *Journal of Magnetic Resonance, 105,* 211–224.

Sheppard, D., Sprangers, R., & Tugarinov, V. (2010). Experimental approaches for NMR studies of side-chain dynamics in high-molecular-weight proteins. *Progress in Nuclear Magnetic Resonance Spectroscopy, 56,* 1–45.

Shilton, B. H., Flocco, M. M., Nilsson, M., & Mowbray, S. L. (1996). Conformational changes of three periplasmic receptors for bacterial chemotaxis and transport: The maltose-, glucose/galactose- and ribose-binding proteins. *Journal of Molecular Biology, 264,* 350–363.

Silva, R. G., Murkin, A. S., & Schramm, V. L. (2011). Femtosecond dynamics coupled to chemical barrier crossing in a Born-Oppenheimer enzyme. *Proceedings of the National Academy of Sciences of the United States of America, 108*, 18661–18665.

Skrynnikov, N. R., Dahlquist, F. W., & Kay, L. E. (2002). Reconstructing NMR spectra of "invisible" excited protein states using HSQC and HMQC experiments. *Journal of the American Chemical Society, 124*, 12352–12360.

Skrynnikov, N. R., Millet, O., & Kay, L. E. (2002). Deuterium spin probes of side-chain dynamics in proteins. 2. Spectral density mapping and identification of nanosecond time-scale side-chain motions. *Journal of the American Chemical Society, 124*, 6449–6460.

Smock, R. G., & Gierasch, L. M. (2009). Sending signals dynamically. *Science (New York, NY), 324*, 198–203.

Sprangers, R., Velyvis, A., & Kay, L. E. (2007). Solution NMR of supramolecular complexes: Providing new insights into function. *Nature Methods, 4*, 697–703.

Tabernero, L., Evans, B. N., Tishmack, P. A., Van Etten, R. L., & Stauffacher, C. V. (1999). The structure of the bovine protein tyrosine phosphatase dimer reveals a potential self-regulation mechanism. *Biochemistry, 38*, 11651–11658.

Tadeo, X., Lopez-Mendez, B., Castano, D., Trigueros, T., & Millet, O. (2009). Protein stabilization and the Hofmeister effect: The role of hydrophobic solvation. *Biophysical Journal, 97*, 2595–2603.

Tang, C., Ghirlando, R., & Clore, G. M. (2008). Visualization of transient ultra-weak protein self-association in solution using paramagnetic relaxation enhancement. *Journal of the American Chemical Society, 130*, 4048–4056.

Tompa, P., Fuxreiter, M., Oldfield, C. J., Simon, I., Dunker, A. K., & Uversky, V. N. (2009). Close encounters of the third kind: Disordered domains and the interactions of proteins. *Bioessays, 31*, 328–335.

Velyvis, A., Yang, Y. R., Schachman, H. K., & Kay, L. E. (2007). A solution NMR study showing that active site ligands and nucleotides directly perturb the allosteric equilibrium in aspartate transcarbamoylase. *Proceedings of the National Academy of Sciences of the United States of America, 104*, 8815–8820.

Venkitakrishnan, R. P., Zaborowski, E., McElheny, D., Benkovic, S. J., Dyson, H. J., & Wright, P. E. (2004). Conformational changes in the active site loops of dihydrofolate reductase during the catalytic cycle. *Biochemistry, 43*, 16046–16055.

Vestergaard, B., Groenning, M., Roessle, M., Kastrup, J. S., van de Weert, M., Flink, J. M., et al. (2007). A helical structural nucleus is the primary elongating unit of insulin amyloid fibrils. *PLoS Biology, 5*, e134.

Vyas, N. K., Vyas, M. N., & Quiocho, F. A. (1983). The 3 A resolution structure of a D-galactose-binding protein for transport and chemotaxis in Escherichia coli. *Proceedings of the National Academy of Sciences of the United States of America, 80*, 1792–1796.

Vyas, N. K., Vyas, M. N., & Quiocho, F. A. (1987). A novel calcium binding site in the galactose-binding protein of bacterial transport and chemotaxis. *Nature, 327*, 635–638.

Vyas, N. K., Vyas, M. N., & Quiocho, F. A. (1988). Sugar and signal-transducer binding sites of the Escherichia coli galactose chemoreceptor protein. *Science (New York, NY), 242*, 1290–1295.

Vyas, N. K., Vyas, M. N., & Quiocho, F. A. (1991). Comparison of the periplasmic receptors for L-arabinose, D-glucose/D-galactose, and D-ribose. Structural and Functional Similarity. *The Journal of Biological Chemistry, 266*, 5226–5237.

Waheed, A., Laidler, P. M., Wo, Y. Y., & Van Etten, R. L. (1988). Purification and physicochemical characterization of a human placental acid phosphatase possessing phosphotyrosyl protein phosphatase activity. *Biochemistry, 27*, 4265–4273.

Watt, E. D., Shimada, H., Kovrigin, E. L., & Loria, J. P. (2007). The mechanism of rate-limiting motions in enzyme function. *Proceedings of the National Academy of Sciences of the United States of America, 104*, 11981–11986.

Wedemeyer, W. J., Welker, E., & Scheraga, H. A. (2002). Proline cis–trans isomerization and protein folding. *Biochemistry*, *41*, 14637–14644.

Weiss, A., & Schlessinger, J. (1998). Switching signals on or off by receptor dimerization. *Cell*, *94*, 277–280.

Williamson, T. E., Craig, B. A., Kondrashkina, E., Bailey-Kellogg, C., & Friedman, A. M. (2008). Analysis of self-associating proteins by singular value decomposition of solution scattering data. *Biophysical Journal*, *94*, 4906–4923.

Xu, H., Xia, B., & Jin, C. (2006). Solution structure of a low-molecular-weight protein tyrosine phosphatase from Bacillus subtilis. *Journal of Bacteriology*, *188*, 1509–1517.

Yun, S., Jang, D. S., Kim, D. H., Choi, K. Y., & Lee, H. C. (2001). 15N NMR relaxation studies of backbone dynamics in free and steroid-bound Delta 5-3-ketosteroid isomerase from Pseudomonas testosteroni. *Biochemistry*, *40*, 3967–3973.

CHAPTER SEVEN

Protein–DNA Electrostatics: Toward a New Paradigm for Protein Sliding

Maria Barbi*,1, Fabien Paillusson†

*Laboratoire de Physique Théorique de la Matière Condensée, Université Pierre et Marie Curie, Paris cedex, France
†Department of Chemistry, University of Cambridge, Cambridge, United Kingdom
1Corresponding author: e-mail address: barbi@lptmc.jussieu.fr

Contents

Advances in Protein Chemistry and Structural Biology, Volume 92
ISSN 1876-1623
http://dx.doi.org/10.1016/B978-0-12-411636-8.00007-9

Abstract

Gene expression and regulation rely on an apparently finely tuned set of reactions between some proteins and DNA. Such DNA-binding proteins have to find specific sequences on very long DNA molecules and they mostly do so in the absence of any active process. It has been rapidly recognized that, to achieve this task, these proteins should be efficient at both searching (i.e., sampling fast relevant parts of DNA) and finding (i.e., recognizing the specific site). A two-mode search and variants of it have been suggested since the 1970s to explain either a fast search or an efficient recognition. Combining these two properties at a phenomenological level is, however, more difficult as they appear to have antagonist roles. To overcome this difficulty, one may simply need to drop the dichotomic view inherent to the two-mode search and look more thoroughly at the set of interactions between DNA-binding proteins and a given DNA segment either specific or nonspecific. This chapter demonstrates that, in doing so in a very generic way, one may indeed find a potential reconciliation between a fast search and an efficient recognition. Although a lot remains to be done, this could be the time for a change of paradigm.

1. INTRODUCTION: THE SEARCH OF TARGET SEQUENCES

The observation of gene regulatory networks made possible by proteomics (the study of the ensemble of proteins in a cell or tissue in given conditions) and transcriptome analysis (the set of messenger RNA resulting from the expression of a portion of the genome of a cell tissue or cell type) reveals the set of interactions between different cellular components. It is then necessary to specify the nature of these interactions, from the structural, energetic, spatial, and temporal point of view, in order to reveal the mechanisms underlying the "cellular timing": how appropriate macromolecules are recruited at the right time and at the right place?

Many proteins indeed have to search and bind specific, relatively short DNA sequences in order to perform their biological task. These specific proteins include not only polymerases and a number of transcription factors involved in the regulation of gene expression but also proteins with different functions as, for example, nucleases. Knowing that the total length of DNA may reach millions or billions of base pairs (bp), one understands that finding the target sequence is a formidable challenge. The problem of this search kinetics has been debated since, in the 1970s, researchers realized that the relatively short time needed for a protein to find its target sequence on DNA cannot be explained by a simple search by 3D diffusion in the cell (according to the Smoluchowski theory) followed by random collisions with the DNA: the actual association constant is approximately two orders of magnitude larger (Richter & Eigen, 1974; Riggs, Bourgeois, & Cohn, 1970). Since then, many people have been interested in the search process, and a large amount of theoretical work has been done (Coppey, Bénichou, Voituriez, & Moreau, 2004; von Hippel & Berg, 1989). Interestingly, despite the fact that the role of electrostatics had been explicitly invoked in the original works (Halford, 2009; Riggs et al., 1970; von Hippel & Berg, 1989), most of the work has been based on a purely kinetic approach. The main results can be summarized by the finding that 3D excursions should be alternated by phases of 1D diffusion, named *sliding*, during which the protein binds DNA and slides along the double helix by thermal 1D diffusion (Berg, Winter, & von Hippel, 1981). This intermittent process has been called "facilitated diffusion."

The existence of 1D diffusion or sliding has then been proved by several experiments (Halford & Marko, 2004; Terry, Jack, & Modrich, 1985; Winter, Berg, & von Hippel, 1981) and, in particular, by fluorescence microscopy (Blainey, van Oijen, Banerjee, Verdine, & Xie, 2006; Bonnet et al., 2008; Elf, Li, & Xie, 2007; Shimamoto, 1999). In this kind of experiments, the two extremities of a DNA molecule are bound on a surface in such a manner that the DNA is softly stretched. The movement of a fluorescent protein moving along the DNA direction can then be recorded and analyzed. These experiments confirm that proteins may slide along DNA and generally display a standard diffusion dynamics. Experiments also show that the sliding lifetime is sensitive to the salt concentration (Blainey et al., 2006; Bonnet et al., 2008; Riggs et al., 1970). This supports the idea that electrostatics is involved to some extent in the intermittent behavior, with a probable role for the dissolved salt ions. Electrostatics plays indeed a major role in the protein–DNA interaction (Carrivain et al., 2012; Jones, van Heyningen, Berman, & Thornton, 1999; Kalodimos et al., 2004; Nadassy, Wodak, & Janin, 1999; Takeda, Ross, & Mudd, 1992; Viadiu & Aggarwal, 2000; Von Hippel, 2007). The reason is that DNA is very strongly negatively charged (-2 charges per bp). On the other hand, DNA-binding proteins are most often positively charged on the surface that faces DNA, so to be attracted onto it (Jones et al., 1999; Nadassy et al., 1999).

Together with its role in the search kinetics, the protein sliding is also supposed to have another crucial role: it allows the protein to read the DNA sequence and therefore to distinguish the target site among all other sequences. This reading can be performed, besides other interactions, by the formation of hydrogen bonds between the protein and the side of the base pairs exposed toward the major groove, without opening the double helix

(Jones et al., 1999; Nadassy et al., 1999). As the patterns of hydrogen bonds that may be formed on each base pair are different, a protein can discriminate precisely a target site by looking for the formation of the good hydrogen bond pattern along the entire sequence visited. However (and independently from the precise reading interaction), this reading mechanism leads to a paradox. An efficient discrimination between sequences implies indeed a rough interaction energy strongly varying as a function of the protein position along DNA, and such an energy profile leads in turn to a trapping of the protein, which reduces considerably its mobility. The mobility of the protein seems therefore to be in contradiction with its specificity, that is, its capability of discriminating the good sequence (Barbi, Place, Popkov, & Salerno, 2004a; Barbi, Popkov, Place, & Salerno, 2004b; Slusky, Kardar, & Mirny, 2004; Slutsky & Mirny, 2004). This paradox is not always taken into account in the literature concerning target search, but some authors have addressed the problem. Intuitively, one solution seems to be the existence of two different states for the protein: one state where the protein slides but cannot recognize the sequence, and another state where it reads but moves in a much slower way. Mirny and coworkers proposed that the protein could undergo conformational changes between a search state and a recognition state in an intermittent way (Slusky et al., 2004; Slutsky & Mirny, 2004). We have proposed an alternative mechanism, where the key parameter will be the distance between the protein and the DNA (Dahirel, Paillusson, Jardat, Barbi, & Victor, 2009; Paillusson, Dahirel, Jardat, Victor, & Barbi, 2011). As the range of H-bonds is rather short, one can guess that this distance can indeed play a crucial role. Our starting point has been the study of the physics of the interaction between protein and DNA, with a focus on the electrostatic interaction.

A second important ingredient, usually neglected in the modeling of protein–DNA system, emerges from this study: the protein shape. We have shown indeed that a charged convex body (like DNA) counterintuitively repels an oppositely charged concave body (like DNA-binding proteins), provided the two bodies do not exactly neutralize each other (Dahirel et al., 2009; Paillusson et al., 2011). In the following, we will describe how to obtain this result and discuss its implications on the search mechanism, a possible solution for the *mobility-specificity paradox*.

2. PROTEIN DIFFUSION IN THE CELL

2.1. Diffusion: A stochastic regulation tool?

The search of a target DNA sequence may have a particularly evident biological importance in cell differentiation as evidenced in some recent theories. Among others, Kupiec rejects the predominant role attributed to the "genetic program" (all information necessary for the development of the organism is encoded in the genome) and stereospecificity (for each cellular function, there is a specific protein that acts through a deterministic "key–lock" recognition mechanism). An alternative model for cellular functioning is proposed, based on evolutionary approach. In this model, in brief, proteins diffuse into the cell and interact randomly with DNA. Gene expression is also random. However, these interactions are statistically regulated by the position of genes in the cell space and along the genome: the probability of interacting with a closer site is higher, and this effect is strong enough to introduce a differentiation in gene expression. This mechanism finally leads to a kinetic competition that allows to set the appropriate gene expression and to stabilize it as best suited to the needs of the cell (Fig. 7.1) (Kupiec, 1997).

Even without adopting this point of view entirely, it is interesting to note that it involves several important elements of the cell functioning. Most

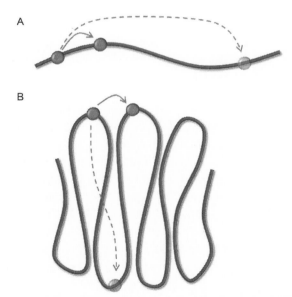

Figure 7.1 In the scenario proposed by Kupiec, the diffusion of proteins is responsible for the activation of different genes. The distance between the site where the transcription factor is synthesized and the gene activation sites determines the speed of search and therefore the efficiency of activation. The distance concerned may be either the linear distance along the molecule (A) or the 3D distance due to the arrangement of the DNA into the nucleus (B). (For color version of this figure, the reader is referred to the online version of this chapter.)

of them are nowadays well substantiated. First, it is clear that the affinity of proteins for their target sequences is relative (see, e.g., Stormo & Fields, 1998). This introduces the problem of obtaining specific recognition while avoiding an excessive competition between slightly different sequences, an effect which can lead to a trapping effect (Gerland, Moroz, & Hwa, 2002; Slusky et al., 2004; Slutsky & Mirny, 2004). On the other hand, it is also clear that there is a stochastic component in the search mechanisms, related to the presence of a diffusive dynamics, which allows proteins to move and meet their specific sequences. It follows that gene regulation depends on a stochastic and complex dynamics, and it is therefore appropriate to propose a statistical physics approach to describe regulation, based on a precise description of diffusion, recognition, and competition mechanisms.

From the point of view of the diffusion dynamics, target sequences search is indeed a very active research field, involving both theoretical and experimental groups. (Halford and Marko (2004) wrote a recent comprehensive review of this literature.) From the pioneering works of Berg and von Hippel (Berg, 1978; Berg & Blomberg, 1976, 1977, 1978; von Hippel & Berg, 1989; Winter et al., 1981), attention has focused on the rate constant of the association reaction between the protein and its target sequence. Then appeared a difficulty: assuming that the protein finds its target by simple random diffusion within the cell leads to reaction times which are too low if compared with those experimentally observed. In 1970, Riggs et al. (1970) showed that the association constant of Lac repressor with the initiation site of the lactose operon was two to three orders of magnitude higher than the theoretical prediction of the Smoluchowski theory for chemical reactions limited by diffusion (Richter & Eigen, 1974) ($k_a \simeq 10^{10}\,\mathrm{M}^{-1}\,\mathrm{s}^{-1}$ against 10^7–10^8 obtained from the theory). In addition, it was noted that the association constant of the Lac repressor with its specific site is also an increasing function of the length of the flanking *nonspecific* DNA present in the sample (Winter et al., 1981). These observations suggest the existence of an additional mechanism, involving the interaction of the protein with nonspecific DNA, and leading to a relevant acceleration of the search for the target sequence.

2.2. 3D versus 1D

It was then proposed that this particular strategy, able to optimize the target search time and called *facilitated diffusion* (von Hippel & Berg, 1989), can be associated with an *intermittent* diffusion, composed of several different displacement modes (Fig. 7.2). The newer idea was to include a mode called

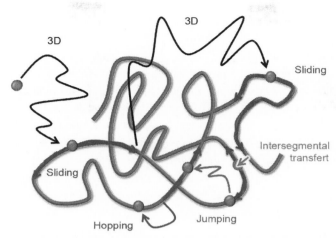

Figure 7.2 The search modes usually considered in literature: 3D diffusion, *sliding* or 1D diffusion along the double helix, *hopping* at a close site, *jumping* to a different DNA stretch, and *intersegmental transfer*, involving simultaneous binding to two distinct DNA stretches. (For color version of this figure, the reader is referred to the online version of this chapter.)

sliding: a 1D, thermal diffusion of the protein along the double helix. The diffusion of the protein during the *sliding* has been initially considered either as a free diffusive movement on a (2D) cylindrical surface surrounding DNA or as a motion along the helical path following one DNA groove. The last hypothesis has the advantage to keep the protein in closest and constant contact with the DNA base pairs, allowing the protein to maintain a specific orientation with respect to the DNA helix. A helical trajectory has been then indirectly proved for the case of some DNA-binding proteins (Blainey, 2009; Dikić, 2012), but the question remains open in general (Kampmann, 2004).

Two other displacement modes, rather similar to each other and called *hopping* and *jumping*, consist of diffusion excursions in the 3D space, allowing the protein to jump to more or less distant sites along the chain. Finally, during *intersegmental transfer*, proteins can transiently bind to two different DNA sites at a time and then directly move from one region to the second one without any intermediate diffusion.

The advantage common to all these mechanisms is to reduce the size of the searched space, thus accelerating the localization of the target sequence. Among them, 1D *sliding* has been soon considered as necessary by most authors. The relative weight of *sliding* with respect to 3D diffusion has then been subjected to debate (Halford, 2009). It is obvious that *pure* sliding would not be very effective if the starting position of the protein on DNA is far from the target sequence, as the protein will then spend too much time in searching remote regions unnecessarily. This effect is of course enhanced dramatically by the slow progression that characterizes diffusion (the visited space scales as the square root of time). Under certain assumptions, it is possible to prove that there exists an optimal choice of the mean times spent in 1D and 3D phases, respectively, that minimize the overall target search time (Coppey et al., 2004). However, the precise mechanisms governing these two types of motion and the transition from one to the other have still not been elucidated.

2.3. Experiments: Biochemistry, atomic force microscopy, and fluorescence microscopy

From the experimental point of view, the possibility to observe 1D diffusion of proteins along DNA has aroused great interest. Biochemical experiments have been performed to measure the average protein–DNA reaction rates as a function of different parameters, and in particular of the lengths of DNA sequences where the target is inserted, were reported (Riggs et al., 1970; Shimamoto, 1999; Winter et al., 1981). A more quantitative and accurate method, but only applicable to certain proteins, is based on the evaluation of the correlation between the activity levels of a protein in two remote sites located at a known distance on a DNA molecule (*processivity*) (Stanford, Szczelkun, Marko, & Halford, 2000). It is interesting to note that, despite its good performances, this experience is open to multiple interpretations (Halford & Marko, 2004), and its results are difficult to reproduce by simple models (Stanford et al., 2000). Alternative techniques, such as atomic force microscopy (AFM; Guthold et al., 1999) (Fig. 7.3) and fluorescence microscopy (Harada, 1999) (Fig. 7.4), allow a direct visualization of the protein movement.

The basic principle of the AFM is to scan the surface of an object by a nanometer-sized tip to reconstruct the geometry of the surface. In the case of protein–DNA systems, protein and DNA can be either fixed adsorbed onto the surface or loosely enough to be able to diffuse on it (Guthold et al., 1999). Despite the very high spatial resolution, this technique was initially limited by a low temporal resolution: tens of seconds between two images. More recently, high-speed AFM allows scanning biological samples in buffer up to 30 frames per second (Ando et al., 2001).

$T = 0$ s $T = 200$ s $T = 400$ s

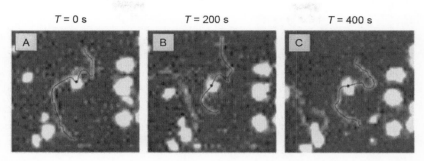

Figure 7.3 Three successive AFM images showing the complex formed by the RNA polymerase of *Escherichia coli*, fixed on a mica surface, and a nonspecific DNA sequence, semi-adsorbed on the same surface. This type of experience can show the relative movement of the protein along DNA, but fails in giving a quantitative description of the diffusion due to geometrical constraints. *Adapted from Guthold et al. (1999).*

However, another limitation, particularly relevant in the study of diffusion, is due to the presence of the surface itself, which limits the free space around the molecules. Double-stranded DNA immobilized on the surface may function as a trap reducing Brownian motion (Sanchez, Suzuki, Yokokawa, Takeyasu, & Wyman, 2011). Similarly, DNA sliding through a fixed protein may induce anomalous diffusion as for the passage of a polymer in a pore (Dubbeldam, Milchev, Rostiashvili, & Vilgis, 2007).

Fluorescence microscopy is used to study processes on large spatial scales and temporal areas (from nanometer to micrometer and from nanosecond to second) (Blainey et al., 2006; Bonnet et al., 2008; Gorman & Greene, 2008; Kim & Larson, 2007; Tafvizi et al., 2008; Wang, Austin, & Cox, 2006). The operating principle is simple: the protein is chemically linked to a fluorescent label (organic fluorophores, fluorescent nanocrystals, fluorescent proteins, quantum dots, . . .) and can therefore be observed optically. In practice, however, the experience is very sensitive and dependent on many details, particularly related to the properties of fluorescence markers (lifetime of the light emission, flashing, . . .).

Moreover, in order to observe the diffusive motion of a protein around a DNA molecule, it is necessary to fix the DNA in an appropriate manner, to immobilize it while leaving the space necessary for the interaction with the protein. Techniques of DNA "combing" have been proposed to this aim. Starting from the DNA molecule in its random-coil configuration (the form in which it is found naturally in solution), one of its ends is first bound on a chemically treated glass surface. Then the surface is slowly withdrawn causing the stretching of the molecules by capillarity. Alternatively, combing can

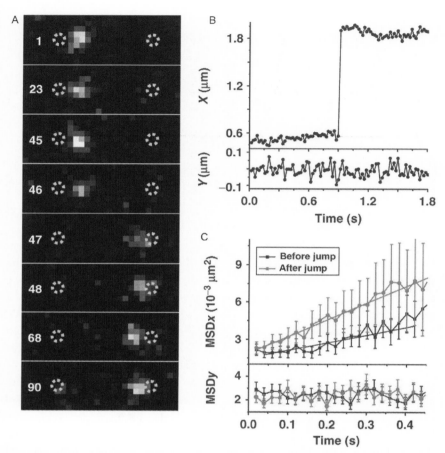

Figure 7.4 Figure of Bonnet et al. (2008) in which *sliding* and *jumping* events are directly observed. (A) Subsequent fluorescent images of the protein (white spot) moving along a stretched DNA (yellow circles on both sides of the figure show the two ends of the DNA segment). Between frames 46 and 47, a *jump* can be observed. (B) Longitudinal (*X*) and transverse (*Y*) displacement of the protein as a function of time. The jump of about 1300 nm is again detected in the *X*-trajectory. (C) The longitudinal MSD calculated before and after the jump display 1D diffusion similar to that observed during events without large jumps. Values of the diffusion constant are between 0.3 and 0.6×10^{-2} $\mu m^2/s$. *Bonnet et al. (2008), figure 3. By permission of Oxford University Press.* (For interpretation of the references to color in this figure legend, the reader is referred to the online version of this chapter.)

be obtained through the application of a hydrodynamic flow of DNA molecules attached at one end: this method enables a more soft stretching, which in addition can be controlled so as to obtain more or less important stretching degrees (Crut, Lasne, Allemand, Dahan, & Desbiolles, 2003).

Similar to any conventional optical microscopy technique, fluorescence microscopy is limited by the diffraction of light. Its resolving power is about 200 nm. However, it is possible to go down to about 30 nm resolution by image analysis techniques for determining the center of the light spot recorded. This gives a good enough resolution to detect the movement of the protein between two successive images, which are usually separated by a few tens of milliseconds.

An example of the results obtained by fluorescence microscopy is represented by the work of Pierre Desbiolles group (Bonnet et al., 2008), an extract of which is given in Fig. 7.4. The registration of the position of the endonuclease *EcoRV* when bound nonspecifically to DNA is decomposed into a longitudinal component and a transverse component. If the latter remains limited, the longitudinal component mean square displacement is proportional to time, consistently with 1D diffusion along DNA. In addition, several dissociation/reassociation events are observed, as indicated by a faster movement leading to the reassociation on a distant DNA position in a single time frame, that is, a *hopping* process following the usual definition (Fig. 7.4A and B).

2.4. Who helps who?

Thanks to fluorescence microscopy experiments, *sliding* has become a reality and its existence as a step in target sequence search is nowadays largely accepted. Nonetheless, the actual role of this searching mechanism is still under discussion. An important element in this discussion has been the Halford's paper (Halford, 2009), where the author contests the need of any mechanism to facilitate the search and affirms that "no known example of a protein binding to a specific DNA site at a rate above the diffusion limit" exists. Indeed, if both 1D and 3D diffusion processes can be observed, the conclusion that facilitated diffusion may greatly enhance DNA–protein association rates is more questionable. The point raised by Halford is that the rapidity of these reactions is instead due primarily to electrostatic interactions between oppositely charged molecules (Halford, 2009). The large association rates reported in the pioneering work (Riggs et al., 1970) were indeed obtained at very low ionic strength, suggesting a role of the electrostatic attraction that becomes negligible, due to screening effects, in higher salt. This conclusion has, however, been overlooked in the following literature, until Halford's work. We emphasize, in particular, the crucial role attributed to electrostatic, a point to which we will come back in the following.

It is also interesting to note that electrostatic should also determine another important feature of the search process, namely the protein–DNA association strength and therefore the lifetime of the 1D diffusion phase, and therefore the relative weight of 1D and 3D processes. This is another important question evoked in discussing the relevance of sliding as an enhancing mechanism in target search. In Gowers, Wilson, and Halford (2005), the same Halford and coworkers showed for the restriction enzyme *Eco*RV that, at low salt, the protein only *slides* continuously on DNA for distances shorter than 50 bp. Transfers of more than 30 bp at *in vivo* salt and over distances of more than 50 bp at any salt always included at least one dissociation step. The authors then conclude that 3D dissociation/reassociation is its main mode of translocation for this protein.

To end this discussion, we would like to point out that that question of the relative role of 3D diffusion and *sliding* can also be seen in an opposite way. Due to the electrostatic attraction, indeed, one can take as reference the weakly bound state where the protein stays along DNA. The question is then whether or not *3D excursions* may help the protein 1D search and reduce the search time. This is the point of view adopted, for example, by the group of Bénichou (Sheinman, Bnichou, Kafri, & Voituriez, 2012).

Whatever the philosophy one adopts, the question of the target sequence search reveals an unexpected richness. Electrostatics seems to be an essential ingredient, and if intermittency is expected to improve the search time in any case, observations and models invoke different *sliding* mechanisms (along the helical path or not), together with *jumps* and *hops*. Moreover, as we will see in Section 3, alternative *sliding modes* have been proposed in order to solve additional difficulties in explaining the protein mobility. It is therefore tempting to ask whether a different "paradigm" for the search, based on a different description (or parameterization) of the whole process, may be more adapted.

3. DIFFUSION ALONG THE DNA: WHAT ROLE FOR THE SEQUENCE?

3.1. Reading the sequence

3.1.1 Direct and indirect interaction

While experiments on *sliding* were multiplying and becoming more refined, this problem was attracting more and more theoreticians, seeking a consistent modeling of the observed phenomena.

Different models have been proposed. However, all models seem to lead to more or less important inconsistencies, and a unified model has not yet been imposed. Some authors (Halford & Marko, 2004; von Hippel & Berg, 1989) consider DNA as a uniform cylindrical space in which the protein is trapped by electrostatic interaction and could slide spontaneously under the effect of thermal agitation. Some models where the protein would even slide along the helical structure of DNA have been envisioned fairly early (Schurr, 1979).

However, as some authors stressed rather soon (Barbi, Place, et al., 2004a; Barbi, Popkov, et al., 2004b; Bruinsma, 2002; Slutsky & Mirny, 2004), the recognition of the target sequence needs a way of *reading* the sequence, which cannot be taken into account by a homogeneous interaction. In order to discriminate the target sequence, it is necessary to introduce a sequence-dependent interaction, albeit small.

To get a concrete picture of this interaction, let us consider as an example a particular protein, the RNA polymerase of T7 virus. The specific complex formed by the T7 RNA polymerase and its target sequence (a gene promoter) has been studied by crystallography (Cheetham, Jeruzalmi, & Steitz, 1999) (Fig. 7.5). The protein–DNA interaction occurs in three regions: in a first region of 5 bp, the double helix is bent by the presence of the protein; in a second region, 5 bp long, a set of hydrogen bonds between the side chains of the protein and the base pairs is made; finally, in correspondence of a third

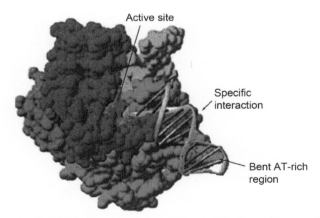

Figure 7.5 Crystallographic reconstruction of the interaction between the RNA polymerase and its T7 target sequence. The three interaction regions mentioned in the main text are indicated. *Adapted from Cheetham et al. (1999).* (For color version of this figure, the reader is referred to the online version of this chapter.)

site, a portion of the protein is inserted between the two helices of DNA causing a local opening of the double helix.

Among the different interactions, some are likely to participate in the target sequence search, and others are probably induced only once the target is reached. The latter interaction, which characterizes the formation of the *open complex* (the preactivated state, ready to start the gene transcription), is most probably absent during the search. The two other modes of interaction are two typical examples of direct (chemical) and indirect (mechanical) interaction (Paillard & Lavery, 2004). The first interaction will include, typically, direct hydrogen bonds to base pairs and Van der Waals interactions (Nadassy et al., 1999; Seeman, Rosenberg, & Rich, 1976). Hydrogen bonds provide the higher level of sequence specificity and may be used to define a simple code to explain sequence reading. In the following, we will precise how this specificity is obtained.

Entropic contributions due either to the loss of degrees of freedom of the protein and DNA or to the expulsion of ions and water molecules from the protein–DNA interface may also contribute to the direct pair of the interaction, but their degree of specificity is less easily quantified.

On the other hand, sequence-dependent changes in DNA structure, or in its mechanical or dynamic properties, can also play a role in recognition (Paillard & Lavery, 2004). Sequence-induced protein deformations may also be considered. Such mechanical effects may be used by the protein as discriminating tools. They may give rise to rather smooth energy profiles (Slutsky & Mirny, 2004), correlated over distances comparable to the length of the target sequence (Fig. 7.6), and have interesting dynamic properties not yet fully explored.

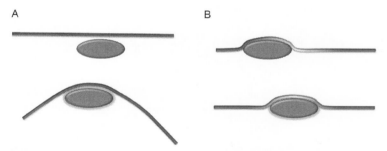

Figure 7.6 Local DNA curvature (A) or flexibility (yellow region, B) can affect the protein–DNA interaction. These physical properties being sequence dependent provide a sequence-dependent contribution to the interaction energy profile. With respect to direct chemical bond, the curvature/flexibility effect is expected to vary in a smoother fashion. (For interpretation of the references to color in this figure legend, the reader is referred to the online version of this chapter.)

3.1.2 Hydrogen bonding

Let us now just take into account the hydrogen bond contribution to the overall interaction and precise its origin. All DNA base pairs expose in the major groove a regular pattern of four chemical groups that can be donors or acceptors of hydrogen bonds (Fig. 7.7). On the other side, a protein like the T7 RNA polymerase presents a reactive site that contains, through the arrangement of its side chains, a recognition pattern containing the information on the correct disposition of donor and acceptor groups in the target. It seems reasonable to assume that the protein looks for this same pattern on any sequence during the search. We also assume that the H–bonds formed in the DNA–protein complex at the recognition site are known (this information can be obtained from crystallographic analysis of the DNA–protein complex). The interaction between the protein and a given sequence can therefore be simply described by counting the number of bonds it can make at that position, that is, the number of DNA groups that are consistent with the protein recognition pattern. Within this model, the protein can be represented by a *recognition matrix* containing the pattern of H-bonds formed by the protein and the DNA at the recognition site (Fig. 7.8).

When the protein is at position n, the sequence that it is visiting can be represented as a list of vectors, $D^{(n)} = b_{n+1}, b_{n+2}, \ldots, b_{n+N}$, where

$$
b_n = \begin{cases} (1, -1, 1, 0)^T & \text{for base A} \\ (0, 1, -1, 1)^T & \text{for base T} \\ (1, 1, -1, 0)^T & \text{for base G} \\ (0, -1, 1, 1)^T & \text{for base C} \end{cases}
$$

and where the number N of vectors corresponds to the length of the visited sequence. The recognition matrix is then a $N \times 4$ matrix containing the "good" pattern of hydrogen bonds, that is, the one that will be made on the target. In the specific case of the T7 DNA polymerase, for example, the recognition matrix reads

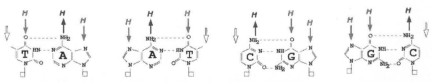

Figure 7.7 Hydrogen bond acceptor (red) and donor (blue) sites on the four base pairs accessible through the major groove. Note that a similar four-site pattern can be defined for each base pair, but associated with a different acceptor/donor order. (For interpretation of the references to color in this figure legend, the reader is referred to the online version of this chapter.)

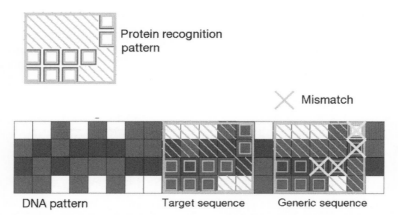

Figure 7.8 While sliding along DNA, the protein applies a recognition pattern to *read* the sequence by counting the number of acceptor or donor groups that corresponds to its own motif. (For color version of this figure, the reader is referred to the online version of this chapter.)

$$R = \begin{pmatrix} 1 & 1 & 0 & 0 \\ 1 & -1 & 0 & 0 \\ 1 & 1 & 0 & 0 \\ 0 & 1/2 & 0 & 0 \\ 0 & 0 & 1/2 & 1 \end{pmatrix} \tag{7.1}$$

where the factors $1/2$ have been introduced in order to reproduce one hydrogen bond shared by 2 bp. The interaction energy for the protein at position n is then given by the sum of all positive matches and can be written as

$$E(n) = -\mathcal{E} \sum_{i=1}^{N} \sum_{j=1}^{4} \max\left(R_{ij} D_j^{(n)}, 0\right) \tag{7.2}$$

where $-\mathcal{E}$ is the net energy gain of a single hydrogen bond (of the order of a fraction of $k_B T$ (Tareste et al., 2002), see discussion below).

Equilibrium measurements (Stormo & Fields, 1998) reveal that the binding energy of a protein to a given sequence can be described, to a good approximation, as the sum of the binding energies to the single base pairs composing the sequence. If the latter can be assumed as independent, then the binding energy can be reasonably described as a Gaussian random variable (Gerland et al., 2002; Sheinman et al., 2012). This is indeed what is measured for some real cases (Sheinman et al., 2012); in the case of the T7 RNA polymerase, the same results have been derived based on a detailed analysis of the protein hydrogen bond pattern (Barbi, Place, et al., 2004a; Barbi, Popkov, et al., 2004b).

3.2. The recognition-mobility paradox

The 1D diffusion along DNA (*sliding*), apparently simple, may hide an unexpected complexity. Most of the authors assume, however, for this diffusive phase a simple diffusive dynamics or *normal diffusion*. In this case, the mean square distance traveled by the protein along DNA after a time t is proportional to time, that is, $\langle r^2 \rangle = 2Dt$, where the only parameter that remains to be fixed is the diffusion constant D. Now, if this model is appropriate when the interaction energy is absolutely uniform along the DNA, it is no longer valid when a sequence-dependent energy profile is taken into account.

Starting from the previous definition of the protein–DNA 1D energy profile,[1] it is easy to model the 1D diffusion. The protein moves by one-site steps on the energy landscape $E(n)$, with rates of translocation between neighboring sites n and $n' = n \pm 1$ defined according to the *Arrhenius law*, that is, proportional to $\exp(-\beta(E(n') - E(n)))$ whenever $E(n') - E(n) > 0$, while it is constant if $E(n') - E(n) \leq 0$. Both expression can be formally written as an identical exponential term of the form $\exp(-\beta \Delta E_{n \to n'})$ by defining $\Delta E(n \to n') = \min(E(n') - E(n), 0)$. If, moreover, we want to include a nonzero probability for the protein to stop at one position, the complete set of translocation rates will reads

$$
\begin{aligned}
r_{n \to n'} &= 1/2 \ \exp(-\beta \Delta E_{n \to n'}), \quad n' = n \pm 1 \\
r_{n \to n} &= 1 - r_{n \to n+1} - r_{n \to n-1}
\end{aligned}
\tag{7.3}
$$

where $\beta = 1/k_B T$.

Note that the case $\Delta E_{n \to n'} = 0 \, \forall n$ corresponds to a constant energy landscape, that is, to a simple 1D diffusion process with diffusion constant $2D = 1$. This limit can also be recovered in the case where $\mathcal{E} = 0$.

The numerical study of this diffusion process gives a predictable result: the trapping effect due to the roughness of the potential gives rise to subdiffusion (Barbi, Place, et al., 2004a; Barbi, Popkov, et al., 2004b). Figure 7.9 shows this effect as a function of the potential roughness $\beta \mathcal{E}$.

In the limit of a $\beta \mathcal{E} = 0$, that is, in the case of a flat underlying potential, the diffusion is of course standard, with $D = 1/2$ and a linear dependence on

---

[1] The energy profile described here may be enriched by adding energy barriers for the translocation from any DNA position to the next one. The results are quantitatively, but not qualitatively, affected.

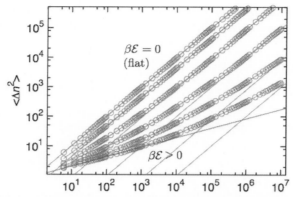

Figure 7.9 Mean squared displacement obtained by simulating the diffusion of a particle on the rough energy profile associated with by hydrogen bonding and defined in the main text. From the upper curve to the bottom: $\beta\mathcal{E}=0$, 0.3, 0.6, 0.9, 1.2, 1.5. Red lines of slope 1 and one blue line of slope 0.3 are reported for comparison. (For interpretation of the references to color in this figure legend, the reader is referred to the online version of this chapter.)

time, so that the corresponding curve is a straight line of slope 1 in the log–log plot. For larger values of $\beta\mathcal{E}$, the dynamics shows initially large deviations from the normal diffusion: in these finite temperature cases, the mean square distance is no longer proportional to time but increases as a power of time which is smaller than unity, according to the law

$$\langle \Delta n^2 \rangle = A t^b, \quad b < 1. \tag{7.4}$$

This effect is transitory: the diffusion becomes normal when one considers long enough time. Accordingly, the exponent b increases with time toward its equilibrium value of 1. This is due to the characteristics of the energy profile, which is rough, but bounded. Roughness thus affects the diffusion for short times, that is, relatively small distances, but it is smoothed out when longer displacements are considered. Overall, on long timescales, it only affects the average.

However, the lifetime of the nonspecific DNA/protein complex can be relatively short: normal diffusion behavior can never be reached, and subdiffusion may be the most appropriate description of the protein motion. Moreover, even if the normal diffusion regime is reached, the transitory subdiffusive phase will significantly change the overall distance traveled by the protein after a given time. By focusing for instance on the time needed to perform a mean squared displacement of 100 bp^2 (therefore, a typical distance of 10 bp), we can see from Fig. 7.9 that this time can be increased by up to three orders of magnitude for the values of $\mathcal{E}$ used.

In conclusion, this deviation from normal diffusion is not a merely academic question: all quantitative estimates made to determine the respective roles of 1D and 3D search would be affected and should be recalculated in view of these results. We stress that it is not easy to obtain a reasonable estimate of the $\mathcal{E}$ parameter. However, rough estimates based on typical hydrogen bond energies (of the order of a fraction of $k_B T$; Tareste et al., 2002) do not seem compatible with the double requirement of a protein which has to be not only free enough to slide along the DNA molecule but also able to bind its target sequence with an energy much higher than for other sequences so as to ensure a good specificity (Barbi, Place, et al., 2004a; Barbi, Popkov, et al., 2004b; Sheinman et al., 2012; Slusky et al., 2004; Slutsky & Mirny, 2004). In this sense, the trapping effect observed in this simple model evidences the existence of a *recognition-mobility* paradox (also called *speed-stability* paradox in the literature). A different way of presenting the paradox, although leading to the same conclusions, is to show that disorder in the binding energy profile on which diffusion takes place leads to an effective diffusion constant that decreases exponentially with the variance of the energy distribution (Sheinman et al., 2012; Slusky et al., 2004; Swanzig, 1988). Essentially, the requirement of a reasonable specificity prohibits the protein to diffuse.

3.3. Two-state models

To solve this paradox, some new mechanisms have been invoked. One of them can be a modified energy distribution where the binding energy at the target is reduced without affecting the energy distribution variance. However, experimental data do not support this hypothesis (Sheinman et al., 2012).

An alternative solution may be associated with protein conformational fluctuations, thus leading to introduce "two-state" models. In brief, the idea is to provide two different 1D *sliding modes*: first, *reading mode*, where the protein is able to *read* the sequence with a reduced mobility, and second, *diffusing mode* where the protein is able to move relatively rapidly along the double helix but is essentially blind to the sequence (Mirny et al., 2009; Murugan, 2010; Winter et al., 1981; Zhou, 2011). The conformation change was initially attributed to a microscopic binding of the protein to the DNA accompanied by water and ion extrusion, but such a transition is usually accompanied by a large heat capacity change (Spolar & Record, 1994)

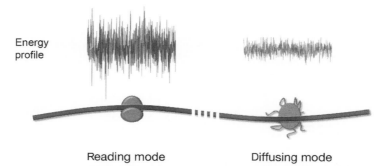

Reading mode   Diffusing mode

Figure 7.10 Slutsky and Mirny (2004) hypothesize that partial denaturation of the protein may be responsible for a significant change in the *effective* energy profile associated with the interaction with DNA. In the *diffusing mode*, the partially denatured protein is much less sensitive to the sequence and its mobility is therefore increased. (For color version of this figure, the reader is referred to the online version of this chapter.)

that in turn needs significant structural changes to be accounted for. Hence, it has been proposed that these two states can be associated to distinct conformational states of the protein–DNA complex (Gerland et al., 2002), eventually associated to a partial protein unfolding (in the *diffusing mode*; Slusky et al., 2004) (Fig. 7.10). However, this mechanism is only efficient if an effective correlation between the transitions between the two modes and the "underlying" energy profile exists. In this way, the transition to the *reading mode* happens mainly when the protein is trapped at a low-energy site of the search landscape, this being related to a mechanism based on residence times (Slusky et al., 2004).

A recent analysis of the efficiency of such mechanism seems to rule out these models, based on quantitative estimates of the relevant parameters (Sheinman et al., 2012). Similarly, it is shown that the presence of a large number of copies of the same protein can resolve the *recognition-mobility* paradox only if the energy profile has a small variance (Sheinman et al., 2012). Instead, a new mechanism which is based on *barrier discrimination* is proposed, which allows to obtain a possible solution for the process (Sheinman et al., 2012). The basic idea is again that the protein has two different conformations, but the additional element is that these conformations are separated by a free energy barrier whose height *depends* on the position along DNA. This implies a difference between transition rates from the *diffusing* to the *reading* mode that finally allows the protein to improve its search time as requested.

But how can this model be justified from a physical point of view? A rationale for this model had already been proposed, based on a more *physical* approach to DNA–protein interaction (Azbel, 1973; Paillusson, Barbi, & Victor, 2009), which we develop in Section 4.

4. ELECTROSTATICS: THE DNA–PROTEIN INTERACTION
4.1. DNA

In the approaches to the study of the kinetics of protein search described until now, the physics of the DNA–protein interaction is only indirectly taken into account. In particular, a description of the *electrostatic* interaction between the two macromolecules in solution was completely missing. In reality, as already mentioned, electrostatics plays a fundamental role in this system.

The mechanical behavior of a DNA molecule of given length can be described, in an effective manner, by different models of polymers (Cocco, Marko, & Monasson, 2002). Different models can be in rather good agreement with experimental results for force–extension experiences, typically performed using optical or magnetic tweezers. In this setup, one end of a DNA molecule is bound to a flat substrate and the other end to a colloidal bead that can be manipulated by an external optical or a magnetic field so to exert a force on the bead and thus on the DNA molecule. The best fit of the resulting data is given by the *worm-like chain* model, describing the DNA as an elastic rod (Fig. 7.11). The torsional rigidity of the rod is accounted for by a given value of the *persistence length* L_p.[2] For DNA, L_p is about 50 nm, that is, approximately 150 bp. This is a quite unusual value for a polymer of ~ 2 nm thickness: we could expect a higher flexibility at a scale much larger than the thickness.

This large persistence length depends on an aspect of DNA that has not yet discussed: it is a polyelectrolyte, that is, a charged polymer. Each phosphate group in the DNA backbone is indeed negatively charged. As there are two phosphate groups per base pair in double-stranded DNA, this corresponds to a linear charge density of the order of $-2\ e$/bp (3.4 nm), or $-6\ e$/nm, or finally a surface charge density of the order of $-1\ e$/nm^2. In comparison, if a power cable in the air had the same surface charge density, the potential difference with respect to the ground would be four orders of magnitude larger than the breakdown voltage in dry air. The DNA molecule

---

[2] Explicitly, the persistence length can be defined as the characteristic length of the exponential decreasing of the angular correlation of the tangent vector to the polymer (see, e.g., Cocco et al., 2002).

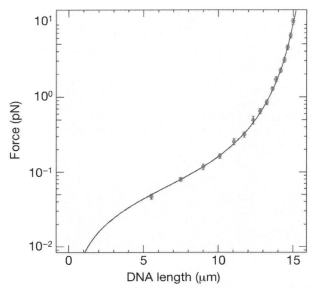

Figure 7.11 Typical experimental results for the extension of DNA when subjected to a constant force, fitted by the *worm-like chain* model. (For color version of this figure, the reader is referred to the online version of this chapter.)

is therefore a highly charged molecule. As a consequence, the phosphate groups strongly repel each other, despite the screening effect due to ions in solution. This adds to the natural rigidity of DNA an additional stiffness, which justifies its large persistence length. At the protein scale, which is of the order of a few tens of nanometers, the DNA molecule can therefore be modeled as a rigid cylinder of radius $R_{DNA} = 1$ nm, carrying a constant surface charge density of $-1\ e/\text{nm}^2$. For simplicity, we can also assume that the dielectric properties of DNA are those of pure water, that is, $\varepsilon_{DNA} = \mathcal{E}_w = 80$.

4.2. Proteins

4.2.1 Charge

Nonspecific interactions between proteins and DNA are poorly documented, but the predominance of electrostatic is undeniable (Jones et al., 1999; Kalodimos et al., 2004; Nadassy et al., 1999; Takeda et al., 1992; Viadiu & Aggarwal, 2000; Von Hippel, 2007). Proteins that bind to DNA are most often positively charged. More precisely, positively charged *patches* are observed in the region which faces the DNA when the specific complex is formed, an effect which can be accounted for by evaluating the *propensity* of positive residues to occur more frequently in a

DNA-binding interface (Ahmad & Sarai, 2004; Dahirel et al., 2009; Jones et al., 1999; Jones, Shanahan, Berman, & Thornton, 2003; Nadassy et al., 1999; Stawiski, Gregoret, & Mandel-Gutfreund, 2003; Szilágyi & Skolnick, 2006).

As an illustration of this effect, we show in Fig. 7.12 an analysis of the large dataset of DNA-binding protein features presented in Jones et al., 1999. Among the proteins analyzed in this work, it is possible to identify a large family of specific proteins, that is, binding to specific sequences: this family includes transcription factors, TATA-binding proteins, and restriction enzymes. Other nonspecific proteins, such as eukaryotic polymerases, repair proteins, and histones, form a second group. We evaluated the surface charge of these proteins in the region of interaction with DNA by counting the charged residues at the interface, and we obtained a very interesting histogram of the charge densities. In all cases, the DNA–protein interface results to be positively charged. Interestingly, in the case of proteins that recognize specific sequences, such as transcription factors and restriction enzymes, we obtained an average density of surface charge $\sigma_{\mathrm{prot}} = (0.17 \pm 0.03)e\ \mathrm{nm}^{-2}$. Besides, we find that nonspecific proteins are more charged: we get $\sigma_{\mathrm{prot}} = (0.27 \pm 0.05)e\ \mathrm{nm}^{-2}$.

Now, the main role of the positive charge of the protein is, of course, to create an electrostatic attraction to DNA. But the difference observed between different classes of proteins and the fact that their charge seems to be rather finely tuned suggest that the surface charge may have a more precise function in the interaction with DNA that it would be interesting to elucidate.

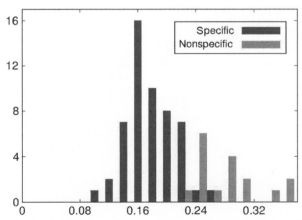

Figure 7.12 Histogram of the surface charge density of the interface binding proteins to DNA. Specific (blue) and nonspecific (orange) proteins are separately considered. (For interpretation of the references to color in this figure legend, the reader is referred to the online version of this chapter.)

4.2.2 Shape

If the charge of the protein immediately appears as one of the main ingredients in an electrostatic model of the protein–DNA interaction, another potentially essential ingredient is less easily recognized. Yet, one of the most characteristic aspects of the DNA-binding proteins is their shape complementarity with DNA. DNA-binding proteins often have a concave shape that fits closely DNA. They can cover the DNA molecule by using up to 35% of their surface (Jones et al., 1999). Averaging over different types of proteins, one obtains for the average surface of the interface a value of S_{prot} ~15 nm^2 (Dahirel et al., 2009; Jones et al., 1999; Nadassy et al., 1999; Stawiski et al., 2003). Generally, and particularly for enzymes, electrostatic patches and significant protein concavities often overlap, so that DNA is "inserted" in this concavities leading to a quite typical *enveloping* or *complementary* shape (Jones et al., 1999; Stawiski et al., 2003) (Fig. 7.13).

This shape complementarity of DNA-binding proteins and DNA enables to maximize the number of direct interactions with DNA base pairs (Jones et al., 1999; Nadassy et al., 1999). Interfaces of DNA-binding proteins have indeed on average more potential hydrogen-bonding groups (more than twice as many) compared with regions that do not bind DNA (Stawiski et al., 2003). In the specific complex, these bonds may closely stack the protein to DNA, so that interfaces exclude solvent molecules from the interstitial space. However, it is tempting to ask whether this particular protein shape may play a role in *nonspecific* protein–DNA interactions, at work during the target sequence search. In this regard, it is

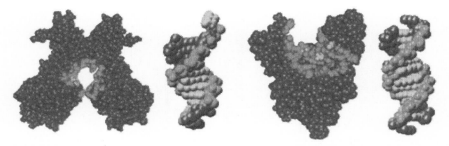

Figure 7.13 Two examples of complementary shape proteins. Left: NF-kB (1nfk); right: *Eco*RI restriction endonuclease (1eri). In blue are represented residues of the protein that do not contact DNA (in red). All protein and DNA groups which come in close contact and form the interface in the protein–DNA complex are shown in green. *Adapted from Jones et al. (1999).* (For interpretation of the references to color in this figure legend, the reader is referred to the online version of this chapter.)

interesting to note that structural studies of some nonspecific protein–DNA complexes show a gap between the two macromolecules, filled with solvent (Jones et al., 1999; Kalodimos et al., 2004; Nadassy et al., 1999; Viadiu & Aggarwal, 2000; Von Hippel, 2007). This observation suggests the existence of a force that counteracts the electrostatic attraction. If this is the case, the question arises as to the physical origin of this repulsive force and how it depends on the precise value of the surface charge of the protein.

4.3. A Monte Carlo study

In order to describe the electrostatic interaction between protein and DNA and the role of the protein charge and shape, we developed a minimal model of DNA–protein system to be studied by Monte Carlo simulation (Dahirel et al., 2009; Paillusson et al., 2011). We modeled the DNA as a regular cylinder, 2 nm in diameter. To compare different protein shapes, we modeled the protein by simple solid bodies: either a sphere, or a cylinder, or a cylinder with a cylindrical cavity. Hollow cubic shapes have been also tested. DNA charges are placed on its axis, and protein charges are placed just below the surface which faces the DNA. The relative orientation between the protein and the DNA was fixed so to orient the charged surface of the protein toward DNA. The distance L between two objects was then varied.

The two bodies are placed in a simulation box with periodic boundary conditions, where water and ions are described by *primitive model* (Hansen & Löwen, 2000): the solvent is treated as a continuum dielectric medium with dielectric constant ε_w, while all ions are modeled by small charged spheres of radius 0.15 nm. Monte Carlo simulation was done in the presence of monovalent salt corresponding to physiological conditions (0.1 mol L^{-1} or 0.06 molecules nm^{-3}). The electrostatic forces acting between protein and DNA can then be calculated and integrated to obtain the free energy profile as a function of the DNA–protein distance L (Dahirel, Jardat, Dufrêche, & Turq, 2007a, 2007b).

Monte Carlo simulations show that while the overall shape of the protein has little influence on the interaction, its complementary with DNA is crucial. The complete comparison of the different protein models have been presented in Dahirel et al. (2009). The main result of this analysis is presented in Fig. 7.14, where the free energy profiles obtained with the spherical and complementary shapes (Dahirel et al., 2009; Paillusson et al., 2011). While in the case of a spherical protein the electrostatic interaction is always attractive, in the case of complementary surfaces a repulsion appears below a distance L of a fraction of nanometer (0.1–0.75 nm, as a function of the protein charge).

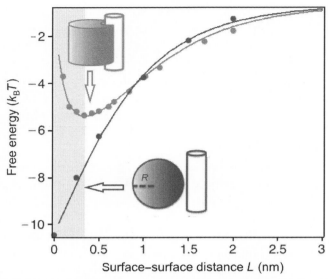

Figure 7.14 Monte Carlo (points) and Poisson–Boltzmann (lines) estimations of the protein–DNA electrostatic interaction for two different protein shapes: a spherical one blue, lower curve and concave, DNA-matching one red, upper curve. In both cases, the results from Poisson–Boltzmann theory applied to the two-plat geometry are adapted to the curved surfaces by mean of a Derjaguin approximation. In the concave case, the osmotic repulsion is clearly observed, while it is absent in the spherical case due to the highly limited area of the interface. (For interpretation of the references to color in this figure legend, the reader is referred to the online version of this chapter.)

A "naive" modeling of the protein as a sphere might be therefore not suitable for the study of the electrostatic interaction. This result is remarkable: above a distance of the order of a nanometer, the protein is *repelled* instead of being attracted by DNA. We will discuss the possible biological role of such an effect in Section 6, but before, we would like to give a closer look at the physical mechanism leading to this rather surprising effect.

5. THEORETICAL APPROACH

What is the physical origin of the repulsion? It is obviously related to the fact that the two charged bodies are immersed in an ionic solution: the physical description of the system will therefore require some notion from colloidal systems physics. On the other hand, Monte Carlo simulations showed that this repulsion is related to the presence of the two

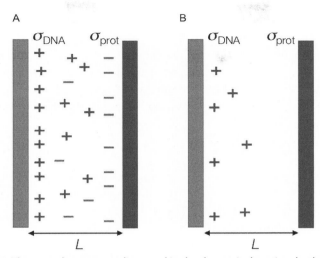

Figure 7.15 The two-plate system discussed in the theoretical section, both in presence of salt (A) or in the counterions only regime (B). (For color version of this figure, the reader is referred to the online version of this chapter.)

complementary surfaces, which create a large interface between the two charged macromolecules. We can then assume that for small distances between the two bodies, the system can be reasonably approximated by two planar charged surfaces approaching one another (e.g., the DNA plate at $x=0$ and the protein one at $x=L$, as in Fig. 7.15). This model is very simplified but, precisely for this reason, can be solved by a semi-analytical approach (Ben-Yaakov, Burak, Andelman, & Safran, 2007; Lau, 2000; Ohshima, 1975; Paillusson et al., 2009; Parsegian & Gingel, 1972) whose physical insights are summarized in this section. We will see that having monovalent ions in solution has two consequences on the attraction between two oppositely charged plates. First, ions generate an osmotic repulsion due to the loss of available space for them to move as the plate-to-plate distance decreases. Second, a screening effect due to the presence of a salt in solution. To gain as much physical insight as possible, we shall introduce these two aspects one at a time, starting with the osmotic repulsion.

5.1. Counterions only

We start considering a protein–DNA system modeled as two plates with only one type of monovalent counterions in between so as to ensure electroneutrality (Fig. 7.15B). On the one hand, if $\sigma_{DNA} < 0$ and $0 < \sigma_{prot} < |\sigma_{DNA}|$, respectively, denote DNA's and protein's surface charge densities, then the *direct* electrostatic force per unit area between them is $\Pi_{elec} \approx -|\sigma_{DNA}\sigma_{prot}|/2\varepsilon$. On the other hand, modeling the ions as an ideal gas in a slit of width L, the corresponding osmotic pressure is $\Pi_{osm} \approx n_c k_B T$ with $n_c = (|\sigma_{DNA}| - \sigma_{prot})/L$. Balancing these two pressures yields an equilibrium distance that reads

$$L_{eq} = |\lambda_{DNA} - \lambda_{prot}| \tag{7.5}$$

where we introduced the Gouy–Chapman (GC) length $\lambda_X = 1/(2\pi l_B |\sigma_X|)$ for a plate with surface charge σ_X (in units of e per unit area) and where $l_B = e^2/(4\pi\varepsilon k_B T)$ is the Bjerrum length. In this first limiting case, we have therefore easily estimated the equilibrium distance between the two plates due to the imbalance between electrostatic attraction and ion osmotic pressure.

A comment on the GC length will be useful. The GC length represents the width of the layer of counterions condensed at the plate of charge σ_X they neutralize. It can be retrieved by seeking at what distance from the plate a condensed counterion would go because of a thermal fluctuation. The counterion density at a distance $x > 0$ from the charged plate[3] reads $n_c(x) = (\lambda_X + x)^{-2}/(2\pi l_B)$ (Lau, 2000). Two things are worth noting from this formula. First, the density at zero is $n_c(0) = \sigma_X/\lambda_X$. This result is easily understandable from a physics point of view, as it could have been obtained by imagining that all the counterions are trapped in a layer of width λ_X. Second, as the charge density is not uniform and actually decays as x increases, the cumulative ionic charge over n GC lengths is $\sigma_X(1 - 1/(n+1))$ so that for $n = 1$, only 50% of the charge of the plate is screened (instead of the 100%, one would have guessed from the density at the plate and with uniform assumption).

5.2. High salt concentration

When salt with bulk concentration n_b is added to the system, each counterion has a screened electrostatic interaction with the others and, at a coarser level, the plates also have a screened electrostatic interaction. This screening effect is accounted for by a unique parameter called the Debye screening parameter, $\kappa \equiv \sqrt{8\pi l_B n_b}$ for a 1:1 symmetric electrolyte. It is more intuitive to look at the inverse Debye parameter, $\lambda_D = \kappa^{-1}$, called the Debye length,

---

[3]The given formula works when one considers a plate and a fully neutralizing solution on its right, that is, there is no electrolyte on the left of the plate.

that can be understood as the effective range of the electrostatic interactions in solution.

The osmotic effect, still related with ions thermal motion, plays two different roles when salt is added. First, trapped counterions tend to repel the plates; second, bulk ions tend to increase their accessible volume at the expense of the volume between the plates, and therefore contribute attractively to the osmotic pressure. At high salt concentration, the resulting positive excess osmotic pressure in between the plates reads (Parsegian & Gingel, 1972) $\delta\Pi_{\text{osm}} = 2n_b(\cosh\psi - 1)k_B T \approx n_b\psi^2 k_B T$ where $\psi(x) = \beta e\phi(x)$ is the dimensionless electrostatic potential at x. If we moreover imagine that at close protein–DNA distances L the dimensionless potential is dominated by the most charged plate (i.e., the DNA plate), then we have at the protein plate $\psi \approx 2\lambda_D e^{-\kappa L}/\lambda_{\text{DNA}}$ and $\delta\Pi_{\text{osm}} \approx 4n_b\lambda_D^2 e^{-2\kappa L}/\lambda_{\text{DNA}}^2$.

As the electrostatic force is screened, we can assume that at the protein plate it equals $\Pi_{\text{elec}} \approx -|\sigma_{\text{DNA}}\sigma_{\text{prot}}|e^{-\kappa L}/2$. As before, equating these two contributions allows one to get an equilibrium distance

$$L_{\text{eq}} \approx \lambda_D \left| \ln\frac{\lambda_{\text{prot}}}{\lambda_{\text{DNA}}} \right|. \tag{7.6}$$

Although the assumptions we used to derive Eq. (7.6) in a simple manner seem very restrictive, this last result is much more robust and holds whenever the salt concentration is high (Ohshima, 1975; Paillusson et al., 2009; Parsegian & Gingel, 1972). It is also worth noting that Eq. (7.6) can be rewritten in a way similar to Eq. (7.5) by introducing an effective counterion cloud size at high salt concentration $\lambda_X^{\text{salt}} \approx \lambda_D(\ln 2 + \ln\kappa\lambda_X)$ so that Eq. (7.6) reads now

$$L_{\text{eq}} \approx \left| \lambda_{\text{DNA}}^{\text{salt}} - \lambda_{\text{prot}}^{\text{salt}} \right|. \tag{7.7}$$

The expression given for λ_X^{salt} cannot be interpreted as simply as the GC length because the presence of salt in the system imposes one to choose explicitly a gauge for the potential ψ (Tamashiro & Schiessel, 2003). In practice, the potential offset is commonly chosen so as to be zero in bulk solution (i.e., far away from the plates). This implies that in a high salt regime the potential $|\psi_0|$ at the plate is of order $\mathcal{O}(1/(\kappa\lambda_X)) \ll 1$ and asking at which distance from the plate a fluctuation $k_B T$ can bring a counterion does not make sense in this context (while it did in the absence of salt). Finding an interpretation is not desperate however and one can check easily that at a distance $n\lambda_X^{\text{salt}}$ away from the plate, the potential is of order

$\mathcal{O}\left(1/(\kappa\lambda_X)^{n+1}\right) \ll |\psi_0| \ll \mathcal{O}$ (Eq. 7.1). Hence, each step λ_X^{salt} away from the plate decreases drastically—by the same proportion—the potential toward zero. Another way to look at this question is to compute the cumulative charge over a width $n\lambda_X^{salt}$ from the plate. This quantity scales as $\sigma(1 - 1/(2\kappa\lambda_X)^n)$: hence, almost 100% of the plate charge is screened by this cumulative charge and we now exactly how far it is from 100%. Finally, note that the cumulative ionic charge in the high salt case is much faster closer to the charge plate σ_X than in the counterion case. This reflects the very different behavior of the charge density in those two cases: in the case of counterions, only the charge density decays algebraically, while in the high salt case it decays exponentially.

5.3. General case

In general, the screening effects do not write as simple exponentials and both electrostatic and osmotic contributions are complicated to assess. Eventually, one can find the exact equilibrium distance within the Poisson–Boltzmann framework (Ben-Yaakov et al., 2007; Ohshima, 1975; Paillusson et al., 2009). We will try to give an intuition for the result by extrapolating the above relations (7.5) and (7.7) to a more general situation. We will assume that if an equilibrium distance exists, then it should take the form of a difference between two effective counterions cloud sizes λ_{DNA}^{eff} and λ_{prot}^{eff}, respectively, brought by the DNA and the protein plates. For each plate of charge density σ_X, this effective length has to be a function of λ_X and λ_D. In addition, in low salt regime (i.e., $\kappa\lambda_X \ll 1$), $\lambda_X^{eff} \to \lambda_X$ while at high salt concentration (i.e., $\kappa\lambda_X \gg 1$), $\lambda_X^{eff} \to \lambda_X^{salt}$. The only form that satisfies these constraints is

$$\lambda_X^{eff} = \lambda_D \arcsin h(\kappa\lambda_X). \tag{7.8}$$

A full physical analysis of this particular lengthscale in the general case of a single plate neutralized by an electrolyte can be done semi-analytically from an exact formula for the potential (see, e.g., Ben-Yaakov et al., 2007) or numerically. Here, we will just emphasize that, after n steps of size λ_X^{eff}, the potential goes as $\sim\gamma/(\gamma + 2\kappa\lambda_X)^n$, where $\gamma > 0$ and for n sufficiently big and therefore tends to zero. Depending on the value of $\kappa\lambda$, the true charge density will lie in between an algebraically decaying form and an exponentially decaying one so that the cumulative ionic charge gotten over a width λ_X^{eff} can take any value in between 50% and 100% of the charge plate.

Now, extrapolating from before we therefore assume that

$$L_{eq} = \left| \lambda_{DNA}^{eff} - \lambda_{prot}^{eff} \right| = \left| \ln \frac{\kappa \lambda_{DNA} + \sqrt{\kappa^2 \lambda_{DNA}^2 + 1}}{\kappa \lambda_{prot} + \sqrt{\kappa^2 \lambda_{prot}^2 + 1}} \right|. \tag{7.9}$$

This last assumption can in fact be retrieved analytically and has been tested extensively in the past (Ben-Yaakov et al., 2007; Ohshima, 1975; Paillusson et al., 2009).

5.4. Energy at the minimum

Although not intuitive, we have tried to give some motivations for the expression (7.9) that takes the equilibrium position at which (excess) osmotic and (screened) electrostatic pressures balance each other in the general case. Now, it so happens that the free energy per unit area at this very equilibrium position can also be derived exactly and reads (Ohshima, 1975; Paillusson et al., 2009)

$$\beta \Delta F_{well} = 4\sigma^* \left[\sqrt{(\kappa \lambda^*)^2 + 1} - \kappa \lambda^* - \arcsin h \left(\frac{1}{\kappa \lambda^*} \right) \right], \tag{7.10}$$

where σ^* and λ^* are, respectively, the smallest surface charge density (in absolute value) and its corresponding GC length. In our case, $\sigma^* = \sigma_{prot}$.

The free energy per unit area of Eq. (7.10) gives the depth of the electrostatic well at equilibrium and is therefore a direct measure of its stability. Akin to Eq. (7.9), expression (7.10) is quite difficult to guess, in particular, because osmotic and electrostatic effects are now completely intertwined. We can still try to give a flavor of what is happening at least in the high salt regime when $\kappa \lambda_{prot} \gg 1$. In this case, we make use of the fact that $\sqrt{x^2 + 1} \sim x + 1/(2x) + \mathcal{O}(1/x^2)$ as $x \to \infty$ and Eq. (7.10) gives thus $\beta \Delta F_{well} \sim -2\sigma_{prot}/(\kappa \lambda_{prot})$. Let us try to derive this result directly, in the high salt regime. To do so, let us assume that only the screened electrostatic part $\Pi_{elec} \approx - |\sigma_{DNA} \sigma_{prot} e^2 | e^{-\kappa x}/(2\varepsilon)$ is working and that we can neglect any osmotic effect. Integrating Π_{elec} term from infinity to L_{eq} should give us an estimate of the depth of the well. We obtain

$$\Delta F_{well} \approx - \int_{\infty}^{L_{eq}} dx \, \Pi_{elec}(x) \approx - \frac{|\sigma_{DNA}| \sigma_{prot} e^2 \lambda_{DNA}}{2\varepsilon \kappa \lambda_{prot}} \quad \text{(high salt regime).}$$

$$\tag{7.11}$$

Doing a little more algebra leads us to the result $\beta \Delta F_{well} \approx - \sigma_{prot}/(\kappa \lambda_{prot})$ which differs from the exact formula in the high salt limit only by a factor 2 (Parsegian & Gingel, 1972). This missing factor 2 comes from the fact that there is an entropy gain from releasing salt into the bulk as the plates are brought closer from infinity and therefore, the interaction is more attractive than with screened electrostatic only (Ben-Yaakov et al., 2007; Dahirel et al., 2009; Parsegian & Gingel, 1972).

In the simple calculation above, we can also get some insights about why does the well depth only depends on one charge density. As we have seen, the electrostatic pressure is symmetric under the operation of exchanging the plates, hence does not prefer one plate over the other. The equilibrium length, however, has to be positive and cares about which charge density is the smallest. This is therefore the evaluation of a symmetric term in charge densities at a position that is an asymmetric function of σ that selects out the smallest charge density to be relevant for the energy at the minimum.

In summary, it is possible to obtain exact expressions for the position and depth of the free energy minimum corresponding to the equilibrium position induced by the balance between electrostatic attraction and osmotic repulsion (Eqs. 7.9 and 7.10). These quantities depend on the plate charge densities as well as on the salt concentration.[4] Note moreover that Eq. (7.10) gives a free energy *per unit area*, hence the total free energy is also proportional to the area of the interface.

6. TOWARD A NEW PARADIGM FOR THE TARGET SEARCH PROCESS

6.1. Redefining hydrogen bonds

Let us now come back to biology. According to our model, if the protein–DNA interface is large enough, the protein is pushed away from DNA until their distance is of the order of a fraction of nanometer. It is then tempting to guess that this effect can have a significant impact on the search mechanism: instead of "sticking" on DNA, proteins might "float" away from it at a very short distance, as if it were sliding on a thin cushion of air—in this case, a "cushion of ions." Might its mobility along DNA be increased? The distance between DNA and protein in the nonspecific complex allows it to slide without being hampered by the roughness associated with the sequence? And if this is the case, how may the protein still be able to distinguish the target sequence from other sequences with sufficient efficiency?

---

[4] A more detailed analysis of the dependence on these quantities (and on the solution pH) can be found in Paillusson (2009, 2011).

As we have discussed, recognition at the specific site is often character-ized by the formation of hydrogen bonds between residues of the protein and the base pairs. We have assumed that the same pattern of "possible" bonds may be used as reading frame during the search phase. In order to check the effect of the osmotic repulsion on this search mechanism, and therefore its balance with the specific part of the interaction, we should extend the model for this latter. While the number of possible hydrogen bonds at each DNA position can still be described as a Gaussian variable, indeed, we now need to add the energy dependence on the new problem variable: the protein–DNA distance L. An usual way to describe a single hydrogen bond interaction as a function of the bond length is by a Morse potential (Kang, 2000). We will therefore write

$$V_{\text{Morse}}(L) = \mathcal{E}\left[\left(1 - \exp\left(-\frac{L}{\lambda_{\text{M}}}\right)\right)^2 - 1\right] \qquad (7.12)$$

where $\mathcal{E} \simeq -0.5 k_{\text{B}} T$ (Tareste et al., 2002) coincides with the same parameter used in the 1D model of Section 3, but represents now more precisely the depth of the potential well corresponding to the bound state. In the previous expression, the parameter $\lambda_{\text{M}} \simeq 0.05$ nm (Chen, Kortemme, Robertson, Baker, & Varani, 2004; Nadassy et al., 1999) is the bond range.

Then, at each position $z = 0.34n$ (nm) along the sequence, we suppose as before that a number $\mathcal{N}(z)$ of hydrogen bonds can be locally formed by pro-tein with bases between n and $n + N - 1$. The interaction energy at position z and at a distance L of DNA, can be thus written as

$$E(z,L) = \mathcal{N}(z) V_{\text{Morse}}(L) \qquad (7.13)$$

In order to have a rather general model without referring to the case of a particular protein, we will model the number $\mathcal{N}(z)$ of hydrogen bonds by introducing reasonable estimates of its statistical parameters and by assuming a Gaussian distribution (Barbi, Place, et al., 2004a; Barbi, Popkov, et al., 2004b). This assumption, as we have discussed, is in agreement with some experimental data (Gerland et al., 2002; Sheinman et al., 2012; Stormo & Fields, 1998). More precisely, we assume to know the number of bonds between the protein and its target sequence $\mathcal{N}_{\text{max}}$, which correspond to the maximum value of $\mathcal{N}$ (highest affinity). We then describe the distribu-tion of $\mathcal{N}$ by a Gaussian with mean $\langle \mathcal{N} \rangle = \mathcal{N}_{\text{max}}/3$ and standard deviation $\sigma_{\mathcal{N}} = \sqrt{\mathcal{N}_{\text{max}}}$, and we furthermore impose, obviously, $\mathcal{N} \geq 0$. These values are chosen so that the probability of $\mathcal{N} = \mathcal{N}_{\text{max}}$ is realistically low. Indeed,

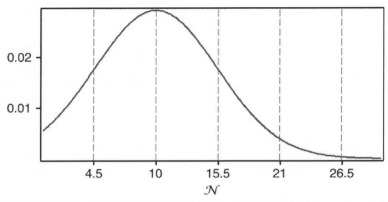

Figure 7.16 Gaussian distribution of the parameter $\mathcal{N}$, corresponding to the number of possible hydrogen bonds between the protein and the DNA, within 0 and $\mathcal{N}_{max}$, using the parameters defined in the main text. The vertical dashed lines are centered on the mean value and are separated by one standard deviation. (For color version of this figure, the reader is referred to the online version of this chapter.)

even for sequences with a high degree of homology to the target one, the number of H-bonds dramatically decreases, as observed, for example, in the crystal structure of noncognate *Bam*HI complex (Viadiu & Aggarwal, 2000).

The maximum number of bonds $\mathcal{N}_{max}$ can be estimated from crystallographic data for specific complexes and gives an average value of about 1.5 hydrogen bonds per nm^2 of DNA–protein interface (Nadassy et al., 1999). For an average surface interaction $S_{prot} = 20\ nm^2$, we obtain $\mathcal{N}_{max} \simeq 30$ and therefore $\langle \mathcal{N} \rangle \simeq 10$ and $\sigma\mathcal{N} \simeq 5.5$. With these choices, the probability of n_{max} bonds is reasonably low (between three and four standard deviations, Fig. 7.16).

6.2. A facilitated sliding

Summing up the two contributions, one coming from the electrostatic interaction, the other associated with hydrogen bonds, we obtain, for the case of a protein surface charge equal to the average value found above for specific proteins ($0.17\ nm\ e^{-2}$), the result presented as a free energy landscape $F(z,L)$ in Fig. 7.17 (Dahirel et al., 2009).

When the protein is precisely at the target, a primary minimum exists almost at the contact with the DNA surface, corresponding to tight binding. Its depth is $\sim 7 k_B T$ with our parameter choice. This primary minimum is separated by an energy barrier of the order of $k_B T$ from a secondary minimum, coming from the electrostatic part of the interaction. A similar scenario will be observed in correspondence with the (rare) sequences that are close to the target sequence and have therefore a high degree of affinity to the protein. On the contrary, for most of the positions along DNA, where the affinity is much lower, the primary minimum practically disappears and only the electrostatic equilibrium position at a distance from the DNA surface remains. Remarkably, the osmotic repulsion between sequence-specific DNA-BPs and DNA dominates along nonspecific sequences: it is almost everywhere strong enough to keep the protein at a distance from DNA, this making it in practice completely *insensitive* to the sequence. Along the equilibrium valley, indeed, the roughness of the sequence-dependent part of the potential is screened out: the protein can therefore easily slide along DNA. At the target site, conversely, the large H-bond interaction significantly reduces the barrier, and the protein can approach the DNA.

Incidentally, the equilibrium gap distance of nearly 0.5 nm that we observe in Fig. 7.17 is in agreement with the distance observed in the complexes of *Eco*RV (0.51 nm; Jones et al., 1999) with nonspecific sequences. This also gives a rational basis to some *ad hoc* protein sizes that had to be put by hand in recent coarse grained simulations of protein sliding on DNA to ensure the protein would not go closer to DNA than the distance observed in the non-specific complex (Florescu & Joyeux, 2009a, 2009b; Givaty & Levy, 2009).

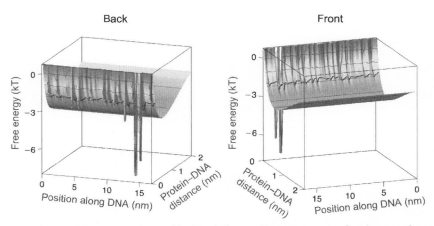

Figure 7.17 Free energy is here calculated along a DNA sequence of 50 bp, as a function of the protein–DNA distance L and of the position z of the protein along the DNA, for $\sigma_{prot} = 0.17\ e\,nm^{-2}$. The distance between the contour lines is $k_B T$. For clarity, we show the same graph from two opposite sides (back and front). Red and blue curves are added as a guide for the eye in the approximate position (along DNA) of the primary minimum and of the barrier, respectively. (For interpretation of the references to color in this figure legend, the reader is referred to the online version of this chapter.)

In other words, what we obtain is a mechanism that we could name *facilitated siding*: the mobility of the protein is guaranteed by the osmotic repulsion, until it reaches a good sequence and can bind it (Dahirel et al., 2009). This mechanism may represent an efficient solution of the mobility-specificity paradox, since it introduces *de facto* a two-mode search: the protein is actually insensitive to the sequence all along nonspecific DNA, except for a few traps, and in the *diffusing mode* evoked in Section 3. However, note that, unlike previous models, the coupling between *diffusing mode* and "wrong" sequences is here explicit and does not require any additional "switching" mechanism. Moreover, in spite of the fact that the effective search obtained in our model can be intuitively described as a combination of *diffusing* and *reading mode*, the real mechanism is in fact different: the protein is no more sensitive to the sequence, whatever the position along DNA, but it is now sensitive to the *free energy barrier* that separates it from the sequence. Therefore, the interaction is always described in a similar way, but it allows for an energy-activated change in the protein–DNA complex state (bringing the two bodies closer) for some special positions. Interestingly, a similar barrier-dependent mechanism is also invoked in Sheinman et al. (2012) as a solution for the mobility-specificity paradox, although the details of the model and notably the correlation between the barrier, the primary minimum, and the sequence are somehow different. This allows the authors to fit the available quantitative data on the search kinetics by a simple and generic kinetic model.

6.3. Toward a different modeling of the protein search

As we have discussed in Section 2, many theoretical models (see Sheinman, 2012, for a good review) have been proposed to catch the essential features of the search mechanism. We note that all these models include sliding (to different extent) and focus on dichotomic views of the search process: sliding versus 3D diffusion, "reading" versus "diffusing" modes, specific versus nonspecific binding at the target, or specific versus nonspecific interaction (all along the DNA).

From a numerical point of view, detailed molecular dynamics simulations seem to suggest a more complicated scenario (Bouvier & Lavery, 2009; Chen & Pettitt, 2011; Dahirel et al., 2009; Shoemaker, Portman, & Wolynes, 2000) where DNA deformations, protein deformation, flexible protein tails behavior, entropic costs participate in defining a complex energy landscape for the protein–DNA complex, with rather continuous and complicated variations as a function of the relative position of protein and DNA, either along the sequence (and therefore on and off the target) or in the radial direction, but also associated with the protein rotation and with the protein and/or DNA deformations (see Zakrzewska & Lavery, 2012, for a more exhaustive discussion). On the other hand, it is known that a significant stabilizing effect of the specific complex is associated with the release of water molecules (Bouvier & Lavery, 2009; Chen & Pettitt, 2011), which implies the presence of a layer of water between proteins and DNA in the nonspecific complex.

Very interestingly, the scenario obtained by our model shares some central features with what is found numerically by some authors. In particular, either Bouvier and Lavery (2009) or Chen and Pettitt (2011) evidence the presence of two distinct free energy minima, one closer to DNA and the other farer from it, separated by a free energy barrier. The relative positions of the three states are smaller but not incompatible with what obtained in our model[5].

These finding suggest an alternative way to describe the search process, by replacing the usual dichotomic view by a more "soft" approach where the interaction is described in terms of *continuous variables*. The protein–DNA distance is indeed a crucial variable, potentially leading to a description of the protein kinetics where the distinction between *sliding, hopping, jumping*, and 3D diffusion becomes somehow obsolete. More concretely, the movement of the protein in the vicinity of DNA can, in our scenario, be treated as a diffusion in the landscape of Fig. 7.17. Unfortunately, *in vitro* experiments, which assess for sliding cannot reach the resolution needed to describe the protein–DNA interaction (and associated kinetics) at the scale involved in this model. However, experimentalists clearly distinguish at least phases where the proteins are "on" DNA (and can therefore be observed) from phases where the protein dissociates from it. Moreover, rapid displacements along a same DNA molecule have been observed (Bonnet et al., 2008) that cannot be compatible with pure 1D diffusion along the double helix. The question therefore arises of how these different protein *states* or *modes of displacement* can be accounted for in the context of a continuous description.

6.4. Defining a physical-meaningful sliding time

By comparing our model to experimental estimates of the chemical rates of protein binding and unbinding, one can in principle get more decisive feedback about the landscape, as binding and unbinding events involve a wide

---

[5] In Bouvier and Lavery (2009), the secondary minimum, barrier, and primary minimum locations are found, respectively, at protein–DNA distances of 0.32, 0.31, and <0.3 nm, in Chen and Pettitt (2011), at 0.26, 0.13, and 0.08 nm, respectively.

range of DNA–protein distances. In the following, for the sake of simplicity, we shall focus on the dissociation rate of a nonspecific protein–DNA complex, although the binding rate can also be considered without too much difficulty following, for example, Berne, Borkovec, and Straub (1988). Moreover, we neglect, here, the effects due to the hydrogen bond interaction, only relevant at very short distances: the aim of this calculation is indeed to evaluate the time needed for the protein to escape from a generic, nonspecific position along DNA, that is, to exit the secondary minimum defined by the electrostatic part of the interaction.

We are interested in the following reaction

$$(\text{Prot}|\text{DNA})_{\text{complex}} \rightarrow \text{Prot} + \text{DNA} \tag{7.14}$$

We will assume that the size of the particles is big enough for the unbinding process to be diffusion dominated (Berne et al., 1988). Considering the energy landscape we derived in the previous parts, it is natural to use the surface-to-surface DNA–protein separation L as the reaction coordinate. Moreover, if the energy landscape displays a well-defined barrier between the two chemical states of reaction (Eq. 7.14), then we can use Kramers' rate theory for a 1D isomerization process (Berne et al., 1988). The dissociation rate k_{diss} reads then

$$k_{\text{diss}} \approx \frac{D}{2\pi} \sqrt{\beta |G''(L_A)| |\beta G''(L_B)|} e^{-\beta \Delta_{AB} G} \tag{7.15}$$

where D is the diffusion constant of the protein, A corresponds to the minimum of the binding well, B is the location of the dividing surface, that is, the top of the energy barrier (cf. Fig. 7.18), $\Delta_{AB} G = G_B - G_A$, and G'' stands for a second derivative of the energy G with respect to L. The total effective interaction $G(L)$ in Eq. (7.15) is defined so that the ratio of the marginal probabilities to be either at L_1 or L_2 reads

$$\frac{p(L_1)}{p(L_2)} = \frac{e^{-\beta G(L_1)}}{e^{-\beta G(L_2)}}. \tag{7.16}$$

for any L_1 and L_2.

On the other hand, it is also possible to state that this same ratio should read

$$\frac{p(L_1)}{p(L_2)} = \frac{2\pi (R_{\text{DNA}} + L_1) e^{-\beta F(L_1)}}{2\pi (R_{\text{DNA}} + L_2) e^{-\beta F(L_2)}} \tag{7.17}$$

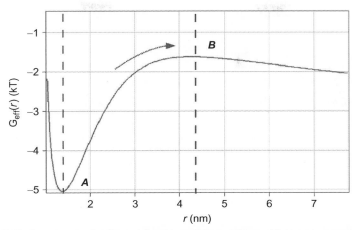

Figure 7.18 Free energy landscape for the radial coordinate. The curve represents the thermodynamic potential G associated with the effective diffusion in the radial direction under physiological conditions (i.e., $c_{rmsalt} = 0.1$ mol L^{-1}). Point A corresponds to the bound state allowing a 1D diffusion along DNA, while point B is the point beyond which the protein can diffuse freely in three dimensions and therefore corresponds to the dividing surface. (For color version of this figure, the reader is referred to the online version of this chapter.)

where the $F(L)$ is the free energy (that we estimated in previous sections) that corresponds to the work, one has to do to bring a protein from infinity to a distance L from a DNA segment for any fixed value of the polar angle that locates the protein within the plane perpendicular to the DNA axis. The $2\pi(R_{DNA} + L)$ factor is a degeneracy term, associated to the probability of being at a particular distance from the axis of the DNA molecule. This probability grows indeed as the circumference of a circle of radius $R_{DNA} + L$.

Note that, unlike $F(L)$, $G(L)$ may present a maximum, that is, an energy barrier between the location of the electrostatic minimum and the region $L \rightarrow \infty$ (see Fig. 7.18). From Eqs. (7.16) and (7.17), we thus find that the total effective interaction $G(L)$ associated to a distance L has to have the form

$$G(L) = F(L) - k_B T \ln\left(\frac{R_{DNA} + L}{R_0}\right) \tag{7.18}$$

where R_0 is some unimportant distance whose purpose is to have a dimensionless argument inside the logarithm. Now that we have understood that we can try to interpret Kramers' formula (Eq. 7.15). To do so, we rewrite (7.15) in a slightly different way

$$k_{\text{diss}} \approx \frac{1}{2\pi} \sqrt{\frac{D}{\delta L_A^2} \frac{D}{\delta L_B^2}} e^{-\beta \Delta_{AB} G} = \frac{\sqrt{v_A v_B}}{2\pi} e^{-\beta \Delta_{AB} G} \qquad (7.19)$$

where $\delta L_A \equiv 1/\sqrt{\beta G''(L_A)}$ and $\delta L_B \equiv 1/\sqrt{\beta G''(L_B)}$ are, respectively, the typical sizes of the bottom of the well and the top of the barrier, and where $v_A^{-1} \equiv \delta L_A^2/D$ and $v_B^{-1} \equiv \delta L_B^2/D$ are the typical times it takes for a diffusive protein to travel over the lengths δL_A and δL_B, respectively. Thus, the prefactor $\sqrt{v_A v_B}$ is nothing but the geometric mean of the natural rates v_A and v_B. To get some insights from Eq. (7.19), we first calculated k_{diss} from the model with parameters used for Fig. 7.18, that is, in the case of a physiological salt concentration $n_b = 0.1$ mol L^{-1}. We found that $\beta \Delta_{AB} G \approx 3.4$ while the prefactor $\sqrt{v_A v_B}/2\pi \approx 10^3 \text{ms}^{-1}$. Overall, the rate is $k_{\text{diss}}[0.1 \text{ M}] \approx 35$ ms^{-1}. It thus means that, on average in physiological conditions, a protein with a landscape as that of Fig. 7.18 will stay less than a millisecond on a given DNA segment before leaving it. This observation seems however in contradiction with measured average sliding times in experiments (Bonnet et al., 2008) where a protein can be bound to a DNA segment for up to few seconds. This discrepancy is without accounting for the fact that the mentioned experiments are done at much lower salt concentration. In fact, as we have seen before, increasing the salt concentration can have a very strong effect on the free energy landscape. We thus recalculated it with the same protein and DNA parameters but with $n_b = 0.01$ M. We got that $\beta \Delta_{AB} G \approx 9$ while the prefactor in Eq. (7.19) is about 10^2 ms^{-1}. Overall, the dissociation rate k_{diss} is $k_{\text{diss}}[0.01 \text{ M}] \approx 10^{-2}$ ms^{-1} which is about three orders of magnitude lower than in physiological conditions! Also, in this particular case, the typical life time of the nonspecific complex is comparable to those observed in Bonnet et al. (2008).

In this part, we were able to relate our continuous description to observable quantities such as the dissociation rates of the nonspecific complex of arbitrary proteins. To apply Kramers' theory, we emphasized the fact that the reaction coordinate is a radial coordinate that gives rise to an entropic repulsive force that allows for a nonambiguous definition of the barrier between the bound state and the unbound one. In absence of the mentioned $2\pi(R_{\text{DNA}} + L)$ degeneracy, however (i.e., in a truly 1D case), there is no consensus on where to put the dividing surface for free energies as those of Fig. 7.15, and one should be careful about this point (Hänggi, Talkner, & Borkovec, 1990).

Evidently, the next step in exploiting the model described here will be to try to predict the features of the protein diffusion along DNA during sliding and to compare them with experiments. Note, however, that although we can in principle estimate the sliding diffusion coefficient D_1 from diffusion properties of the protein in bulk and get an estimate of the typical sliding length $\left(\sim \sqrt{D_1/k_{\mathrm{diss}}} \right)$ that is measured in many experiments (*in vitro* but also *in vivo*, see, e.g., Hammar, 2012), it is in fact more subtle than expected. Indeed, as it was imagined by Schurr (1979), some DNA-binding proteins slide with a helical motion along DNA (Blainey, 2009; Dikić, 2012). The resulting effective diffusion coefficient then depends on the DNA–protein distance in the bound state (Bagchi, Blainey, & Xie, 2008; Blainey, 2009; Dikić, 2012), and we have seen that the latter depends on the salt concentration; the sliding diffusion coefficient therefore depends on the salt concentration. This additionally supports a potential need for the change of paradigm that has been stressed throughout this chapter in order to understand fully what are the relevant parameters to describe the observed binding kinetics of proteins to their specific sites on DNA.

REFERENCES

Ahmad, S., & Sarai, A. (2004). Moment-based prediction of DNA-binding proteins. *Journal of Molecular Biology*, *341*(1), 65–71.

Ando, T., Kodera, N., Takai, E., Maruyama, D., Saito, K., & Toda, A. (2001). A high-speed atomic force microscope for studying biological macromolecules. *Proceedings of the National Academy of Sciences of the United States of America*, *98*(22), 12468–12472.

Azbel, M. Y. (1973). Random two–component one–dimensional Ising model for heteropolymer melting. *Physical Review Letters*, *31*, 589.

Bagchi, B., Blainey, P., & Xie, X. (2008). Diffusion constant of a nonspecifically bound protein undergoing curvilinear motion along DNA. *The Journal of Chemical Physics B*, *112*, 6282–6284.

Barbi, M., Place, C., Popkov, V., & Salerno, M. (2004a). Base–sequence–dependent sliding of proteins on DNA. *Physical Review E*, *70*, 041901.

Barbi, M., Popkov, V., Place, C., & Salerno, M. (2004b). A model of sequence dependent protein diffusion along DNA. *Journal of Biological Physics*, *30*, 203–226.

Ben-Yaakov, D., Burak, Y., Andelman, D., & Safran, S. A. (2007). Electrostatic interactions of asymmetrically charged membranes. *Europhysics Letters*, *79*, 48002–48008.

Berg, O. G. (1978). On diffusion-controlled dissociation. *The Journal of Chemical Physics*, *31*, 47–57.

Berg, O., & Blomberg, C. (1976). Association kinetics with coupled diffusional flows. *Biophysical Chemistry*, *4*, 367–381.

Berg, O., & Blomberg, C. (1977). Association kinetics with coupled diffusion. An extension to coiled–chain macromolecules applied to the lac repressor-operator system. *Biophysical Chemistry*, *7*, 33–39.

Berg, O., & Blomberg, C. (1978). Association kinetics with coupled diffusion. III. Ionic-strength dependence of the lac repressor-operator association. *Biophysical Chemistry, 8*, 271–280.

Berg, O. G., Winter, R. B., & von Hippel, P. H. (1981). Diffusion driven mechanism of protein translocation on nucleic acids. I. Models and theory. *Biochemistry, 20*, 6929.

Berne, B., Borkovec, M., & Straub, J. (1988). Classical and modern methods in reaction rate theory. *The Journal of Physical Chemistry, 92*, 3711–3725.

Blainey, P., van Oijen, A., Banerjee, A., Verdine, G., & Xie, X. (2006). A base–excision DNA–repair protein finds intrahelical lesion bases by fast sliding in contact with DNA. *Proceedings of the National Academy of Sciences of the United States of America, 103*, 5752–5757.

Blainey, P. C. (2009). Nonspecifically bound proteins spin while diffusing along dna. *Nature Structural & Molecular Biology, 16*, 1224–1229.

Bonnet, I., Biebricher, A., Porté, P.-L., Loverdo, C., Bénichou, O., Voituriez, R., et al. (2008). Sliding and jumping of single EcoRV restriction enzymes on noncognate DNA. *Nucleic Acids Research, 36*(12), 4118–4127.

Bouvier, B., & Lavery, R. (2009). A free energy pathway for the interaction of the SRY protein with its binding site on DNA from atomistic simulations. *Journal of the American Chemical Society, 131*, 9864–9865.

Bruinsma, R. F. (2002). Physics of protein-DNA interaction. *Physica A, 313*, 211–237.

Carrivain, P., Cournac, A., Lavelle, C., Lesne, A., Mozziconacci, J., Paillusson, F., et al. (2012). Electrostatics of DNA compaction in viruses, bacteria and eukaryotes: Functional insights and evolutionary perspective. *Soft Matter, 8*, 9285–9301.

Cheetham, J. M. T., Jeruzalmi, D., & Steitz, T. A. (1999). Structural basis for initiation of transcription from an RNA polymerase–promoter complex. *Nature, 399*.

Chen, Y., Kortemme, T., Robertson, T., Baker, D., & Varani, G. (2004). A new hydrogen–bonding potential for the design of protein–RNA interactions predicts specific contacts and discriminates decoys. *Nucleic Acids Research, 32*, 5147–5162.

Chen, C., & Pettitt, B. M. (2011). The binding process of a nonspecific enzyme with DNA. *Biophysical Journal, 101*(5), 1139–1147.

Cocco, S., Marko, J. F., & Monasson, R. (2002). Theoretical models for single-molecule DNA and RNA experiments: From elasticity to unzipping. *Comptes Rendus Physique, 3*(5), 569–584.

Coppey, M., Bénichou, O., Voituriez, R., & Moreau, M. (2004). Kinetics of target site localization of a protein on DNA: A stochastic approach. *Biophysical Journal, 87*, 1640–1649.

Crut, A., Lasne, D., Allemand, J.-F., Dahan, M., & Desbiolles, P. (2003). Transverse fluctuations of single DNA molecules attached at both extremities to a surface. *Physical Review E, 67*, 051910.

Dahirel, V., Jardat, M., Dufrêche, J. F., & Turq, P. (2007a). How the excluded volume architecture influences ion–mediated forces between proteins. *Physical Review E, 76*, 040902.

Dahirel, V., Jardat, M., Dufrêche, J. F., & Turq, P. (2007b). Toward the description of electrostatic interactions between globular proteins: Potential of mean force in the primitive model. *The Journal of Chemical Physics, 127*, 095101.

Dahirel, V., Paillusson, F., Jardat, M., Barbi, M., & Victor, J.-M. (2009). Nonspecific DNA-protein interaction: Why proteins can diffuse along DNA. *Physical Review Letters, 102*, 228101.

Dikić, J. (2012). The rotation-coupled sliding of ecorv. *Nucleic Acids Research, 40*(9), 4064–4070.

Dubbeldam, J. L. A., Milchev, A., Rostiashvili, V. G., & Vilgis, T. A. (2007). Polymer translocation through a nanopore: A showcase of anomalous diffusion. *Physical Review E, 76*, 010801.

Elf, J., Li, G., & Xie, X. (2007). Probing transcription factor dynamics at the single–molecule level in a living cell. *Science, 316*, 1191–1194.

Florescu, A.-M., & Joyeux, M. (2009a). Description of nonspecific DNA-protein interaction and facilitated diffusion with a dynamical model. *The Journal of Chemical Physics, 130*, 015103.

Florescu, A.-M., & Joyeux, M. (2009b). Dynamical model of DNA-protein interaction: Effect of protein charge distribution and mechanical properties. *The Journal of Chemical Physics, 131*, 105102.

Gerland, U., Moroz, J. D., & Hwa, T. (2002). Physical constraints and functional characteristics of transcription factor-DNA interaction. *Proceedings of the National Academy of Sciences of the United States of America, 99*, 12015–12020.

Givaty, O., & Levy, Y. (2009). Protein sliding along DNA: Dynamics and structural characterization. *Journal of Molecular Biology, 385*, 1087–1097.

Gorman, J., & Greene, E. (2008). Visualizing one-dimensional diffusion of proteins along DNA. *Nature Structural & Molecular Biology, 15*, 5752–5757.

Gowers, D., Wilson, G., & Halford, S. (2005). Measurement of the contributions of 1D and 3D pathways to the translocation of a protein along DNA. *Proceedings of the National Academy of Sciences of the United States of America, 102*, 15883–15888.

Guthold, M., Zhu, X., Rivetti, C., Yang, G., Thomson, N. H., Kasas, S., et al. (1999). Direct observation of one-dimensional diffusion and transcription by Escherichia coli RNA polymerase. *Biophysical Journal, 77*, 2284–2294.

Halford, S. E. (2009). An end to 40 years of mistakes in DNA-protein association kinetics? *Biochemical Society Transactions, 37*, 343–348.

Halford, S. E., & Marko, J. F. (2004). How do site-specific DNA–binding proteins find their targets? *Nucleic Acids Research, 32*, 3040–3052.

Hammar, P. (2012). The lac repressor displays facilitated diffusion in living cells. *Science, 336*, 1595–1598.

Hänggi, P., Talkner, P., & Borkovec, M. (1990). Reaction-rate theory: Fifty years after Kramers. *Reviews of Modern Physics, 62*, 251–341.

Hansen, J.-P., & Löwen, H. (2000). Effective interactions between electric double layers. *Annual Review of Physical Chemistry, 51*, 209.

Harada, Y. (1999). Single-molecule imaging of RNA polymerase-dna interactions in real time. *Biophysical Journal, 76*, 709–715.

Jones, S., van Heyningen, P., Berman, H., & Thornton, J. (1999). Protein–DNA interactions: A structural analysis. *Journal of Molecular Biology, 287*, 877–896.

Jones, S., Shanahan, H., Berman, H., & Thornton, J. (2003). Using electrostatic potentials to predict DNA–binding sites on DNA–binding proteins. *Nucleic Acids Research, 31*, 7189–7198.

Kalodimos, C., Biris, N., Bonvin, A. M. J. J., Levandoski, M. M., Guennuegues, M., Boelens, R., et al. (2004). Structure and flexibility adaptation in nonspecific and specific protein–DNA complexes. *Science, 305*, 386–389.

Kampmann, M. (2004). Obstacle bypass in protein motion along DNA by two-dimensional rather than one-dimensional sliding. *The Journal of Biological Chemistry, 279*(37), 38715–38720.

Kang, Y. (2000). Which functional form is appropriate for hydrogen bond of amides? *The Journal of Physical Chemistry. B, 104*, 8321–8326.

Kim, J., & Larson, R. (2007). Single–molecule analysis of 1d diffusion and transcription elongation of T7 RNA polymerase along individual stretched DNA molecules. *Nucleic Acid Research, 35*, 3848–3858.

Kupiec, J. J. (1997). A Darwinian theory for the origin of cellular differentiation. *Molecular & General Genetics, 255*, 201–208.

Lau, A. W.-C. (2000). Fluctuation and correlation effects in electrostatics of charged membranes. PhD thesis, University of California at Santa Barbara.

Mirny, L., Slutsky, M., Wunderlich, Z., Tafvizi, A., Leith, J., & Kosmrlj, A. (2009). How a protein searches for its site on DNA: The mechanism of facilitated diffusion. *Journal of Physics A: Mathematical and Theoretical, 42*(43), 434013.

Murugan, R. (2010). Theory of site-specific DNA-protein interactions in the presence of conformational fluctuations of DNA binding domains. *Biophysical Journal*, *99*(2), 353–359.

Nadassy, K., Wodak, S. J., & Janin, J. (1999). Structural features of protein–nucleic acid recognition sites. *Biochemistry*, *38*, 1999–2017.

Ohshima, H. (1975). Diffuse double layer interaction between two parallel plates with constant surface charge density in an electrolyte solution III: Potential energy of double layer interaction. *Colloid and Polymer Sciences*, *253*, 150–157.

Paillard, G., & Lavery, R. (2004). Analyzing protein-DNA recognition mechanisms. *Structure*, *12*, 113–122.

Paillusson, F., Barbi, M., & Victor, J.-M. (2009). Poisson-Boltzmann for oppositely charged bodies: An explicit derivation. *The Journal of Chemical Physics*, *107*, 1379–1391.

Paillusson, F., Dahirel, V., Jardat, M., Victor, J.-M., & Barbi, M. (2011). Effective interaction between charged nanoparticles and dna. *Physical Chemistry Chemical Physics*, *13*, 12603–12613.

Parsegian, V. A., & Gingel, D. (1972). On the electrostatic interaction across a salt solution between two bodies bearing unequal charges. *Biophysical Journal*, *12*, 1192–1204.

Richter, P., & Eigen, M. (1974). Diffusion controlled reaction rates in spheroidal geometry. Application to repressor–operator association and membrane bound enzymes. *Biophysical Chemistry*, *2*, 255–263.

Riggs, A. D., Bourgeois, S., & Cohn, M. (1970). The lac repressor–operator interaction. 3. Kinetic studies. *Journal of Molecular Biology*, *53*, 401–417.

Sanchez, H., Suzuki, Y., Yokokawa, M., Takeyasu, K., & Wyman, C. (2011). Protein-DNA interactions in high speed AFM: Single molecule diffusion analysis of human rad54. *Integrative Biology*, *3*, 1127–1134.

Schurr, J. M. (1979). The one dimensional diffusion coefficient of protein absorbed on DNA. *Biophysical Chemistry*, *9*, 413–414.

Seeman, N. C., Rosenberg, J. M., & Rich, A. (1976). Sequence–specific recognition of double helical nucleic acids by proteins. *Proceedings of the National Academy of Sciences of the United States of America*, *73*, 804.

Sheinman, M., Bnichou, O., Kafri, Y., & Voituriez, R. (2012). Classes of fast and specific search mechanisms for proteins on DNA. *Reports on Progress in Physics*, *75*(2), 026601.

Shimamoto, N. (1999). One–dimensional diffusion of proteins along DNA. Its biological and chemical significance revealed by single–molecule measurements. *The Journal of Biological Chemistry*, *274*, 15293–15296.

Shoemaker, B. A., Portman, J. J., & Wolynes, P. G. (2000). Speeding molecular recognition by using the folding funnel: The fly-casting mechanism. *Proceedings of the National Academy of Sciences of the United States of America*, *97*(16), 8868–8873.

Slusky, M., Kardar, M., & Mirny, L. A. (2004). Diffusion in correlated random potentials, with applications to dna. *Physical Review E*, *70*, 049901.

Slutsky, M., & Mirny, L. A. (2004). Kinetics of protein–DNA interaction: Facilitated target location in sequence–dependent potential. *Biophysical Journal*, *87*, 4021–4035.

Spolar, R., & Record, M., Jr. (1994). Coupling of local folding to site–specific binding of proteins to DNA. *Science*, *263*, 777.

Stanford, N. P., Szczelkun, J., Marko, N. P., & Halford, S. (2000). One- and three-dimensional pathways for proteins to reach specific DNA sites. *The EMBO Journal*, *19*, 6546–6557.

Stawiski, E. W., Gregoret, L. M., & Mandel-Gutfreund, Y. (2003). Annotating nucleic acid-binding function based on protein structure. *Journal of Molecular Biology*, *326*(4), 1065–1079.

Stormo, G. D., & Fields, D. S. (1998). Specificity, free energy and information content in protein-DNA interactions. *Trends in Biochemical Sciences*, *23*, 109–113.

Swanzig, R. (1988). Diffusion in a rough potential. *Proceedings of the National Academy of Sciences of the United States of America, 85,* 2029.

Szilágyi, A., & Skolnick, J. (2006). Efficient prediction of nucleic acid binding function from low-resolution protein structures. *Journal of Molecular Biology, 358*(3), 922–933.

Tafvizi, A., Huang, F., Leith, J. S., Fersht, A. R., Mirny, L. A., & van Oijen, A. M. (2008). Tumor suppressor p53 slides on DNA with low friction and high stability. *Biophysical Journal, 95*(1), L01–L03.

Takeda, Y., Ross, P. D., & Mudd, C. P. (1992). Thermodynamics of Cro protein–DNA interactions. *Proceedings of the National Academy of Sciences of the United States of America, 89,* 8180–8184.

Tamashiro, M. N., & Schiessel, H. (2003). Where the linearized Poisson-Boltzmann cell model fails: The planar case as a prototype study. *Physical Review E, 68,* 066106.

Tareste, D., Pincet, F., Perez, E., Rickling, S., Mioskowski, C., & Lebeau, L. (2002). Energy of hydrogen bonds probed by the adhesion of functionalized lipid layers. *Biophysical Journal, 83,* 3675–3681.

Terry, B. J., Jack, W. E., & Modrich, P. (1985). Facilitated diffusion during catalysis by EcoRI endonuclease. Nonspecific interactions in EcoRI catalysis. *The Journal of Biological Chemistry, 260*(24), 13130–13137.

Viadiu, H., & Aggarwal, A. (2000). Structure of BamHI bound to nonspecific DNA: A model for DNA sliding. *Molecular Cell, 5,* 889.

von Hippel, P. H. (2007). From simple DNA–protein interaction to the macromolecular machines of gene expression. *Annual Review of Biophysics and Biomolecular Structure, 36,* 79–105.

von Hippel, P. H., & Berg, O. (1989). Facilitated target location in biological systems. *The Journal of Biological Chemistry, 264,* 675–678.

Wang, Y., Austin, R., & Cox, E. (2006). Single molecule measurements of repressor protein 1D diffusion on DNA. *Physical Review Letters, 97,* 048302.

Winter, R. B., Berg, O. G., & von Hippel, P. H. (1981). Diffusion driven mechanism of protein translocation on nucleic acids. III. The E. coli lac repressor–operator interaction: Kinetic measurements and conclusions. *Biochemistry, 20,* 6961–6977.

Zakrzewska, K., & Lavery, R. (2012). Towards a molecular view of transcriptional control. *Current Opinion in Structural Biology, 22*(2), 160–167.

Zhou, H.-X. (2011). Rapid search for specific sites on DNA through conformational switch of non-specifically bound proteins. *Proceedings of the National Academy of Sciences of the United States of America, 108,* 8651–8656.

CHAPTER EIGHT

Structural and Dynamical Aspects of HIV-1 Protease and Its Role in Drug Resistance

Biswa Ranjan Meher[*,1,2], Seema Patel[†,2]
[*]Department of Chemistry, Gottwald Center for the Sciences, University of Richmond, Richmond, Virginia, USA
[†]Center for Bioinformatics and Medical Informatics, San Diego State University, San Diego, California, USA
[1]Corresponding author: e-mail address: brmeher@gmail.com
[2]Both the authors contributed equally in preparing the manuscript.

Contents

Abstract

Acquired immunodeficiency syndrome (AIDS) caused by the retrovirus human immunodeficiency virus (HIV) has become a major epidemic afflicting mankind. The Joint United Nations Program on HIV/AIDS (UNAIDS) projection shows the existence of millions of AIDS patients at the end of 2012. All the Food and Drug Administration (FDA)-approved drugs are getting ineffective due to resistance offered by the mutation-prone HIV. Hence, there is an urgent need for developing new drugs with greater potential. HIV life cycle is controlled by the activities of its essential proteins like glycoproteins (gp41 and gp120), HIV reverse transcriptase (HIV-RT), HIV integrase (HIV-IN), and HIV-1 protease (HIV-pr). This chapter focuses on the protein HIV-pr, which is important for the cleavage

Advances in Protein Chemistry and Structural Biology, Volume 92
ISSN 1876-1623
http://dx.doi.org/10.1016/B978-0-12-411636-8.00008-0

of Gag and Gag-Pol polyproteins to form mature, structural, and functional virions. The conformation and dynamics of the protein HIV-pr play a pivotal role in ligand binding and the catalytic process, which is affected by the rapid point mutations and various physiological parameters. The effect of the mutations and the varied simulation protocols on conformational dynamics and drug resistance of HIV-pr is discussed.

1. INTRODUCTION

1.1. Acquired immunodeficiency syndrome

Acquired immunodeficiency syndrome (AIDS) transmitted by human immunodeficiency virus (HIV) has emerged as a pandemic (Sanou, De Groot, Murphey-Corb, Levy, & Yamamoto, 2012). The deadly virus kills the helper T cells (T_H cell) and renders the host immunity compromised (Hatziioannou & Evans, 2012). The loss of immunity exposes the patient to Kaposi's sarcoma, non–Hodgkin's lymphoma, cervical cancer, *Pneumocystis carinii* pneumonia, and other opportunistic infections (Winstone, Man, Hull, Montaner, & Sin, 2013). Several HIV-positive patients show an earlier onset of aging-related chronic conditions such as cardiovascular disease, osteoporosis, diabetes, and neoplasia with respect to uninfected individuals (Gibellini et al., 2012). Despite robust public health initiatives and rigorous research efforts, AIDS continues to be a fatal disease. Though antiretroviral therapy has effectively tackled HIV replication and treated AIDS to some extent, absolute triumph is still remote. HIV mutation-induced drug resistance has nullified the clinical potency of contemporary drugs.

1.2. HIV

HIV is an enveloped, spherical, or pleomorphic retrovirus, containing two copies of single-stranded, positive-sense RNA (Ganser-Pornillos, Yeager, & Pornillos, 2012) (Fig. 8.1). HIV attaches to the host T_H cell receptor CD4 and coreceptor beta-chemokine (either CCR5 or CXCR4) by its envelope antireceptors gp41 and gp120 (Tran et al., 2012). RNA is injected into the host cell while the nucleocapsid is discarded. Upon entrance, the RNA is reverse transcribed to viral DNA by the enzyme reverse transcriptase (Le Grice, 2012). The integration of viral DNA into host genome is followed by transcription and translation culminated by assembly and budding of HIV proteins (Bukrinskaya, 2007). Viral inhibition by targeting the key regulators is exploited in drug designing. Till now, more than 20 antiretroviral drugs

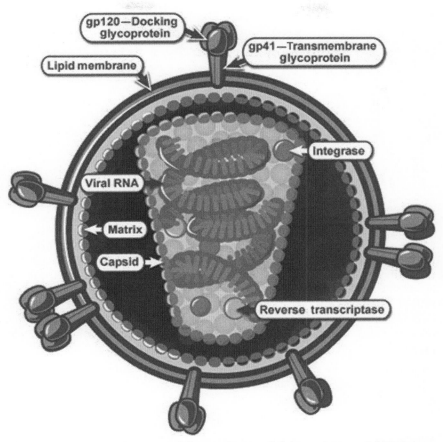

Figure 8.1 A schematic structure of a human immunodeficiency virus type 1 (HIV-1) showing its cellular contents. *Credit: National Institute of Allergy and Infectious Diseases (NIAID).* (For color version of this figure, the reader is referred to the online version of this chapter.)

have been formulated, targeting reverse transcriptase, protease (HIV-pr), integrase, fusion, and cellular CCR5 (Miyamoto & Kodama, 2012).

2. HIV PROTEASE

HIV protease (HIV-pr), a member of aspartyl protease family, has been a major target of most of antiviral therapy strategies. It cleaves Gag and Gag-Pol polyproteins generating mature infectious virions (Fun, Wensing, Verheyen, & Nijhuis, 2012). The assumption that incomplete

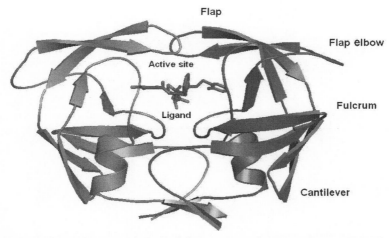

Figure 8.2 Schematic representation of the structure of HIV-1 protease (HIV-pr) bound to the ligand in the active site. Important regions of the protein are labeled. The homodimeric protein is shown in magenta and cyan ribbon structures. (For interpretation of the references to color in this figure legend, the reader is referred to the online version of this chapter.)

processing of these polyproteins will produce immature, noninfectious viral particles has fuelled interest in designing HIV-pr inhibitors.

HIV-pr is a 99-amino acid protease which functions as a homodimer with only one active site which is C2 symmetric in the free form (Brik & Wong, 2003). Each polypeptide chain has a molecular weight of ∼22 kDa (Cardinale et al., 2010). The residues of HIV-pr are numbered as 1–99 for chain A and 100–198 (or 1′–99′) for chain B. Flap (residues 43–58 and 142–157), flap elbow (residues 35–42 and 134–141), fulcrum (residues 11–22 and 110–121), cantilever (residues 59–75 and 158–174), and ligand-binding regions constitute different parts of the enzyme. The active site of the enzyme where the ligand binds is conserved and characterized by the catalytic triad sequence Asp-Thr-Gly. The Asp residues are vital for catalysis while the Thr and Gly residues are buried in the active site (Mager, 2001). The active site is capped by two identical flexible glycine-rich β-hairpins or flaps which control the size of the active site regulating the access of ligands into it (Fig. 8.2).

3. SIDE EFFECTS AND RESISTANCE OF PROTEASE INHIBITORS

Though the HIV-pr-based therapeutic strategies have achieved considerable success, they are plagued by the challenges of serious side effects and resistance. Conventional protease inhibitors cause side effects such as dyslipidemia, characterized by increased plasma levels of triglycerides, low-density lipoprotein cholesterol, and total cholesterol. The abnormal lipid profile predisposes the patients to coronary diseases (Overton, Arathoon, Baraldi, & Tomaka, 2012). Also, certain inhibitors result in insulin resistance, the metabolic dysfunction contributing to cognitive impairment (Gupta et al., 2012). The administration of several drugs has put selective pressure on HIV leading to frequent mutations and subsequent evolution of resistant forms (Chen & Lee, 2006). Multidrug-resistant HIV has emerged as a cause of treatment failure, morbidity, and mortality. U.S. Food and Drug Administration (FDA)-approved HIV-pr inhibitor drugs face various degrees of resistance (Fig. 8.3). There have been a large number of computer simulation studies to understand the HIV-pr dynamics and drug resistance behavior primarily using molecular dynamics (MD) simulations (Collins, Burt, & Erickson, 1995; Hornak, Okur, Rizzo, & Simmerling, 2006; Meagher & Carlson, 2005; Ode, Neya, Hata, Sugiura, & Hoshino, 2006; Perryman, Lin, & McCammon, 2004; Piana, Carloni, & Parrinello, 2002; Piana, Carloni, & Rothlisberger, 2002; Scott & Schiffer, 2000; Toth & Borics, 2006). The affinity of amprenavir to HIV-pr variants V32I, I50V, I54V, I54M, I84V, and L90M decreased 3- to 30-fold compared to the wild type (Kar & Knecht, 2012). The mutation V82T greatly enhances drug resistance of HIV-pr toward darunavir and tipranavir, by altering the hydrophobicity of the binding pocket (Wang et al., 2011).

4. PROTEIN DYNAMICS

Biological processes critically depend on protein structures and functions. For proper functionality, the protein molecules need to be highly dynamic (Kokkinidis, Glykos, & Fadouloglou, 2012). Their conformational changes are vital for muscle contraction, signal transduction, immune function, transportation, etc. Also, their flexibility is fundamental for enzymatic processes such as catalysis, regulation, and substrate recognition. Protein dynamics can be motion as mild as a side-chain displacement or large-scale rearrangements of entire domains (Ho & Agard, 2009). Side-chain and backbone fluctuations largely occur on different timescales. Typical side-chain fluctuations range from picoseconds to nanoseconds, while backbone fluctuations range from nanoseconds to seconds or longer (Petsko & Ringe, 2004). Proteins assume their native states through stepwise dynamic transitions (Breuker & McLafferty, 2008).

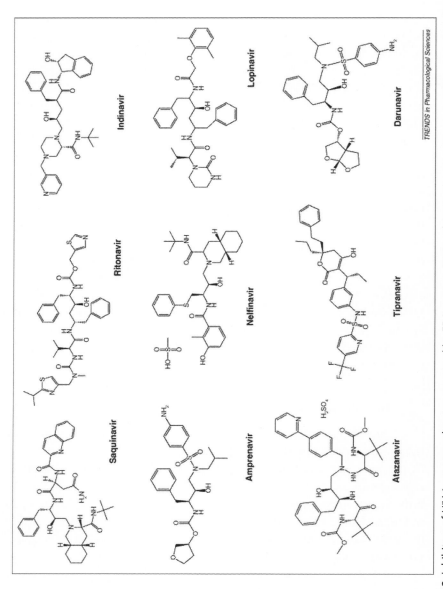

Figure 8.3 Inhibitors of HIV-1 protease that are approved by FDA to treat AIDS and AIDS-related malignancies (from a review (Menendez-Arias & Tozser, 2008)). *Reprinted from Menendez-Arias and Tozser (2008). Copyright Trends in Pharmacological Sciences 2008 Elsevier.*

5. COMPUTATIONAL SIMULATION STUDIES

Several biophysical techniques exist to monitor protein conformational changes at the single-molecule level. The timescales and magnitudes of protein dynamics have been interpreted by NMR spectroscopy, small angle X-ray scattering, electron microscopy, and single-molecule fluorescence (Henzler-Wildman & Kern, 2007). NMR has revealed a lot about the viral protein mobility (Lakomek et al., 2013). However, it falls short in tracking the path of conformational change (Masterson, Cembran, Shi, & Veglia, 2012). Prediction by simulation is expected to be of considerable help in this regard.

At the current scenario, computer simulation technique is a very powerful tool to derive information about protein dynamics at atomic resolution (Dodson, Lane, & Verma, 2008). MD simulations provide insights into the conformational changes of proteins and nucleic acids at an atomic level and millisecond timescale (Pierce et al., 2012). Further, it is crucial for molecular structure refinement in X-ray crystallography and NMR experiments (Kohn, Afonine, Ruscio, Adams, & Head-Gordon, 2010). MD simulation generates a trajectory that describes how the system evolves through phase space as a function of time (Perryman et al., 2004). Currently, several computationally efficient algorithms exist for integrating the equations of motion. It is interesting to recruit computational simulation tools to trace the conformational dynamics of HIV-pr at atomic level.

6. HIV-PR CONFORMATION AND DYNAMICS

The dynamics of both unliganded and liganded HIV-pr have been analyzed. In the liganded form, the flaps are pulled in toward the bottom of the active site, causing flaps to lose mobility and rendering terminal regions of the monomers more flexible. In the unliganded form, flaps are shifted away from the active site (Hornak et al., 2006). The stabilizing effect of the residues in the active site region and flaps is more pronounced in the liganded form than in the unliganded form (Kurt, Scott, Schiffer, & Haliloglu, 2003).

The mutations in HIV-pr are of two types, one at or near the active site and the others far from the active site (Clemente et al., 2004). The mutations near the active sites are crucial as they lead to reduced affinity between enzyme and ligand. The nonactive site mutations influence ligand binding from various distal locations and their mechanism of action is yet to be deciphered (Mao, 2011). The binding affinities and consequent inhibition by drugs (saquinavir, ritonavir, indinavir, and nelfinavir) are weaker for mutants than for the wild-type HIV-pr. Double mutation results in more intense changes (Mosebi, Morris, Dirr, & Sayed, 2008). The mutations within the binding cavity are very conservative and operate by distorting the shape of the cavity.

6.1. Flap dynamics

The flexibility of the flap tips (Gly48–Gly52 and Gly48'–Gly52') is known as flap dynamics. It opens and closes the flaps determining the cavity size. The conformational alteration in the flaps is correlated with structural reorganization of residues in the active site (Torbeev et al., 2011). The mutations in flap region result in alteration of the nonbonding interactions (van der Waals and electrostatic interactions) between the drugs and protein, consequently promoting drug resistance (Cai, Yilmaz, Myint, Ishima, & Schiffer, 2012; Galiano et al., 2009). Thus, several computational studies have attempted to understand flap dynamics. The effects of protein backbone mutations and simulation protocols (polarization of the system and the force fields) on the conformational dynamics of HIV-pr have been studied.

6.2. Mutation effect on HIV-pr conformation
6.2.1 I47V mutation effect
JE-2147 is an allophenylnorstatine-containing dipeptide HIV-pr inhibitor developed by Agouron, Pfizer (Fig. 8.4). It was considered to be more effective than other existing HIV-pr inhibitors. Also, its resistance profile makes it interesting (Yoshimura et al., 1999). Two major mutations which reduce the efficacy of JE-2147 by altering the binding site are I84V and I47V. While I84V is common for other similar ligands, I47V seems to be specific for JE-2147 (Bandyopadhyay & Meher, 2006). The arsenal of existing drugs falls prone to resistance owing to different mutations on HIV-pr. The effect of I47V mutation is investigated using MD simulation which showed higher mobility of the side chain of mutant Val47 than that of WT Ile47 in chain B of the enzyme. The packing of the inhibitor to the residue 47 was affected leading to this deviation in motion. Further, motion was reported in the flap region, which was more conspicuous in the apo form (Bandyopadhyay & Meher, 2006).

6.2.1.1 Comparing the apo form of protein WT versus mutant
The difference of isotropic temperature (B-factor) furnishes idea about the structural fluctuation between different regions of the WT and the mutant.

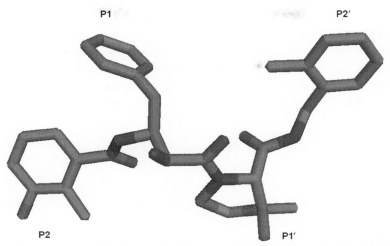

Figure 8.4 Structure of the experimental inhibitor JE-2147. Four sites of interactions (P1, P2, P1′, and P2′) to the protein are labeled. Atoms are shown in color as carbon: green, oxygen: red, nitrogen: blue, and sulfur: brown. (For interpretation of the references to color in this figure legend, the reader is referred to the online version of this chapter.)

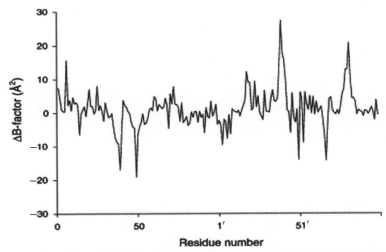

Figure 8.5 Difference of B-factor values from molecular dynamics (MD) simulation for WT and mutant HIV-pr simulation of the apo protein (mutant B-factor–WT B-factor). *Reprinted from Bandyopadhyay and Meher (2006) with permission from* Chem. Biol. Drug Des. *and publisher John Wiley & Sons, Inc.*

The maximum changes occurred for the residues in the flap elbows of the two chains (34–35, 37, 35′–41′), flaps of the two chains (48–51, 46′–54′), and part of the cantilever region (65–70, 65′–68′). Also, there were other changes too, but of lesser degree (Fig. 8.5). Several mutations affect dynamics of flap. For instance, L90M, G48V, and V82F/I84V mutations open the flap more in the mutant than the wild type. On the other hand, M46I mutation makes the flap more closed. The simulation results by Bandyopadhyay and Meher (2006) suggest that, for chain A, the average flap tip to active site distance is less in the case of the mutant. Similar findings have been reported earlier by Wittayanarakul et al. (2005). The distance between Ile50 Cα to that of Ile149 Cα fluctuates more in the case of the WT than that of the mutant (Fig. 8.6).

6.2.1.2 Comparing the complexed form of protein WT versus mutant

The parameters viz. B–factor, Asp25(25′)–Ile50(50′) distance and Ile50–Ile50′ distance, protein–ligand distance, and dihedral angle reflecting the orientation of protein and ligand were investigated to distinguish between result of mutation on complexed form of both the WT and the mutant. Compared to apo form, the complexed form has reduced variation in B–factor between the WT and the mutant state. The Asp25(25′)–Ile50 (50′) distances were monitored, and it was found that in the ligand–bound state, the distance between flap tips and the active site does not differ

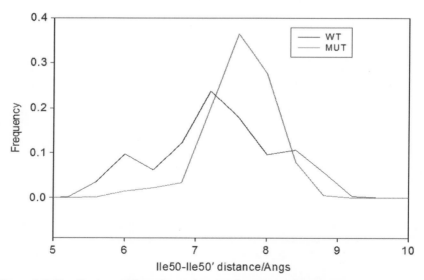

Figure 8.6 Distributions of Ile50-Ile149 distance for WT and mutant HIV-pr simulation of the apo protein. *Reprinted from Bandyopadhyay and Meher (2006) with permission from* Chem. Biol. Drug Des. *and publisher John Wiley & Sons, Inc.* (For color version of this figure, the reader is referred to the online version of this chapter.)

significantly on mutation for chain A, but is remarkable for chain B. The Ile50–Ile50′ distance was also found similar in the complex state, yet, there is difference in the Asp–Ile distance in chain B. Calculation of the Asp25 and Asp25′ distance to the ligand showed that the distributions are essentially same for the WT and the mutant, which indicates that the ligand is bound strongly to the catalytic aspartates and mutation does not have any pronounced effect (Fig. 8.7). The dihedral angle for mutant HIV–pr showed a greater fluctuation compared to that of the WT. The results showed that the relative orientation of residue 47′ changes more rapidly and with a wider range of angles relative to the P2′. It causes the loss of hydrophobic interactions between Ile47′ and the ligand.

6.2.1.3 The molecular mechanisms of drug resistance

In complexed HIV-pr, the difference in dynamic motion of flaps and flap elbows between the WT and the mutant was less than that of the apo protein. The most distinct motion for the HIV-pr complex was the movement of the side chain of Val47′ about the inhibitor which was greater than that of Ile47′ in the WT. It was attributed to the mutation–caused loss of one —CH2 group in the side chain and the resultant decrease in hydrophobic interactions of ligand and enzyme. The larger volume of the mutant

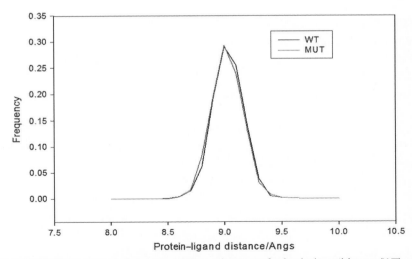

Figure 8.7 Distributions of protein–inhibitor distances for both the wild-type (WT) and the mutant HIV-pr simulation. *Reprinted from Bandyopadhyay and Meher (2006) with permission from* Chem. Biol. Drug Des. *and publisher John Wiley & Sons, Inc.* (For color version of this figure, the reader is referred to the online version of this chapter.)

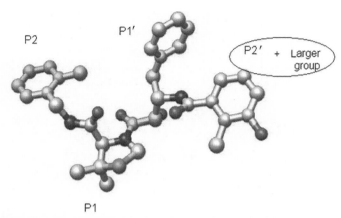

Figure 8.8 Region showing the P2′ position in JE-2147, where the larger group can be added to minimize the mutation-induced loss of interaction between the ligand and the protein. (For color version of this figure, the reader is referred to the online version of this chapter.)

cavity reduces the packing density leading to diminished strength of the binding energy (Ohtaka, Velazquez-Campoy, Xie, & Freire, 2002). Positioning a larger group at the P2′ position of JE-2147 is assumed to counteract the weakened interaction (Fig. 8.8).

6.2.2 I50V and I50L/A71V mutations effect

Meher and Wang investigated the binding of inhibitor TMC114 (Fig. 8.9) to WT, single (I50V) as well as double (I50L/A71V) mutant HIV-pr with all-atom MD simulations as well as MM-PBSA calculation. In double mutant apo HIV-pr, flap–flap distance and the distance from the active site to the flap residues were smaller than wild-type or single mutant form (Fig. 8.10). In double mutant form, complexed HIV-pr shows a less curling of the flap tips and less flexibility than WT and the single mutant I50V. The single mutant I50V decreases the binding affinity of I50V–HIV-pr to inhibitor, resulting in a drug resistance, whereas the double mutant I50L/A71V increases the binding affinity (Fig. 8.11). The increase of the binding affinity for the double mutant I50L/A71V–HIV-pr can be attributed to the increase in electrostatic energy and van der Waals force (Meher & Wang, 2012).

6.2.2.1 Comparing the apo form of protein WT versus mutant

The difference in isotropic temperature factor between the mutant and WT HIV-pr for each residue shows that the highest changes in B-factor occur in

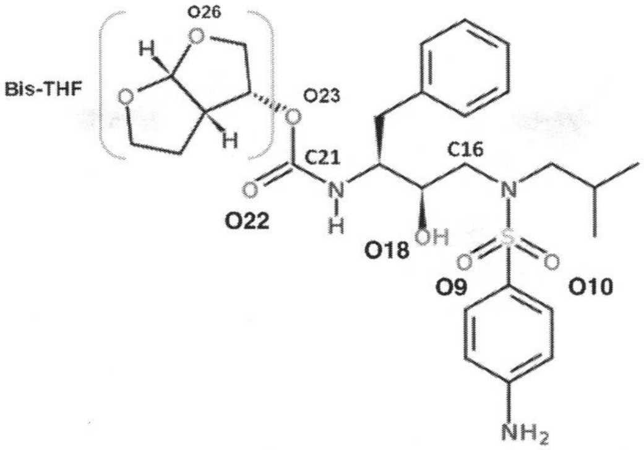

Figure 8.9 Molecular structure of the inhibitor TMC114. The moiety bis-THF is labeled with a square bracket in color green. Important atoms like O9, O10, O18, and O22 which are involved in the interactions between the inhibitor and protein are also labeled in black bold letters. *Reprinted from Meher and Wang (2012) with permission from J. Phys. Chem. B. and publisher American Chemical Society (ACS).* (For interpretation of the references to color in this figure legend, the reader is referred to the online version of this chapter.)

the dimer interface region (6, 8 and 6′, 8′), flap elbow of the chain A (35, 37, 39–41), and flap of the chain A (49–52). Simulation results from Meher and Wang (2012) show that the distance between the flap tips were recognized to fluctuate more in the case of the WT and the single mutant I50V than in the double mutant I50L/A71V (Fig. 8.12). Hence, the mean of the double mutant structure is significantly less (1.15 Å) than the WT, suggesting that there is a close movement of flaps in I50L/A71V as compared to WT and I50V. The reduced active site conformations of I50L/A71V due to the flap dynamics behavior may help in better binding of the inhibitor to the active site.

6.2.2.2 Comparing the complexed form of protein WT versus mutant

Compared to the apo protein, the difference of B-factor between complexed WT and mutant is lesser for most of the residues. It was observed that, four regions around 17 (17′), 41 (41′), 53 (53′), and 70 (70′) show the highest dynamic fluctuations. The relatively smaller B-factor of the double mutant I50L/A71V complex may be explained by the relatively lesser conformational fluctuations and stronger binding. To explore the relative motion of the flap tips, the Ile50–Ile50′ distance was examined. The difference between the complexed WT, I50V, and I50L/A71V HIV-pr was found to be less and narrower than that of the apo HIV-pr (Fig. 8.13). The results indicate that although the Ile50–Ile50′ distance was similar in the complexed HIV-pr, difference exists in the Asp25′–Ile50′ distance. It implies the different behavior of the two chains of HIV-pr.

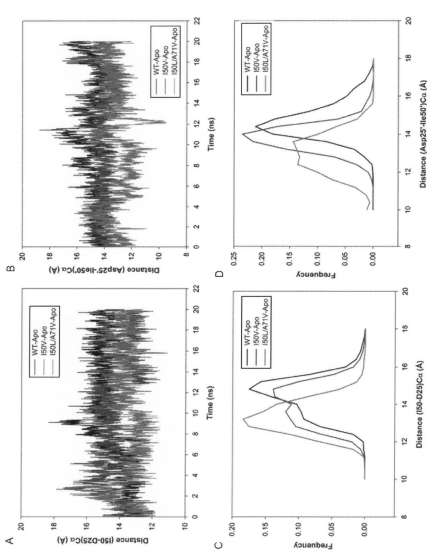

Figure 8.10 Variability of (A) the Ile50-Asp25 Cα distances; (B) the Ile50′-Asp25′ Cα distances of the apo WT, I50V mutant, and I50L/A71V double mutant HIV-pr; (C) histogram distributions of Ile50-Asp25 distance; and (D) histogram distributions of Ile50′-Asp25′ distance for WT and all mutant HIV-pr simulation of the apo protein. *Reprinted from Meher and Wang (2012) with permission from J. Phys. Chem. B. and publisher American Chemical Society (ACS).* (For color version of this figure, the reader is referred to the online version of this chapter.)

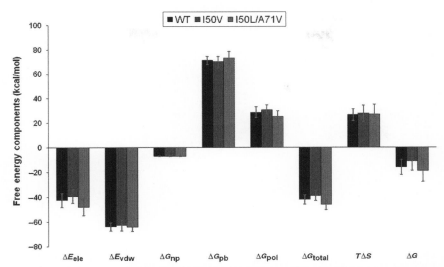

Figure 8.11 Energy components (kcal/mol) for the binding of TMC114 to the WT, I50V, and I50L/A71V: ΔE_{ele}, electrostatic energy in the gas phase; ΔE_{vdw}, van der Waals energy; ΔG_{np}, nonpolar solvation energy; ΔG_{pb}, polar solvation energy; ΔG_{pol}, $\Delta E_{ele} + \Delta G_{pb}$; $T\Delta S$, total entropy contribution; $\Delta G_{total} = \Delta E_{ele} + \Delta E_{vdw} + \Delta E_{int} + \Delta G_{pb}$; $\Delta G = \Delta G_{total} - T\Delta S$. Error bars in green solid line indicates the difference. *Reprinted from Meher and Wang (2012) with permission from* J. Phys. Chem. B. *and publisher American Chemical Society (ACS).* (For interpretation of the references to color in this figure legend, the reader is referred to the online version of this chapter.)

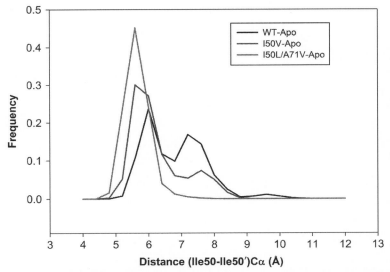

Figure 8.12 Histogram distribution of distance between the flap tips in the apo form WT, I50V, and I50L/A71V proteins. *Reprinted from Meher and Wang (2012) with permission from* J. Phys. Chem. B. *and publisher American Chemical Society (ACS).* (For color version of this figure, the reader is referred to the online version of this chapter.)

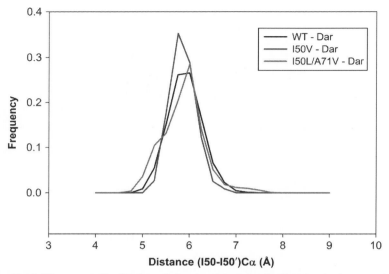

Figure 8.13 Histogram distribution of distance between the flap tips in the complexed form WT, I50V, and I50L/A71V proteins. *Reprinted from Meher and Wang (2012) with permission from* J. Phys. Chem. B. *and publisher American Chemical Society (ACS).* (For color version of this figure, the reader is referred to the online version of this chapter.)

6.2.2.3 Molecular mechanisms of drug resistance

In the I50V mutant HIV-pr, the replacement of isoleucine with valine results in a loss of $-CH_2$ group, which apparently decreases the interaction with the central phenyl of TMC114 through $C-H...\pi$, the size of the hydrophobic side chain, and possible increase of the size of the active site with a reduced binding affinity to TMC114. This change results in a decrease of van der Waals energy between Val50 and TMC114 by about 0.24 kcal/mol relative to the WT. However, for Val50', the change shows a significant decrease in van der Waals energy by 0.54 kcal/mol, which may be due to the lessening of $C-H...O$ interactions between the Val50' side chains and the O22 of TMC114. Figure 8.14 confirms that the C...O22 for I50V-HIV-pr is longer than those for WT and I50L/A71V-HIV-pr, which might be the reason for the less binding affinity and drug resistance.

6.3. Effect of simulation protocols

6.3.1 Effect of polarization

The foundation for any MD simulation is an accurate force field (Dommert, Wendler, Berger, Delle Site, & Holm, 2012). For expressing the interaction between the atoms of a macromolecule, simple potential functions (also known as force fields) consisting of nonbonded (electrostatics and van der Waals

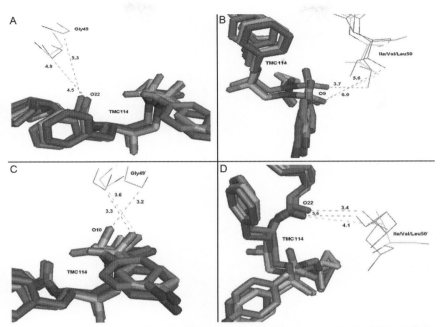

Figure 8.14 C—H. . .O interactions between the TMC114 and the flap residues (Gly49, Gly40′, Ile/Val/Leu50, and Ile/Val/Leu50′). TMC114 in sticks is colored by the atom type and residues are shown as lines (green, WT; blue, I50V; purple, I50L/A71V). *Reprinted from Meher and Wang (2012) with permission from* J. Phys. Chem. B. *and publisher American Chemical Society (ACS).* (For interpretation of the references to color in this figure legend, the reader is referred to the online version of this chapter.)

interactions) and bonded (bond, angle, torsion) terms are used. Polarization plays critical role in dynamic stability of macromolecules (Li, Ji, Xu, & Zhang, 2012). The effect of AMBER force fields viz. ff99 (nonpolarizable) and ff02 (polarizable) were investigated on HIV-pr using MD simulation (Meher, Satish Kumar, & Bandyopadhyay, 2009). The starting structure for both the simulations is a semi-open structure of 2.7 Å resolution (pdb ID 1HHP). TIP3P (Jorgensen, Chandrasekhar, Madura, Impey, & Klein, 1983) and POL3 (Caldwell & Kollman, 1999) are used as the water models for the ff99 and ff02 simulations, respectively. The protein is solvated in a water box containing more than 8000 water molecules. The total charge of the whole system was made neutral by addition of chloride ions. The results showed that for both types of force fields, water count and radial distribution function differ significantly near the charged residue catalytic Asp25. More water molecules are found around that residue in the nonpolarizable force field, ff99 (Fig. 8.15). However, the water movement is quite similar near the polar (Ser37) and hydrophobic (Ile85) residues. Polarization is likely to influence both global and specific local

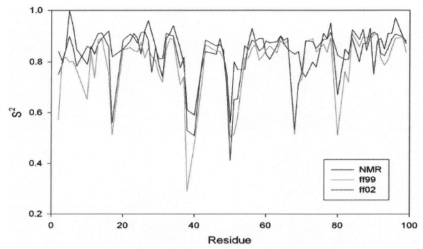

Figure 8.15 Comparison of NMR order parameters for the N—H bond vector with that calculated from ff99 and ff02 trajectories. *Reprinted from Meher et al. (2009) with permission from 2009 Indian Association for Cultivation of Sciences (IACS) and Springer India.* (For color version of this figure, the reader is referred to the online version of this chapter.)

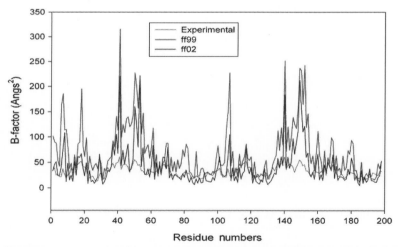

Figure 8.16 Comparison of B-factors for the X-ray structure, nonpolarizable (ff99) and polarizable (ff02) HIV-pr simulations. (For color version of this figure, the reader is referred to the online version of this chapter.)

motions of protein and solvent. Comparison of the calculated S^2 order parameter with the NMR S^2 order parameter results showed that flexibility is more in case of ff99 than ff02 (Fig. 8.16). Also, the comparison of the B–factor values adds additional support with regard to the higher flexibility for ff99 (Fig. 8.17). The

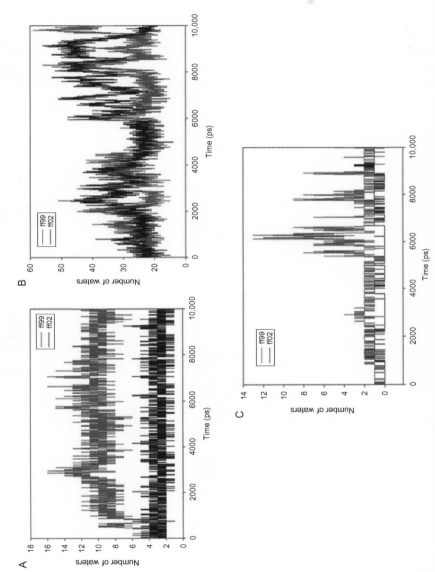

Figure 8.17 Number of waters within 8 Å from (A) Asp25, (B) Ser37, and (C) Ile85 for the ff99 and ff02 trajectories. *Reprinted from Meher et al. (2009) with permission from 2009 Indian Association for Cultivation of Sciences (IACS) and Springer India.* (For color version of this figure, the reader is referred to the online version of this chapter.)

flap–active site distance, a measure of flap opening, is distinctly more in the non-polarizable simulation (Fig. 8.18). The overall structural fluctuation of HIV-pr is reduced in the polarizable simulation making it rigid than the nonpolarizable simulation (Meher et al., 2009). Few polar interactions and hydrogen bonds involving the flap residues were found stronger with polarizable (ff02) force field. Interchain hydrophobic cluster formation (between flap tip of one chain and active site wall of another chain) was found to be prevailed in the semi–open conformations sampled from the simulations regardless of the force field used. It was proposed that an inhibitor proficient of stimulating this interchain hydrophobic cluster may make the flaps stiffer and the effect of mutations on the ligand could be minimized (Meher, Satish Kumar, & Bandyopadhyay, 2013).

6.3.2 Effect of force fields

The effect of AMBER nonpolarizable force fields ff99SB and ff03 was studied on HIV-pr. Two different 30-ns MD trajectories were used to check the ff03 and ff99SB S^2 values and the parameters were found to agree reasonably well with NMR S^2 values. S^2 values for the loop residues were lower in ff03 force field (Fig. 8.19). Thus, the loops were more flexible in the protein models with the ff03 force field resulting in a larger active site cavity.

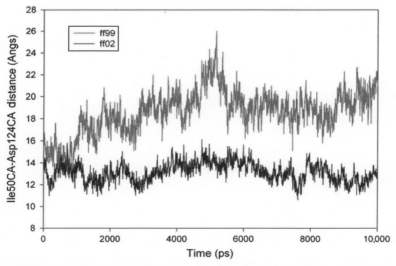

Figure 8.18 Flap-active site (Ile50Ca-Asp25Ca) distance from the ff99 and ff02 trajectories. *Reprinted from Meher et al. (2009) with permission from 2009 Indian Association for Cultivation of Sciences (IACS) and Springer India.* (For color version of this figure, the reader is referred to the online version of this chapter.)

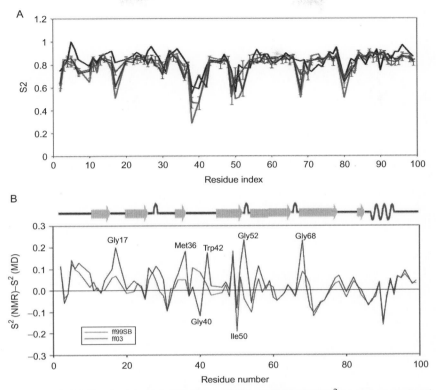

Figure 8.19 (A) Comparison of experimental (black line) NMR S^2 N—H-generalized order parameter values with the calculated values from ff99 (cyan line), ff99SB (red line), and ff03 (blue line) force field simulations. Order parameters (S^2) averaged for both monomers, with error bars reflecting the difference in case of ff99SB and ff03. (B) The difference between ff99SB and ff03 calculated order parameters from the NMR S^2 values. *Reprinted from Meher, Satish Kumar, Sharma, and Bandyopadhyay (2012) with permission from 2012 Imperial College Press (ICP) and J. Bioinforma. Comp. Biol. (For interpretation of the references to color in this figure legend, the reader is referred to the online version of this chapter.)*

Analyzing the flap–active site distance shows significant difference between the two force fields ff99SB and ff03. The flap–active site distance is longer in ff03 simulation; thus, the cavity opening is more (Fig. 8.20). A combination of different factors (such as H-bonding, different torsion parameter) is assumed to contribute to this difference in fluctuation. The superimposition of two representative structures from ff03 and ff99SB force fields shows larger cavity size in the ff03 simulation (Fig. 8.21).

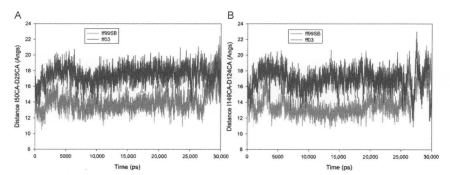

Figure 8.20 Flap-active site distance for (A) chain A and (B) chain B from the ff99SB and ff03 force field simulations. *Reprinted from Meher et al. (2012) with permission from 2012 Imperial College Press (ICP) and* J. Bioinforma. Comp. Biol. (For color version of this figure, the reader is referred to the online version of this chapter.)

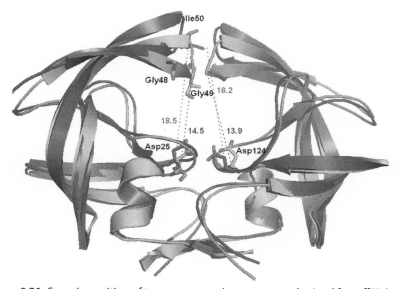

Figure 8.21 Superimposition of two representative structures obtained from ff03 (cyan) and ff99SB (magenta) simulations. Residues (Asp25 (124), Gly48, Gly49, and Ile50) involved in the distance and angle calculations are labeled in the figure. However, for clarity reason, we have not shown the residues for flap region in case of chain B. Structure with ff03 force fields shows larger cavity size compared to ff99SB. The relative distances between the active site to the flap {Asp25 (124) Cα — Ile50 (149) Cα} are more in case of ff03 (18.5 and 18.2) compared to ff99SB (14.5 and 13.9) force field. The red dashed line and numbers indicates the distance for ff03, whereas blue is for ff99SB. *Reprinted from Meher et al. (2012) with permission from 2012 Imperial College Press (ICP) and* J. Bioinforma. Comp. Biol. (For interpretation of the references to color in this figure legend, the reader is referred to the online version of this chapter.)

7. FUTURE PERSPECTIVES

Probing the relation between mutation and conformational dynamics has elucidated the mechanisms of drug resistance. The selection of the correct setups and force fields can help carry out effective MD simulation of other variants of HIV-pr. In *vitro* and *in silico* drug designing, effect of polarization, and importance of force fields can be explored further. The specific inference from the study carried out was that (i) positioning a larger group at the P2' position of JE-2147 may counteract resistance issue, (ii) polarization should be included in the system setup to influence protein rigidity and differential motion of water molecules around a charged residue, and (iii) force fields with greater stability for loop residues can be developed.

8. CONCLUSION

The MD simulation provides ample insights into both biological and technical aspects of macromolecular conformation and dynamics at atomistic level. It sheds light on the mechanisms of drug action against and resistance toward HIV-pr. The investigation led to the finding that conformational dynamics of HIV-pr is affected by simulation setups and force fields difference. Mutations have important role on HIV-pr conformation and dynamics. These results are expected to be vital for designing new promising inhibitors against HIV. The implication of MD simulations can be ramified to an array of mutations and inhibitors, including nanotube-based drugs, to garner profound knowledge of drug–enzyme interactions.

ACKNOWLEDGMENT

The authors gratefully acknowledge the Indian Institute of Technology Guwahati (IITG), India, Albany State University (ASU), Albany, Georgia, USA, and the Pittsburgh Supercomputing Center (PSC), Pittsburgh, USA, for provision of computing facility to accomplish this study.

REFERENCES

Bandyopadhyay, P., & Meher, B. R. (2006). Drug resistance of HIV-1 protease against JE-2147: I47V mutation investigated by molecular dynamics simulation. *Chemical Biology & Drug Design, 67*, 155–161.

Breuker, K., & McLafferty, F. W. (2008). Stepwise evolution of protein native structure with electrospray into the gas phase, 10^{-12} to 10^{2}s. *Proceedings of the National Academy of Sciences of the United States of America, 105*, 18145–18152.

Brik, A., & Wong, C.-H. (2003). HIV-1 protease: Mechanism and drug discovery. *Organic & Biomolecular Chemistry*, *1*, 5–14.

Bukrinskaya, A. (2007). HIV-1 matrix protein: A mysterious regulator of the viral life cycle. *Virus Research*, *124*, 1–11.

Cai, Y., Yilmaz, N. K., Myint, W., Ishima, R., & Schiffer, C. A. (2012). Differential flap dynamics in Wild-type and a drug resistant variant of HIV-1 protease revealed by molecular dynamics and NMR relaxation. *Journal of Chemical Theory and Computation*, *8*, 3452–3462.

Caldwell, J. W., & Kollman, P. A. (1999). Structure and properties of neat liquids using non-additive molecular dynamics: Water, methanol, and N-methylacetamide. *Journal of Physical Chemistry*, *99*, 6208–6219.

Cardinale, D., Salo-Ahen, O. M., Ferrari, S., Ponterini, G., Cruciani, G., Tochowicz, A. M., et al. (2010). Homodimeric enzymes as drug targets. *Current Medicinal Chemistry*, *17*, 826–846.

Chen, L., & Lee, C. (2006). Distinguishing HIV-1 drug resistance, accessory, and viral fitness mutations using conditional selection pressure analysis of treated versus untreated patient samples. *Biology Direct*, *1*, 14.

Clemente, J. C., Moose, R. E., Hemrajani, R., Whitford, L. R. S., Govindasamy, L., Reutzel, R., et al. (2004). Comparing the accumulation of active- and nonactive-site mutations in the HIV-1 protease. *Biochemistry*, *43*, 12141–12151.

Collins, J. R., Burt, S. K., & Erickson, J. W. (1995). Flap opening in HIV-1 protease simulated by 'activated' molecular dynamics. *Nature Structural Biology*, *2*, 334–338.

Dodson, G. G., Lane, D. P., & Verma, C. S. (2008). Molecular simulations of protein dynamics: New windows on mechanisms in biology. *EMBO Reports*, *9*, 144–150.

Dommert, F., Wendler, K., Berger, R., Delle Site, L., & Holm, C. (2012). Force fields for studying the structure and dynamics of ionic liquids: A critical review of recent developments. *Chemphyschem*, *13*, 1625–1637.

Fun, A., Wensing, A. M., Verheyen, J., & Nijhuis, M. (2012). Human Immunodeficiency Virus Gag and protease: Partners in resistance. *Retrovirology*, *9*, 63.

Galiano, L., Ding, F., Veloro, A. M., Blackburn, M. E., Simmerling, C., & Fanucci, G. E. (2009). Drug pressure selected mutations in HIV-1 protease alter flap conformations. *Journal of the American Chemical Society*, *131*, 430–431.

Ganser-Pornillos, B. K., Yeager, M., & Pornillos, O. (2012). Assembly and architecture of HIV. *Advances in Experimental Medicine and Biology*, *726*, 441–465.

Gibellini, D., Borderi, M., Clo, A., Morini, S., Miserocchi, A., Bon, I., et al. (2012). Antiretroviral molecules and cardiovascular diseases. *New Microbiology*, *35*, 359–375.

Gupta, S., Knight, A. G., Losso, B. Y., Ingram, D. K., Keller, J. N., & Bruce-Keller, A. J. (2012). Brain injury caused by HIV protease inhibitors: Role of lipodystrophy and insulin resistance. *Antiviral Research*, *95*, 19–29.

Hatziioannou, T., & Evans, D. T. (2012). Animal models for HIV/AIDS research. *Nature Reviews. Microbiology*, *10*, 852–867.

Henzler-Wildman, K., & Kern, D. (2007). Dynamic personalities of proteins. *Nature*, *450*, 964–972.

Ho, B. K., & Agard, D. A. (2009). Probing the flexibility of large conformational changes in protein structures through local perturbations. *PLoS Computational Biology*, *5*, e1000343.

Hornak, V., Okur, A., Rizzo, R. C., & Simmerling, C. (2006). HIV-1 protease flaps spontaneously open and reclose in molecular dynamics simulations. *Proceedings of the National Academy of Sciences of the United States of America*, *103*, 915–920.

Jorgensen, W. L., Chandrasekhar, J., Madura, J. D., Impey, R. W., & Klein, M. L. (1983). Comparison of simple potential functions for simulating liquid water. *The Journal of Chemical Physics*, *79*, 926.

Kar, P., & Knecht, V. (2012). Energetic basis for drug resistance of HIV-1 protease mutants against amprenavir. *Journal of Computer-Aided Molecular Design*, *26*, 215–232.

Kohn, J. E., Afonine, P. V., Ruscio, J. Z., Adams, P. D., & Head-Gordon, T. (2010). Evidence of functional protein dynamics from X-ray crystallographic ensembles. *PLoS Computational Biology*, *6*, e1000911.

Kokkinidis, M., Glykos, N. M., & Fadouloglou, V. E. (2012). Protein flexibility and enzymatic catalysis. *Advances in Protein Chemistry and Structural Biology*, *87*, 181–218.

Kurt, N., Scott, W. R., Schiffer, C. A., & Haliloglu, T. (2003). Cooperative fluctuations of unliganded and substrate-bound HIV-1 protease: A structure-based analysis on a variety of conformations from crystallography and molecular dynamics simulations. *Proteins*, *51*, 409–422.

Lakomek, N. A., Kaufman, J. D., Stahl, S. J., Louis, J. M., Grishaev, A., Wingfield, P. T., et al. (2013). Internal dynamics of the homotrimeric HIV-1 viral coat protein gp41 on multiple time scales. *Angewandte Chemie (International Edition in English)*, http://dx.doi.org/10.1002/anie.201207266.

Le Grice, S. F. (2012). Human immunodeficiency virus reverse transcriptase: 25 years of research, drug discovery, and promise. *The Journal of Biological Chemistry*, *287*, 40850–40857.

Li, Y., Ji, C., Xu, W., & Zhang, J. Z. (2012). Dynamical stability and assembly cooperativity of β-sheet amyloid oligomers—Effect of polarization. *The Journal of Physical Chemistry. B*, *16*, 13368–13373.

Mager, P. P. (2001). The active site of HIV-1 protease. *Medicinal Research Reviews*, *21*, 348–353.

Mao, Y. (2011). Dynamical basis for drug resistance of HIV-1 protease. *BMC Structural Biology*, *11*, 31.

Masterson, L. R., Cembran, A., Shi, L., & Veglia, G. (2012). Allostery and binding cooperativity of the catalytic subunit of protein kinase A by NMR spectroscopy and molecular dynamics simulations. *Advances in Protein Chemistry and Structural Biology*, *87*, 363–389.

Meagher, K. L., & Carlson, H. A. (2005). Solvation influences flap collapse in HIV-1 protease. *Proteins: Structure, Function, and Bioinformatics*, *58*, 119–125.

Meher, B. R., Satish Kumar, M. V., & Bandyopadhyay, P. (2009). Molecular dynamics simulation of HIV-protease with polarizable and non-polarizable force fields. *Indian Journal of Physics*, *83*, 81–90.

Meher, B. R., Satish Kumar, M. V., & Bandyopadhyay, P. (2013). Inter-chain hydrophobic clustering promotes rigidity in HIV-1 protease flap dynamics: New insights from molecular dynamics. *Journal of Biomolecular Structure & Dynamics*, http://dx.doi.org/10.1080/07391102.2013.795873.

Meher, B. R., Satish Kumar, M. V., Sharma, S., & Bandyopadhyay, P. (2012). Conformational dynamics of HIV-1 protease: A comparative molecular dynamics simulation study with multiple amber force fields. *Journal of Bioinformatics and Computational Biology*, *10*, 150018.

Meher, B. R., & Wang, Y. (2012). Interaction of I50V mutant and I50L/A71V double mutant HIV-protease with inhibitor TMC114 (Darunavir): Molecular dynamics simulation and binding free energy studies. *The Journal of Physical Chemistry. B*, *116*, 1884–1900.

Menendez-Arias, L., & Tozser, J. (2008). HIV-1 protease inhibitors: Effects on HIV-2 replication and resistance. *Trends in Pharmacological Sciences*, *29*(1), 42–49.

Miyamoto, F., & Kodama, E. N. (2012). Novel HIV-1 fusion inhibition peptides: Designing the next generation of drugs. *Antiviral Chemistry & Chemotherapy*, *22*, 151–158.

Mosebi, S., Morris, L., Dirr, H. W., & Sayed, Y. (2008). Active-site mutations in the South African human immunodeficiency virus type 1 subtype C protease have a significant impact on clinical inhibitor binding: Kinetic and thermodynamic study. *Journal of Virology*, *82*, 11476–11479.

Ode, H., Neya, S., Hata, M., Sugiura, W., & Hoshino, T. (2006). Computational simulations of HIV-1 proteases multi-drug resistance due to non-active site mutation L90M. *Journal of the American Chemical Society*, *128*, 7887–7895.

Ohtaka, H., Velazquez-Campoy, A., Xie, D., & Freire, E. (2002). Overcoming drug resistance in HIV-1 chemotherapy: The binding thermodynamics of Amprenavir and TMC-126 to wild-type and drug-resistant mutants of the HIV-1 protease. *Protein Science*, *11*, 1908–1916.

Overton, E. T., Arathoon, E., Baraldi, E., & Tomaka, F. (2012). Effect of darunavir on lipid profile in HIV-infected patients. *HIV Clinical Trials*, *13*, 256–270.

Perryman, A. L., Lin, J.-H., & McCammon, J. A. (2004). HIV-1 protease molecular dynamics of a wild-type and of the V82F/I84V mutant: Possible contributions to drug resistance and a potential new target site for drugs. *Protein Science*, *13*, 1108–1123.

Petsko, G. A., & Ringe, D. (2004). *Protein structure and function*. London, UK: New Science Press Ltd.

Piana, S., Carloni, P., & Parrinello, M. (2002). Role of conformational fluctuations in the enzymatic reaction of HIV-1 protease. *Journal of Molecular Biology*, *319*, 567–583.

Piana, S., Carloni, P., & Rothlisberger, U. (2002). Drug resistance in HIV-1 protease: Flexibility assisted mechanism of compensatory mutations. *Protein Science*, *11*, 2393–2402.

Pierce, L. C., Salomon-Ferrer, R., de Augusto, F., Oliveira, C., McCammon, J. A., & Walker, R. C. (2012). Routine access to millisecond time scale events with accelerated molecular dynamics. *Journal of Chemical Theory and Computation*, *8*, 2997–3002.

Sanou, M. P., De Groot, A. S., Murphey-Corb, M., Levy, J. A., & Yamamoto, J. K. (2012). HIV-1 vaccine trials: Evolving concepts and designs. *Open AIDS Journal*, *6*, 274–288.

Scott, W. R., & Schiffer, C. A. (2000). Curling of flap tips in HIV-1 protease as a mechanism for substrate entry and tolerance of drug resistance. *Structure*, *8*, 1259–1265.

Torbeev, V. Y., Raghuraman, H., Hamelberg, D., Tonelli, M., Westler, W. M., Perozo, E., et al. (2011). Protein conformational dynamics in the mechanism of HIV-1 protease catalysis. *Proceedings of the National Academy of Sciences of the United States of America*, *108*, 20982–20987.

Toth, G., & Borics, A. (2006). Flap opening mechanism of HIV-1 protease. *Journal of Molecular Graphics & Modelling*, *24*, 465–474.

Tran, E. E., Borgnia, M. J., Kuybeda, O., Schauder, D. M., Bartesaghi, A., Frank, G. A., et al. (2012). Structural mechanism of trimeric HIV-1 envelope glycoprotein activation. *PLoS Pathogens*, *8*, 1002797.

Wang, Y., Liu, Z., Brunzelle, J. S., Kovari, I. A., Dewdney, T. G., Reiter, S. J., et al. (2011). The higher barrier of darunavir and tipranavir resistance for HIV-1 protease. *Biochemical and Biophysical Research Communications*, *412*, 737–742.

Winstone, T. A., Man, S. F., Hull, S. F., Montaner, J. S., & Sin, D. D. (2013). Epidemic of lung cancer in patients with HIV infection. *Chest*, *143*, 305–314.

Wittayanarakul, K., Aruksakunwong, O., Saen-oon, S., Chantratita, W., Parasuk, V., Sompornpisut, P., et al. (2005). Insights into saquinavir resistance in the G48V HIV-1 protease: Quantum calculations and molecular dynamic simulations. *Biophysics Journal*, *88*, 867–879.

Yoshimura, K., Kato, R., Yusa, K., Kavlick, M. F., Maroun, V., Nguyen, A., et al. (1999). JE-2147: A dipeptide protease inhibitor (PI) that potently inhibits multi-PI-resistant HIV-1. *Proceedings of the National Academy of Sciences of the United States of America*, *96*, 8675–8680.

AUTHOR INDEX

Note: Page numbers followed by "*f*" indicate figures, and "*t*" indicate tables, and *np* indicate footnote.

A

Abagyan, R., 74
Abdelnoor, M., 201
Abel, R., 82–83
Abu-Abed, M., 220–221
Achari, A., 11–12, 29
Adams, P. D., 305
Adcock, S. A., 180–181, 221
Afonine, P. V., 305
Agard, D. A., 303
Agashe, V. R., 112–113
Aggarwal, A., 255, 274–275, 276–277, 285–286
Agha-Amiri, K., 190–191
Ahmad, S., 274–275
Aigaki, T., 181–182
Ainavarapu, S. R., 122
Ainsworth, C. F., 137–138
Akerud, T., 240, 242, 243–244
Akke, M., 234–235, 240, 242, 243–244
Alcala, M., 14
Alden, C. J., 6–7, 28–29
Alden, R. A., 236–238
Alder, B., 180–181
Alemany, A., 109
Alessandro, L., 37, 38–39
Alexander, M., 5–6
Allison, R. D., 22
Amadei, A., 74
Ames, J. B., 182, 185–186
Andersson, U., 204–206
Ando, T., 260
Andrea, B., 190–191
Andreas, 255, 261, 262*f*, 263, 289, 291–292
Andrew, R., 201
Anfinsen, C. B., 110–111
Antonini, I., 33
Antosiewicz, J. M., 182–183
Aqvist, J., 236
Arathoon, E., 302–303
Argos, P., 74

Armitage, J. P., 233
Arnott, S., 6–7, 28–29, 193
Arseniev, A. S., 226
Aruksakunwong, O., 306–308
Arumugam, S., 234–235
Ashkin, A., 95
Astoul, C. H., 143–145, 149–150, 151–153
Atanasov, B., 191
Athappilly, F., 122
Atkins, P. W., 22
Austin, R., 261
Austin, S. E., 190–191
Aymami, J., 4–5
Azbel, M. Y., 273
Azuaga, A. I., 241–242

B

Baase, W. A., 100–101
Babcock, H. P., 240–241
Bachhawat, K., 138–139
Bachmann, T. T., 2
Badilla, C. L., 100, 112–113, 122
Badt, D., 100
Bagchi, B., 41–42, 43*f*, 48–50, 110–111, 293
Baginski, M., 36
Bahadur, R. P., 192
Bahar, I., 221
Bailey-Kellogg, C., 241–242
Baker, J. K., 109
Balamurali, M. M., 122
Balci, H., 95
Baldini, G., 9, 16, 17*f*
Balzarini, J., 136
Bandyopadhyay, P., 306–308, 307*f*, 308*f*, 309*f*, 314–318, 316*f*, 317*f*, 318*f*, 319*f*
Banerjee, R., 137–138, 139–141, 146–147, 173
Bansal, S., 241–242
Banumathi, S., 146–147
Banuprakash Reddy, G., 146–147
Baptista, A. M., 182–183

SUBJECT INDEX

Note: Page numbers followed by "*f*" indicate figures, and "*t*" indicate tables.

Printed and bound by CPI Group (UK) Ltd, Croydon, CR0 4YY

15/05/2025

01872389-0001